HUMAN BIOLOGICAL AGING

HUMAN BIOLOGICAL AGING
From Macromolecules to Organ Systems

Glenda Bilder
Gwynedd Mercy University, Gwynedd Valley, PA, USA

WILEY Blackwell

Copyright © 2016 by John Wiley & Sons, Inc. All rights reserved

Published by John Wiley & Sons, Inc., Hoboken, New Jersey
Published simultaneously in Canada

No part of this publication may be reproduced, stored in a retrieval system, or transmitted in any form or by any means, electronic, mechanical, photocopying, recording, scanning, or otherwise, except as permitted under Section 107 or 108 of the 1976 United States Copyright Act, without either the prior written permission of the Publisher, or authorization through payment of the appropriate per-copy fee to the Copyright Clearance Center, Inc., 222 Rosewood Drive, Danvers, MA 01923, (978) 750–8400, fax (978) 750–4470, or on the web at www.copyright.com. Requests to the Publisher for permission should be addressed to the Permissions Department, John Wiley & Sons, Inc., 111 River Street, Hoboken, NJ 07030, (201) 748–6011, fax (201) 748–6008, or online at http://www.wiley.com/go/permission.

Limit of Liability/Disclaimer of Warranty: While the publisher and author have used their best efforts in preparing this book, they make no representations or warranties with respect to the accuracy or completeness of the contents of this book and specifically disclaim any implied warranties of merchantability or fitness for a particular purpose. No warranty may be created or extended by sales representatives or written sales materials. The advice and strategies contained herein may not be suitable for your situation. You should consult with a professional where appropriate. Neither the publisher nor author shall be liable for any loss of profit or any other commercial damages, including but not limited to special, incidental, consequential, or other damages.

For general information on our other products and services or for technical support, please contact our Customer Care Department within the United States at (800) 762–2974, outside the United States at (317) 572–3993 or fax (317) 572–4002.

Wiley also publishes its books in a variety of electronic formats. Some content that appears in print may not be available in electronic formats. For more information about Wiley products, visit our web site at www.wiley.com.

Library of Congress Cataloging-in-Publication Data
Names: Bilder, Glenda, author.
Title: Human biological aging : from macromolecules to organ systems / Glenda Bilder.
Description: Hoboken, New Jersey : John Wiley & Sons, Inc., [2016] | Includes index.
Identifiers: LCCN 2015035901 | ISBN 9781118967027 (paperback)
Subjects: LCSH: Aging–Physiological aspects. | Macromolecules. | BISAC:
  SCIENCE / Life Sciences / Human Anatomy & Physiology.
Classification: LCC QP86 .B513 2016 | DDC 612.6/7–dc23 LC record available at http://lccn.loc.gov/2015035901

Cover image supplied by: gettyimages.com/178807611/SergeyNivens, gettyimages.com/178807611/hxdbzwy, gettyimages.com/ 466327944/YukiL4a, gettyimages.com/ 469299034/3quarks

Printed in Singapore by C.O.S. Printers Pte Ltd

10 9 8 7 6 5 4 3 2 1

CONTENTS

Preface vii

About the Companion Website ix

Section I   THE FOUNDATION 1

1 ORIENTATION 3

2 MEASUREMENTS AND MODELS 17

3 EVOLUTIONARY THEORIES OF AGING 35

Section II   BASIC COMPONENTS 47

4 AGING OF MACROMOLECULES 53

5 AGING OF CELLS 77

Section III   ORGAN SYSTEMS: OUTER COVERING AND MOVEMENT: INTEGUMENTARY, SKELETAL MUSCLES, AND SKELETAL SYSTEMS 101

6 AGING OF THE INTEGUMENTARY SYSTEM 103

7 AGING OF THE SKELETAL MUSCLE SYSTEM 123

8 AGING OF THE SKELETAL SYSTEM 143

Section IV   INTERNAL ORGAN SYSTEMS: CARDIOVASCULAR, PULMONARY, GASTROINTESTINAL, AND URINARY SYSTEMS 163

9 AGING OF THE CARDIOVASCULAR SYSTEM 165

10 AGING OF THE PULMONARY SYSTEM   193

11 AGING OF THE GASTROINTESTINAL AND URINARY SYSTEMS   207

Section V REGULATORY ORGAN SYSTEMS: CENTRAL NERVOUS SYSTEM, SENSORY, ENDOCRINE, AND IMMUNE SYSTEMS   223

12 AGING OF THE CENTRAL NERVOUS SYSTEM   225

13 AGING OF THE SENSORY SYSTEM   255

14 AGING OF THE ENDOCRINE SYSTEM   275

15 AGING OF THE IMMUNE SYSTEM   303

Index   323

PREFACE

My first objective in writing *Human Biological Aging: From Macromolecules to Organ Systems* is to provide an introductory textbook for non-science majors interested in learning about the biological aging process in man. This would include college students with majors in gerontology, allied health, psychology, and sociology. Since biological aging builds on an understanding of basic scientific principles, my second objective is to craft a biology of aging textbook that incorporates sufficient basic biological science to render the aging process more comprehensible. Thus, this textbook seeks to present to students with modest to minimal science education, the essentials of human biological aging: descriptions; mechanisms and theories of aging; strategies to extend the health span and aging-related disease vulnerabilities. It is hoped that the intertwined and supplemental basic science material will facilitate a successful avenue to the appreciation of the aging process.

In an endeavor to achieve these goals, the book predominately discusses results from studies of human aging and presents the aging process from macromolecules to organ systems. In particular, the reader will learn the principal theories of aging, study designs / models of aging, and age changes in the structure and function of macromolecules, cells, skin, muscles, bone, lungs, heart and blood vessels, brain, kidney, gastrointestinal tract, endocrine glands, sensory organs, and the immune system.

To aid understanding, several learning tools have been employed. Subdivisions of every chapter are introduced with a phrase or one-sentence header (in bold type) that summarizes the essences of the material to follow. Within each discussion, important reinforcing or supportive data and information are highlighted with italics. Both aging and related scientific background information are managed in this fashion. Additionally, each chapter contains a list of key terms, a formal summary of age changes, numerous illustrations and tables, and side boxes with supplemental material. Questions to inspire exploratory thinking relevant to chapter content and associated controversial issues accompany each chapter. Use of a select bibliography for each chapter is appropriate for a textbook of this size and focus. My expectation is that the chosen references will serve as a starting point of future inquiry by the interested student.

The study of human aging is a fertile arena for new discoveries. The rapid growth of the biogerontology disciple is witness to this. Not surprisingly, there is no shortage of discrepancies and controversies. Some of these are introduced in this textbook. However, my prime effort has been to capture the current understanding of aging at the various biological levels and to organize it for assimilation by future gerontology and allied health workers who will serve the increasing number of elderly in our society.

I am grateful to my colleague Dr. Camilo Rojas at Johns Hopkins University for his critique of portions of this text. His insightful suggestions on presentation and content have been invaluable. I am appreciative of the computer and editing expertise of my son Dr. Patrick Bilder at Albert Einstein College of Medicine. His assistance has aided this work significantly. I am thankful for the repeated opportunity provided by Gwynedd Mercy University to teach the Biology of Aging course. Student comments and support from GMU Natural Science chairman, Dr. Michelle McEliece, were helpful to this project. I am also sincerely indebted to my husband Chuck for his unwavering encouragement.

GLENDA BILDER

ABOUT THE COMPANION WEBSITE

This book is accompanied by a companion website:

   www.wiley.com\go\Bilder\HumanBiologicalAging

The website includes downloadable photographs, illustrations and tables from the book.

SECTION I

THE FOUNDATION

ESSENTIAL PREPARATORY MATERIAL

Chapters 1–3 are foundation chapters that present an overview of the field of biological aging. Chapter 1 provides a description of aging from the perspectives of established biogerontologists, introduces the theories of aging, and sets out a working model to understand aging in relation to other phases of the life cycle. Chapter 2 reviews the scientific method, assets and pitfalls of study designs used to evaluate aging in man, and the numerous animal models of aging that provide invaluable insights into conserved preservation mechanisms. Chapter 3 presents the evolutionary theory of aging, a persuasive theory that offers a convincing explanation as to why organisms age and consequently positions aging as a legitimate biological entity.

Human Biological Aging: From Macromolecules to Organ Systems, First Edition. Glenda Bilder.
© 2016 John Wiley & Sons, Inc. Published 2016 by John Wiley & Sons, Inc.

1

ORIENTATION

One of many interpretations of Samuel Clemen's (Mark Twain) famous quote, "Age is an issue of mind over matter, if you don't mind, it doesn't matter," is that although it is easy enough to ignore aging, it may not be a wise approach. The reason is that aging, unlike other stages of the lifespan (prenatal, birth, infancy, childhood, adolescence, and adulthood) is uniquely different. Distinct and nearly opposite from that observed in other life stages, the contribution of heredity (genes) to aging is modest, barely reaching 30%. This allows a larger contribution (near 70%) from the environment and its interaction with heredity. Thus, the lesser role of genes in aging allows for a substantial influence of the environment, for example, life style choices and societal improvements, on the expression of individual aging. The more one learns of the aging process, the greater will be the understanding of what choices will make a difference in both quality and quantity of life.

BEGINNINGS OF BIOGERONTOLOGY

Multiple Disciplines Come Together to Study Biological Aging

Historically, gerontology was the only scientific discipline devoted to research on the aging process and for many years received little attention or research funding. As the demand to comprehend the aging process mounted, energized by insights from evolutionary biology, plus societal needs engendered by an expanding class of senior citizens, scientists from diverse disciplines, for example, molecular/cellular biology, biochemistry, neuroscience, vertebrate/invertebrate genetics, comparative/

Human Biological Aging: From Macromolecules to Organ Systems, First Edition. Glenda Bilder.
© 2016 John Wiley & Sons, Inc. Published 2016 by John Wiley & Sons, Inc.

evolutionary biology, endocrinology, and physiology, found the aging process to be worthy of serious evaluation. Their collective effort has dramatically accelerated the accumulation of novel observations in biological aging. Not surprisingly, it prompted the introduction of the term biogerontology to replace gerontology.

This shared effort yielded the following insights:

1. The aging process is understandable in the context of established biological principles.
2. The aging process is distinct from the disease process; nevertheless, aging remains a risk factor for disease.
3. The aging process is considered a worthwhile research arena in all scientific disciplines, a change that encourages persistent critique of theories of aging and a greater potential to establish reliable guidelines for a healthier life.

The goal of biogerontologists in their scientific endeavors is to prolong life in a way that preserves physical and mental health. Legitimate and feasible research goals are to

1. elucidate the biological mechanisms necessary for a long healthy life (longevity),
2. identify and eliminate factors that cause premature death, and
3. develop strategies to minimize degenerative and devastating diseases.

POPULATION AGING

Dramatic Increase in Life Expectancy Due to Public Health Advancements: Sanitation, Clean Water, Vaccines, and Antibiotics

Data collected worldwide by the United Nations Department of Economic and Social Affairs indicate that as of 2012, there are 810 million individuals 60 years of age and older in the world and by 2050 this number is expected to increase to 2 billion. These numbers are significant because they predict a global population where for the *first time in history, the number of older individuals will exceed the number of younger ones (0–14 years of age)*. Currently, one out of every nine individuals in the world is 60 years or older. By 2050, this will change to one out of every five. The oldest old or those 80 years of age and older now account for 14% of the world population. By 2050, this will increase to 20%. In the United States, the number of individuals aged 60 years and older stands at slightly over 60 million (19% of the population), a number projected to increase to over 100 million (27% of the population) by 2050.

Today, individuals generally live twice as long as those born at the turn of the twentieth century. This comparison is expressed in terms of a mathematical calculation called *life expectancy*. Life expectancy is computed from mortality data of a population (demographic information). As defined by Murphy et al. (2013): "Life expectancy at birth represents the average number of years that a group of infants would live if the group was to experience throughout life the age-specific death rate present in the year of birth". Commonly, life expectancy is expressed relative to a birth year, for example, 2011, or alternatively to a particular age in a specified year, for example, 65 in 2011. If *birth is the reference point, life expectancy equates to the average or mean lifespan for the particular population under study*. For example, an individual born in 1900 in the United States could expect to live to 47.3 years compared to an individual born in 2010 in the United States who could expect to live to 76.2 years (males) and 81 years (females). The mean lifespan nearly doubled over the last century.

Several key societal advancements contributed to the increase in life expectancy. Public innovations that occurred in the *first half of the twentieth century and slightly before* improved life expectancy to the greatest extent. One important development was the acceptance of the validity of the Germ Theory of Infectious Disease proposed by Louis Pasteur, Robert Koch, and others. This enlightenment propelled reformations in sewage handling, sanitation, and clean water and led to the availability of vaccines against diphtheria, whooping cough, and tetanus. Moreover, the launch of sulfa drugs and antibiotics, for example, penicillin, significantly reduced the number of infant and childhood deaths from infectious diseases, thereby allowing more individuals to survive to older ages (Figure 1.1).

Life expectancy increased in the second half of the century for *different* reasons. There was a minor reduction in infant mortality brought about by a trend favoring hospital births over traditional home births. For older individuals, mortality rate declined as a result of access to successful management of chronic diseases, especially cardiovascular disease, a major cause of death in the elderly. This came about with the development of safer drugs, for example, antihypertensive drugs, cholesterol-lowering drugs, implantable devices, for example, pacemakers, stents, and defibrillators, and surgical procedures, for example, coronary artery bypass grafts in conjunction with the establishment of Medicare and Medicaid insurance to pay for this care.

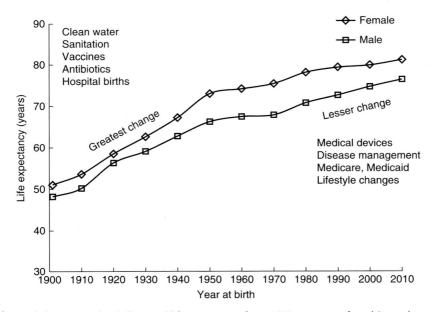

Figure 1.1. Factors that influenced life expectancy from 1900 to present for white males and females of the United States. (Data obtained from Arias (2014).)

In sum, multiple and diverse advances in public health and medical science decreased the mortality rate of our society. These developments initially benefited infants and adolescents by allowing more of them to survive infectious diseases. More recently, older individuals have profited from novel therapies and evidenced-based medicine for the treatment of chronic diseases, but the impact in terms of additional life expectancy years is generally smaller with only a few years added to those 65 years and older.

Does Living Longer Assure Living Healthier?

It is generally concluded that the remarkable increase in life expectancy in the industrialized world over the past 100 years signifies that these societies have become progressively healthier. Since health is generally defined as the "absence of disease," one could argue that the increase in life expectancy comes about in the presence of chronic but managed disabilities and diseases; hence, the question remains unanswered as to whether industrialized societies are indeed any healthier. In the United States, according to 2007–2008 data from the Centers for Disease Control and Prevention (CDC), the percentage of elderly (% male to % female) that manages chronic conditions such as hypertension (55:57), arthritis (42:55), heart disease (37:26), diabetes (20:18), chronic bronchitis/emphysema (9:9), and stroke (9:9) is fairly high. Almost half of elderly men and a third of elderly women admit hearing difficulties. Some elderly (13–15%) have visual problems and up to 25% have no natural teeth. On the other hand, according to the National Long-Term Care Surveys (NLTCS), 1982–2004/2005, the number of individuals with chronic disability has in fact declined compared to the start of data collection, two decades earlier. Furthermore, trends assessed from the 2000–2008 data from the NLTCS and four other national surveys, for example, National Health and Nutrition Examination Survey and Health and Retirement Survey, show that while "personal care and domestic activity" of the oldest old (>85 years of age) have declined they remain unchanged for those 65–84 years of age (Freedman et al., 2013) compared to earlier data. Given the current pace of biogerontological research and society's demand for reliable health-promoting choices, it is reasonable to expect that disabilities and degenerative diseases in the future will affect fewer elderly for shorter periods of time. This remains to be proven.

CHARACTERISTICS OF AGING

The Fundamentals of Physics Describe Aging as the Loss of "Molecular Fidelity" That Exceeds Repair and Replacement

Aging is the *last stage of the life cycle* during which the organism experiences a *gradual loss of organ function and systemic regulation that eventually leads to death*. The complex interaction of known and unknown factors that underlies the aging process influences the onset, the rate (speed), and the anatomical/physiological extent of change.

In humans, senescence or the *senescence span* refers to a deteriorative state with reduced ability to endure stress. Importantly, the time prior to senescence is the *health span* or period of organ maintenance and good health, despite the presence of aging. The *health span extends from peak reproductive years (about 30 years of age or earlier) to the onset of senescence*. During the health span, cellular functions and integrative activities may be suboptimal, but they are below the critical threshold for noticeable dysfunction.

Several useful definitions of aging are presented in Table 1.1. Although there is disagreement with regard to the commencement of aging (discussed below), a summary consensus equates aging to a process that is multifactorial in origin, stochastic (random) in progression and depth of change, dependent on maintenance processes, modulated by the environment, distinct from disease, and deleterious to the point of death.

Hayflick's definition in Table 1.1 is a particularly insightful characterization of aging. He describes aging as "the stochastic process that occurs systemically after reproductive maturity in animals that reach a fixed size in adulthood and is caused by the escalating *loss of molecular fidelity that ultimately exceeds repair capacity* thus increasing vulnerability to pathology or age-related diseases."

Aging may occur by one of two pathways: "By a purposeful program driven by genes or by random accidental events" (Hayflick, 2007). Although genes are key components of longevity determinants (see below), *data are lacking for a gene-driven program of aging*. Instead, aging follows the *fundamental law of physics (Second Law*

TABLE 1.1. Characteristics of Aging

"Aging represents an informational loss . . . one of noise accumulation in homeostatic and copying processes or . . . irreversible switching off of synthetic capacities."
(Comfort, 1974)

"Stochastic process that occurs systemically after reproductive maturity ... caused by the escalating loss of molecular fidelity that ultimately exceeds repair capacity ... vulnerability to pathology." (Hayflick, 2004)

"Evolutionary considerations suggest aging is caused not by active gene programming but by evolved limitations in somatic maintenance, resulting in a build-up of damage."
(Kirkwood, 2005)

"Rate at which aging changes take place can be altered, either by nature or through intervention." (Carrington, 2005)

"Time-independent series of cumulative, progressive, intrinsic, and deleterious functional and structural changes that usually begin to manifest themselves at reproductive maturity and eventually culminate in death." (Arking, 2006)

"Eventual failure of maintenance," "aging is multicausal"; "evolved design of many components of complex animals is incompatible with indefinite survival." (Holliday, 2006)

"Aging is nothing more than the unprogrammed result of selection for early reproductive success." (Faragher et al., 2009)

"Aging is a complex multifactorial process characterized by accumulation of deleterious changes in cells and tissues, progressive deterioration of structural integrity and physiological function across multiple organ systems, and increased risk of death." (Semba et al., 2010) *of Thermodynamics*) that describes "the universal tendency for things to become disordered" (Alberts et al., 2002). This spontaneous disorder or energy dissipation (quantified as *entropy*) is the cause of the random "loss of molecular fidelity" that typifies aging. Cells (hence, the organism) are constantly addressing entropy by taking energy from the environment to create internal order and through chemical reactions dissipating some of the energy (as heat) back into the environment as disorder. Since molecules are characteristically unstable and will change through passive (energy dissipation) or active (attack by oxidants) means, *higher order mechanisms of surveillance, repair, and replacement are absolutely essential*. For a time (to achieve reproductive success, see Chapter 3), entropy is thwarted with excellent repair and restorative processes. Eventually these processes too succumb to disorder and the organism fails and dies. Viewed through a universal law of nature, aging is the "loss of molecular fidelity that exceeds repair and replacement" (Hayflick, 2007). This culminates in a decrement of homeostatic (normalizing) mechanisms and an increased vulnerability to disease, all of which lead to an increased probability of death.

The Commencement of Aging Is Debated

The precise onset of aging is unknown. It is reasoned that since the biological mechanisms that dictate growth, development, and reproductive maturity differ substantially from those implicated in the aging process, biogerontologists have proposed that aging begins somewhere in early adulthood or perhaps slightly before. Mortality data roughly support this (CDC National Center for Health Statistics, 2005) and show that the mortality rate (the inverse of fitness and health) is the lowest around ages 5–14, that is, the time of peak fitness and health. Thereafter, mortality rate doubles approximately every eight years (Arking, 2006).

Rates of Aging Among Different Species May Be Rapid, Gradual, or Negligible

If it is assumed that the period of early reproductive maturity represents an approximate commencement of aging, organisms may be classified as exhibiting *rapid, gradual,* or *negligible rates of aging*. These rates are associated with a lifespan that is short (days or months, for example, fruit fly), intermediate (several years to many decades, for example, humans), or long (hundreds of years, for example, bristlecone pine). Aging in man is gradual (maximal lifespan of 122 years). Humans experience a developmental phase of about 7–10 years, a reproductive phase of 30 years, and a postreproductive adulthood and aging (health span plus senescence span) of about 50 or more years.

The Senescence Phenotype Is Highly Variable

Among species that reproduce sexually, aging is universal. Members of a species or those organisms sharing specific genetic traits generally age in a similar fashion and express an aging or *senescence phenotype*. Phenotype is the sum total of the biology of an organism excluding its genes. Phenotype encompasses all observable traits of an organism, dictated by genes to include structure, function, behavior, and regulation and modified by gene–environment interaction. *The senescence phenotype, therefore, describes all of the observable and measureable traits (structural and functional) that embody aging.* The similarity of aging traits among members of a species is attributed to the expression of the *same vulnerable molecules* randomly affected by entropy that eventually cannot be repaired or replaced.

Despite sharing many basic similarities, senescence phenotypes especially *among humans differ significantly from one individual to another at any one time point*. This accounts for the inability of chronological age to accurately define age changes because *the rate of aging is highly variable, and thus characterized by heterogeneity (nonuniformity)*. Whether in reference to the onset and extent of graying of the hair or appearance of facial fine wrinkles or more serious changes of reduced respiratory function, specific measurements among individuals of the same age, for example, 65 years of age, vary widely. The standard deviation or the variability around the average of a measurement obtained from a group of elderly subjects is larger than the variability observed for the same measurement acquired in young subjects.

Heterogeneity in study measurements arises in part from the inability of biogerontologists to exclude not only elderly with overt disease (obvious disease risk factors, medication use, and smoking) but also those with covert disease (revealed by scans, x-rays). However, even in elderly declared "disease-free" (as best as can be assessed), *heterogeneity of aging persists and is accredited to the interplay between stochastic events that destabilize molecules and the efficacy of survival or maintenance mechanisms* that restabilize them. Irrespective of the similarity of "vulnerable molecules" and also the considerable overlap of maintenance mechanisms among individuals, the profusion of stochastic events including, as in the case of man, life style choices, continues to ensure significant variability or heterogeneity in aging.

COMPONENTS OF LONGEVITY

Longevity Is in Part Heritable Through Expression of Longevity Determinants: Mechanisms of Maintenance, Repair, and Replacement

All life stages depend on the following factors: (i) Genes (hereditary material carried by the DNA) expressed in the organism, (ii) environmental/stochastic (random) influences, and (iii) interaction of genes with environmental and stochastic events termed *epigenetic* effects (Figure 1.2). Several studies based on data from Twin Registry (lifespan information of twins living in Denmark and Sweden from 1870 onward) concluded that gene expression accounts for 25–30% of one's lifespan. Thus, *lifespan is partially heritable*. Similarly, a genetic contribution to lifespan of ~30% has been reported for laboratory animals. Consequently, the environment and its interaction with hereditary material (genes or, generally, genome) contribute prominently (65–70%) to lifespan determination in both man and related animals.

The genetic contribution to lifespan is expression through longevity determinants (longevity assurance genes) and gerontogenes. Longevity assurance genes are defined

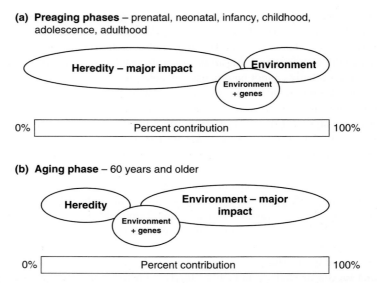

Figure 1.2. Relative influence of genes, environment, and interaction on life stages.

as the genes that contribute to a long life and, in essence, are directly or indirectly involved in upkeep, restoration, and replacement functions. In particular, longevity determinants embody protective cellular mechanisms, antioxidant enzymes and associated proteins, specific lipid constituents of membranes, repair programs for DNA and proteins, homeostatic (stress normalizing) mechanisms, and innate/adaptive immunity. These systems optimize molecular structure and function essential for reproductive success; their *continued presence and efficiency* determine, in part, longevity postreproduction.

Numerous *longevity assurance genes* have been identified in lower animals and are associated with lifespan shortening in animals lacking these genes or, conversely, lifespan extension in animals expressing multiple copies of such genes. An example of a longevity assurance gene is superoxide dismutase (SOD), an antioxidant enzyme capable of suppressing oxidative damage. Manipulation of this gene, for example, addition of multiple copies, lengthens the lifespan of the fruit fly, a popular model of aging. In man only one longevity assurance gene has been identified thus far. It is the apolipoprotein E (APOE) gene that produces a multifunctional protein involved in lipid metabolism, proliferation, and repair. Certain variants of APOE gene confer increased longevity.

Longevity determinants expressed in all members of a species are deemed *conserved* (or *public*). Longevity determinants that are *unique (not shared) confer on acquiring members an advantage of a longer life*. Exclusive determinants are considered private and may be variants of the public determinants, although little is known about private determinants.

In contrast to longevity assurance genes, *gerontogenes are genes whose absence is associated with a 25% or more increase in lifespan* in animal models of aging (see

Chapter 2). Gerontogenes influence energy homeostasis, cell maintenance, stress responses, DNA repair, and inflammatory effects. For example, deletion of the Daf-2 gene in the roundworm or the related (homologous) insulin growth factor-1 receptor (IGFR-1) gene in the mouse increases lifespan by 80 and 30%, respectively. Daf-2 and IGFR-1 are committed to nutrient sensing and regulation of associated metabolic pathways. In their absence, life extension mechanisms are enhanced, possibly because more efficient metabolic pathways are utilized in their absence.

Environmental influences are defined as (i) internal or intrinsic factors within organisms such as chemical reactions and their products and (ii) external or extrinsic factors such as diet, air quality, exposure to ultraviolet radiation, stress, and behavior. Extrinsic factors are modified by lifestyle choices.

Longevity of the Centenarians and Supercentenarians Reveals Few Common Threads

Long-lived species compared to those with shorter lifespans appear to possess a greater abundance of efficient repair or maintenance mechanisms, pathways, programs and systems, lipids and proteins that resist oxidation, and a more sophisticated immune system. In the human population, individuals who reach 100 or more years of age are envied. As subjects of intense study, a *common thread to their longevity remains to be identified*. The only human gene linked to longevity is the APOE gene. Therefore, individuals with the epsilon 2 APOE variant are long lived. Those with the epsilon 4 APOE variant have a higher risk for the development of neurotoxicity, dementia, and other diseases, and hence have a shorter lifespan. Observational studies conclude that centenarians have few factors in common, although they are generally not heavy smokers, not severely overweight, and have a relatively high educational background with reasonably good coping skills. Interestingly, these individuals are not disease-free, suggesting they may have "private" longevity determinants that enable better compensation in the face of disease. As yet there is no "longevity-assuring lifestyle."

Stochastic Events Exert Major Impact on Lifespan

Biogerontologists emphasize the stochastic or random nature of the environmental component of the aging process. Environmental factors are UV radiation (sun), X-rays, pollution, inactivity, lack of nutrition, toxins from smoking, isolation, and mental/physical stress. The mechanisms whereby the environment specifically interacts with organisms are poorly understood. Some environmental factors influence the lifespan through an epigenetic effect. Epigenetic effects change gene expression (but not the gene itself) by one of several processes: (i) DNA methylation (physical addition of a chemical group, in this case, a methyl group, to a gene); (ii) histone modification (alteration of proteins packed around a gene), or (iii) microRNA expression (expression of small nucleic acids called RNA that influence gene expression). Epigenetic-driven molecular changes determine whether a particular gene will be active or silent and in doing so, may lengthen or shorten the lifespan. Epigenetic effects are described more fully in Chapter 4.

THEORIES OF AGING OVERVIEW

There is no shortage of theories of aging (see Table 1.2 for a select few). The plethora of theories is attributed to the complex nature of the aging process. Theories tend to focus on one or more of the numerous maintenance mechanisms or conversely on the multitude of stochastic stresses. Generally, aging theories are sorted into categories of (1) programmed versus stochastic or damage theories; (ii) "how" versus "why" theories; (iii) molecular versus cellular versus systemic theories, and (iv) evolutionary

TABLE 1.2. Theories of Aging

Evolutionary (mutation accumulation; antagonistic pleiotropy; disposable soma)
- Evolutionary pressure is reduced or zero after sexual maturity—allows mutation accumulation (expression of late-acting deleterious genes), antagonistic pleiotropy (expression of mechanisms beneficial to the young but harmful to the old), and disposable soma (limitations of repair and maintenance mechanisms evident after sexual maturity)
(Medawar, 1952; Williams, 1957; Kirkwood, 1977)

Free radical
- Reactive oxygen species (ROS) generated intrinsically and extrinsically are neutralized with antioxidative enzymes and related mechanisms. Excessive accumulation of ROS damage DNA, proteins, and membranes and if unrepaired, lead to cell and organ dysfunction typical of aging
(Gerschman et al., 1954; Harman, 1956)

Redox stress hypothesis
- ROS regulate signal molecules; as redox potential (oxidative state) of cell increases, signally becomes dysfunctional, and homeostasis and response to stress decline
(Sohal and Orr, 2012)

Rate of living
- "Preset limit" on metabolic energy determines lifespan; the faster the metabolic rate, the shorter the lifespan
(Pearl, 1928)

Mitochondrial; lysosomal–mitochondrial axis of postmitotic cells
- Lysosomal and mitochondrial functions are essential to cellular health. Dysfunction of these organelles allows for expression of specific signals that induce cellular suicide. Disappearance of nonreplicating cells, for example, muscle cells, neurons, and cardiac cells reduces organ function and accelerates aging
(Wallace, 2005; Terman et al., 2006)

Cellular senescence (replicative senescence)
- Unrepaired DNA (and many other factors) convert normal cell to replicative senescent cell. Senescent cell contributes to cancer formation and degenerative inflammatory conditions
(Campisi, 2013)

Mitotic clock
- Replicating cells divide a limited number of times (Hayflick's number). Loss of renewal produces organ dysfunction
(Hayflick, 1975)

Immunosenescence
- Progressive alteration in innate/adaptive immunity gives rise to increased susceptibility to "new" but not previously encountered microbes, increased risk for cancers, poor response to vaccines, and a proinflammatory state
(Walford, 1979; Fulop et al., 2011)

Neuroendocrine
- Loss of control of neuroendocrine function reduces homeostasis (metabolism, adaptive responses, reproduction, and immune response) and increases susceptibility to disease/disability leading to death
(Dilman 1986)

theories. The theories summarized in Table 1.2 were selected based on their relative contribution to the understanding of aging.

One of the most important theories is the evolutionary theory of aging, presented in Chapter 3. The remaining theories are discussed in association with the appropriate biological system. For example, the free radical–oxidant theory of aging, redox stress theory, and rate of living theories are discussed within the context of the structure/function of macromolecules (Chapter 4). The mitochondrial, lysosomal–mitochondrial dysfunction, mitotic clock, and senescence cell theories are examined relative to aging in cells (Chapter 5), the endocrine theory of aging is given in relation to aging of the neuroendocrine system (Chapter 14), and the immunosenescence theory is addressed in discussion of the immune system (Chapter 15). All theories must explain species-specific characteristics such as the length of the species-specific lifespan. For example, why the maximal lifespan of man is 122 years, while that of the mouse is only 5 years and additionally why delayed reproduction or prolonged caloric restriction are interventions that consistently lengthen the species-specific lifespan.

SUMMARY

The dramatic increase in life expectancy that has given rise to a larger than ever population of older individuals (65 years of age and older) resulted from effects of several public health and safety improvements of the early twentieth century that impacted mostly infants and children.

The senescence phenotype characterized by severely reduced organ function, inadequate stress and homeostatic responses, and increased vulnerability to disease is expressed as the last stage in life. It is further characterized by considerable heterogeneity.

Although heredity (genes) contributes ~25% to determination of the human lifespan, the contribution from environmental stochastic events that push molecules into disorder is enormous. The genetic contribution to lifespan is defined in terms of longevity assurance genes (negated by gerontogenes) that affect homeostasis, metabolism, stress response, and a multitude of antioxidant and repair mechanisms. Collectively termed maintenance mechanisms, their eventual failure driven by environmental effects underlies age changes that culminate in deterioration and death.

CRITICAL THINKING

Why is chronological age an unreliable indicator of biological age?

What is life expectancy and why has it changed over time?

Why are there so many definitions of aging? Which one is the most convincing?

What are longevity determinants?

What is the influence of heredity on aging? What is the role of random damage?

What is entropy and what role does it play in aging?

KEY TERMS

Biogerontology the study of biology of aging from the perspective of all scientific disciplines.

Demography the study of populations, their size, and change over time; gives insight into health of a population.

Entropy in thermodynamic terms, a tendency for systems to move toward disarray or chaos. Effort or energy required to maintain order.

Epigenetics the process whereby environmental factors influence the expression of genes.

Gene a segment of DNA that codes for (directs) the production of a unique protein. Collectively genes are the blueprint for inherited characteristics.

Gerontogenes genes that have been identified in lower organisms to accelerate aging. Their removal increases the lifespan of the organism.

Gerontology the study of the biology of aging; differs from geriatrics that is the study of diseases of the elderly.

Health span the part of the lifespan from reproductive maturity to overt deterioration (or senescence).

Homeostasis consistency of the internal environment; maintenance of normal function within an optimal range.

Life expectancy a statistical prediction of longevity reflecting in part the health of a population; assuming a constant death rate, the number of years one may statistically expect to live if born in a specific year; or if one has attained a specific age, and death rates are constant, the number of additional years one may expect to live.

Longevity length of life; how long an individual or organism lives.

Longevity determinants genetic programs of maintenance, repair, and replacement that evolutionarily developed to ensure fitness and reproductive success.

Maximal lifespan lifespan of the verifiable longest-lived organism of a particular species. Example: human—122 years (Jeanne Calment); rats: 5–6 years.

Mean lifespan average lifespan of a species; equal to life expectancy of an organism at birth.

Phenotype all observable characteristics of an organism's biological structure and function, excluding the genes or genotype.

Senescence the phase of aging preceding death; phase of marked decline and deterioration.

Senescent phenotype observable changes in an organism characterized as reduced function and deterioration.

Species organisms with similar genetic backgrounds capable of interbreeding. Provides a means of classifying organisms.

Stochastic random or by chance.

BIBLIOGRAPHY

Review

Alberts B, Johnson A, Lewis J, Raff M, Roberts K, Walter P. 2002. *Molecular Biology of the Cell*, 4th ed. New York: GS Garland Science.

Arking R. 2006. *Biology of Aging: Observations and Principles*. 3rd ed. New York: Oxford University Press.

Bonanni, P. 1999. Demographic impact of vaccination: a review. *Vaccine* **17**(Suppl. 3): S120–S125.

Brooks-Wilson AR. 2013. Genetics of healthy aging and longevity. *Hum. Genet.* **132**(12):1323–1338.

Campisi J. 2013. Aging, cellular senescence, and cancer. *Annu. Rev. Physiol.* **75**:685–705.

Carrington JL. 2005. Aging bone and cartilage: cross-cutting issues. *Biochem. Biophys. Res. Commun.* **328**(3):700–708.

Comfort A. 1974. The position of aging studies. *Mech. Ageing Dev.* **3**(1):1–31.

Dilman VM., Revskoy SY, Golubev AG. 1986. Neuroendocrine-ontogenic mechanism of aging: toward an integrated theory of aging. *Int. Rev. Neurobiol.* **28**:89–156.

Faragher RG, Sheerin AN, Ostler EL. 2009. Can we intervene in human ageing? *Expert. Rev. Mol. Med.* **11**:e27.

Finch CE. 1998. Variations in senescence and longevity include the possibility of negligible senescence. *J. Gerontol. Biol. Sci.* **53**A(4):B235–B239.

Finch CE, Tanzi RE. 1997. Genetics of aging. *Science* **278**:407–411.

Fulop T, Larbi A, Kotb R, de Angelis F, Pawelec G. 2011. Aging, immunity, and cancer. *Discov. Med.* **11**(61):537–550.

Gerschman R, Gilbert DL, Nye SW, Dwyer P, Fenn WO. 1954. Oxygen poisoning and X-irradiation: a mechanism in common. *Science* **119**(3097):623–626.

Harman D. 1956. Aging: a theory based on free radical and radiation chemistry. *J. Gerontol.* **11**(3):298–300.

Hayflick L. 1975. Current theories of biological aging. *Fed. Proc.* **34**(1):9–13.

Hayflick L. 2004. Debates: The not-so-close relationship between biological aging and age-associated pathologies in humans. *J. Gerontol. Biol. Sci. Med. Sci.* **59**(6):547–550.

Hayflick L. 2007. Biological aging is no longer an unsolved problem. *Ann. NY Acad. Sci.* **1100**:1–13.

Holliday R. 2006. Aging is no longer an unsolved problem in biology. *Ann. NY Acad. Sci.* **1067**:1–9.

Kinsella KG. 1992. Changes in life expectancy 1900–1990. *Am. J. Clin. Nutr.* **55**(Suppl. 6):1196S–1202S.

Kirkwood TB. 1977. Evolution of ageing. *Nature* **270**(5635):301–304.

Kirkwood TBL. 2005. Understanding the odd science of aging. *Cell* **120**(4):437–447.

McDonald RB, Ruhe RC. 2011. Aging and longevity: why knowing the difference is important to nutrition research. *Nutrients* **3**(3):274–284.

Medawar PB. 1952. *An Unsolved Problem of Biology*. London: H.K. Lewis.

Montesanto A, Dato S, Bellizzi D, Rose G, Passarino G. 2012. Epidemiological, genetic and epigenetic aspects of the research on healthy ageing and longevity. *Immun. Aging* **9**(1):6–18.

Pearl R. 1928. *Rate of Living Theory*. New York: Alfred A. Knopf.

Semba RD, Nicklett EJ, Ferrucci L. 2010. Does accumulation of advanced glycation end products contribute to the aging phenotype? *J. Gerontol. A Biol. Sci. Med. Sci.* **65**(9): 963–975.

Sohal R, Orr WC. 2012. The redox stress hypothesis of aging. *Free Radic. Biol. Med.* **52**(3):539–555.

Terman A, Gustafsson B, Brunk UT. 2006. Mitochondrial damage and intralysosomal degradation in cellular aging. *Mol. Aspects Med.* **27**(5–6):471–82.

Walford RL. 1979. Multigene families, histocompatibility systems, transformation, meiosis, stem cells, and DNA repair. *Mech. Ageing Dev.* **9**(1–2):19–26.

Wallace DC. 2005. A mitochondrial paradigm of metabolic and degenerative diseases, aging, and cancer: a dawn for evolutionary medicine. *Annu. Rev. Genet.* **39**:359–407.

Williams GC. 1957. Pleiotropy, natural selection, and the evolution of senescence. *Evolution* **11**(4):398–411.

Experimental

Arias, E. 2014. United States life tables, 2010. *Natl. Vital Stat. Rep.* **63**(7):1–62 (Table 21).

CDC National Center for Health Statistics. 2005. Worktable 23R, http://www.cdc.gov/nchs/data/statab/MortFinal2005_Worktable23R.pdf

Christensen K, Kyvik KO, Holm NV, Skytthe A. 2011. Register-based research on twins. *Scand. J. Public Health* **39**(Suppl. 7):185–190.

Freedman VA, Spillman BC, Andreski PM, Cornman JC, Crimmins EM, Kramarow E, Lubitz J, Martin LG, Merkin SS, Schoeni RF, Seeman TE, Waidmann TA. 2013. Trends in late-life activity limitations in the United States: an update from five national surveys. *Demography* **50**(2):661–671.

Manton KG, Gu X, Lamb VL. 2006. Change in chronic disability from 1982 to 2004/2005 as measured by long-term changes in function and health in the U.S. elderly population. *Proc. Natl. Acad. Sci. USA* **103**:18374–18379.

Murphy SL, Xu J, Kochanek KD. 2013. Deaths: final data for 2010. *Natl. Vital Stat. Rep.* **61**(4):1–118.

Rajpathak SN, Liu Y, Ben-David O, Reddy S, Atzmon G, Crandall J, Barzilai N. 2011. Lifestyle factors of people with exceptional longevity. *J. Am. Geriatr. Soc.* **59**(8): 1509–1512.

United Nations, Department of Economic and Social Affairs. 2015. www.unpopulation.org

U.S. Census Bureau Statistics. 2010. *Older Americans 2010: key indicators of well being.* Document available at www.agingstats.gov

2

MEASUREMENTS AND MODELS

THE SCIENTIFIC METHOD

Scientists are compelled to adhere to the tenets of the scientific method. The path of inquiry includes systematic observations, hypothesis generation, experimentation, data analysis, and acceptance or rejection of the hypothesis. In practice, a scientist makes a unique observation and in conjunction with review of the published literature determines whether a scientific investigation is needed. If so, a hypothesis is generated. This is an original explanation for the novel observation. The hypothesis guides the scientist in the design of appropriate experiments to be performed under "controlled" conditions for the purpose of hypothesis validation. On completion of each experiment, data are analyzed, the hypothesis is accepted or rejected, and a relevant conclusion is formulated.

An accepted hypothesis by one scientist is subjected to future scrutiny by other scientists. Replication and extension of the data by independent laboratories may support the development of a more comprehensive hypothesis or theory. Confirmed by an abundance of reproducible data, a compelling theory has the capacity to *predict* future results. Predictions can subsequently be confirmed by experimentation. As new discoveries are made, a theory may be further (a) supported, (b) modified, or (c) refuted and discarded. Results of studies that do not comply with the scientific method are suspect and are unlikely to be replicated.

Types of Data

Not All Data are of Equal Value Correlative data differ from *cause/effect* data. Correlative data establish a statistical (mathematical) relationship between

Human Biological Aging: From Macromolecules to Organ Systems, First Edition. Glenda Bilder.
© 2016 John Wiley & Sons, Inc. Published 2016 by John Wiley & Sons, Inc.

variables (the parameters that are measured or are expected to change) and assign a degree of probability regarding the strength of the association between the variables. It is an association and no more. It does not establish a cause and effect relationship where (i) the cause (inducer) produces (ii) the effect (result or change). A cause and effect connection ensues from experiments in which variables are manipulated in a way that enables the scientist to assess the cause and effect relationship with a high degree of assurance. Sadly, the majority of studies on the aging process are correlative and a change in a selected variable is *statistically* associated (or correlated) with an increase in chronological age. Because these variables are linked statistically, their reliability is modest at best and data from cause/effect studies are needed for verification.

Issues with Aging Studies in Man

Studies of Human Aging Encounter Difficulties: Heterogeneity, Organizational Level, and Others

Although age changes in man are the main theater of interest, human heterogeneity assures that studies in man are a formidable challenge. This *heterogeneity* results in part from individual variations in the human genome (hereditary information contained in the DNA, the genotype) and in larger part from environmental influences that affect the genome in multiple and poorly understood ways (see Chapter 1). Consequently, there exist numerous differences in physical aspects of each person's biological structure and function (referred to as the phenotype). Thus, to measure a change that could be attributed to aging with any degree of confidence, *very large numbers of participants are required to study aging in man*. Accordingly, a study in man with a sufficient number of subjects requires not only hundreds of healthy subjects that comply with study demands but also considerable research funding and effort. These are grave obstacles to research progress.

Other concerns among biogerontologists relate to (i) selection of the most relevant biological organizational level to study, for example, molecules, cells, tissues, organ systems, organism, and populations, (ii) characterization of aging mechanisms devoid of covert (hidden) life-shortening factors, for example, alcohol abuse, and (iii) a suitable animal model (if not man). The expected difference in the rate of aging among diverse cell types, tissues, and organs, among animals, and within the human population adds additional complexity.

Aging Assessed from Demographic or Individual Perspective

Analysis of census data provides demographic information on populations, yielding a summary of the entire group and assessment of trends in select aspects of health (only those queried) over time. Demographic information fails to provide information on biological processes in an individual's aging experience. *Cross-sectional or longitudinal study designs are used to reveal individual biological aging*. Additional understanding of aging is obtained from mechanistic studies with animal models. These models are amenable to genetic and environmental manipulations and are the *only* experiments at present that have the potential to establish cause and effect.

MEASUREMENT OF THE AGING PROCESS

Study Designs Are Mainly Cross-Sectional and Longitudinal

The two experimental designs most frequently used to study aging in man and other organisms are the cross-sectional and the longitudinal study designs. Each design has its advantages and disadvantages.

Cross-Sectional Study Design Infers Aging A variable, for example, blood pressure or respiratory capacity, is selected and *measured in groups of individuals of several different ages (each age group referred to as a cohort)*. The response of individuals in each cohort is averaged and a comparison of the average from all cohorts depicts the age change of the selected variable or parameter. The *age change using the cross-sectional study design is inferred* from the average generated from cohorts of different ages. The age effect is *assumed to be identical to the normal age change* in all individuals during the ages represented by the cohorts.

A major disadvantage of the cross-sectional design, especially when used in humans, is that it assumes humans live in a *constant environment*. In the United States, the environment has continuously improved over time and higher standards for air and water quality, work place safety, and medical assistance are achieved on a regular basis. As a result, a 20 year old today lives in an environment *different* from that experienced by an 80 year old when he (she) was 20. Since the aging process is a composite of gene expression, environmental effects, and an interaction between them, a cohort of 20 year olds and a cohort of 80 year olds *differ not only in age* but also with regard to many other external factors, known and unknown. To conclude that kidney function declines with age because kidney function is lower in the cohort of 80 year olds compared to 20 year olds *could possibly be* incorrect. Clearly, many factors other than age contribute to reduced organ function and significantly, those factors *differ* among the various cohorts. *In the cross-sectional design, the effect of birth and period confound normal age changes*. In animal studies, the environmental and pathological influences are managed and presumed to be constant. However, "controlled" environments are too often elusive and assumed rather than verifiably achieved. This restricts interpretation of results from animal studies using the cross-sectional design.

Another limitation of the cross-sectional study design is that cohorts of older individuals represent a select group, that is, survivors of a particular age. Consequently, this study design bears a bias toward survivability and again findings may not represent the aging process of most individuals (if in fact most have died earlier).

Despite the fact that results from cross-sectional studies may yield misleading conclusions, this study design unfortunately is the design of choice. This is because studies employing the cross-sectional design are easier to manage, research expenses are generally modest, and time investments are less.

Longitudinal Study Design Measures Aging Directly The longitudinal study design measures a select variable(s) at specified time points over the lifespan of the organism. This design *measures the rate of change of a parameter* as it actually occurs and gives a more accurate picture of the effect of aging on any particular parameter. The ongoing Baltimore Longitudinal Study of Aging (BLSA) conducted by the National Institute on Aging (NIA) is an example of this study design.

One limitation of the longitudinal design is the possibility that with repeat measures of a function, for example, cognitive function tests, the measured values may inadvertently improve by the "practice" of test repetition. The degree of bias varies with the particular measurement or test. Many tests now have multiple formats to reduce bias, but none eliminate it completely.

The longitudinal design is employed less frequently than the cross-sectional design. The reasons are the inherent lengthy time commitment and associated higher research expenses. In man, compliance (completing a study, once enrolled) tends to decline as study duration increases. Noncompliance adds to expense and errors in data interpretation. This is minimized with use of individuals living in institutions, but this too is problematic as these residents may not represent a "normal" population. Even with the longitudinal study design that studies short-lived laboratory mouse and rat models (lifespan ~3–5 years), the expense and time requirements significantly exceed that of the cross-sectional study design.

To minimize research expenses and time commitment, longitudinal studies of 3–7 years are often employed using cohorts of different ages, achieving a blend of cross-sectional and longitudinal study designs. This combination still suffers from the same disadvantages as noted above, but in some cases it is the only available approach.

Both the cross-sectional and longitudinal designs must develop criteria for subject exclusion. In other words, each must define a "normal" population of subjects. To some extent, this has been addressed with the use of consensus prestudy exclusion criteria developed at least to eliminate individuals with known diseases and/or risk factors for known diseases.

Randomized Controlled Trials and Meta-Analysis are Additional Formats for the Study of Aging in Man

The randomized controlled trial (RCT) is a rigorous study design employed to determine the efficacy of a new drug or medical intervention. The RCT design requires that carefully selected participants be *randomly assigned* to one of two groups: treatment or control (placebo). The group assignment is unknown to the subjects (termed blind) and frequently additionally unknown to the study administrator (double blind). The RCT is considered a study design that yields reliable data. It has been applied to aging studies in cases where a proposed age-modulating intervention, for example, brain exercises, is compared to no intervention (control being something that simulates the intervention). Some RCTs extend out for years and have identified beneficial interventions.

A second type of study frequently encountered in biogerontology is the meta-analysis. This is a statistical analysis of similar quantitative studies. The data from carefully (specified criteria) selected *comparable studies are pooled* and *appropriate statistics are applied* to gain significant probability strength in support of a conclusion, only weakly apparent with a single study. Although there are disadvantages to this approach, for example, study inclusion and choice of statistical methods, it is used in biogerontological studies, especially in areas where many small studies exist and no one conclusion can be drawn with any certainty. The advantage is that the statistical analysis of data merged from multiple parallel studies generates a conclusion potentially worthy of additional investigation. An example is the meta-analysis of the effect of exercise in bone loss prevention (Chapter 8).

CALORIC RESTRICTION: LIFE EXTENSION EXPERIMENT

The phrase, "caloric restriction" (CR), has special meaning in biogerontology that extends beyond the general understanding of dieting by reducing food consumption usually with the goal of shedding pounds. CR refers to an experimental *approach in which caloric consumption (intake of food) is reduced by approximately 30% for an extensive portion of an organism's life*. It is only the content of calories and *not* the nutrients, for example, protein content, vitamins, and minerals, that is reduced. There is no malnutrition.

One of the most important consequences of CR is a measurable and statistically significant extension of the species-specific *maximum lifespan (MLS) of the CR organism or in other words life extension beyond that of the longest lived organism of a species*. The seminal work on CR was published in 1935 by the Cornell University professor of nutrition, Clive McCay and his associates. Using white rats, McCay et al. (1935) set out to observe over a 4-year period the effect of a reduction in calories (energy intake) in the presence of adequate essential nutrients on body size and lifespan. Specifically, lifespan in CR males was 30% longer compared to controls given free access to food (*ad libitum*). It was additionally observed that fur changes paralleled CR; the fur of CR rats was thinner (determined by hair shaft diameter) and finer compared to *ad libitum* controls (with thicker and coarser fur). The fur of CR rats appeared similar to that of young rats, that is, hinting that a longer life could be accompanied by renewed vitality. Many studies over the years have confirmed McCay's original observation of MLS extension with CR. CR protocols have been successful in extending MLS in single-cell organisms (yeast), invertebrates (fruit flies, roundworms), and in mammals (mice, rats, dogs, monkeys).

Physiological Changes with Caloric Restriction

CR produces many exceptional effects. In mammals including the monkey, the following have been reported: (i) a decrease in body weight and change in body composition with loss of fat mass, (ii) multiple endocrine changes to include increased sensitivity to insulin (and lower blood glucose and insulin levels), sustained "youthful" levels of growth hormone, and the purported "rejuvenating" androgen—dehydroepiandrosterone (DHEA), (iii) a reduction in free radical production, for example, less oxidative damage in skeletal muscle mitochondria, (iv) maintenance of immune function, and (v) a delay in the onset of major pathologies, for example, cancers and cardiovascular disease. In nonmammalian organisms, CR produces enhanced cellular activities that include more efficient metabolism, improved cellular maintenance, and greater resistance to stressors.

The observation that CR consistently extends the MLS in organisms as diverse as yeast and mice supports the conclusion that basic and conserved age-related mechanism(s) can be manipulated by an environmental stressor, in this case, lack of food. Furthermore, the establishment by NIA of a colony of CR rats available to researchers emphasizes the collective recognition of the importance of this intervention as a worthwhile approach to explore mechanisms of aging.

CR in Man is Underway Sponsored by NIA, controlled pilot studies of CR in man were initiated in 2002 and designated the CALERIE Study (*C*omprehensive

Assessment of Long-term Effects of Reducing Intake of Energy). Three pilot studies ranging in duration from 6 to 12 months with 26–48 volunteers compared the effects of CR or CR plus exercise with a healthy diet (nonrestricted) or pre-CR values. In particular, CR (where measured ~17%) resulted in decreases in fasting insulin level, fasting glucose level, energy expenditure, DNA fragmentation, core temperature, visceral and body fat mass; an improvement in muscle mitochondrial function; and trend toward elevation in oxidative repair. Although there is no effort to measure longevity in these or future studies, lower values of fasting glucose, body temperature, and oxidative damage evident with CR in man are predictive of longevity in CR exposed animal models.

A phase II study was initiated in 2007 with enrollment of 225 nonobese volunteers. The objective is to measure several physiological parameters, including resting metabolic rate, core body temperature, select functions of the neuroendocrine and immune systems, and cognition in individuals consuming 25% less calories over a 2-year period compared to those eating their regular diet. Results are eagerly awaited.

Mechanisms of Caloric Restriction The molecular mechanisms of CR are currently under intense investigation. The pathway is complex but appears to start with enzymes/receptors that sense nutrients. Nutrient deprivation sets off a cascade of changes that ends at the gene level to "turn-on or turn-off" certain genes. Box 2.1 details these pathways.

Box 2.1. Molecular Mechanisms of Caloric Restriction

There exist an abundance of molecular explanations for the life extension effects of CR. Two intensely studied pathways are those mediated by inhibition of insulin/IGF-1 signaling and inactivation of an enzyme named mTOR (mammalian target of rapamycin). Both are considered nutrient-sensing pathways (insulin for glucose and mTOR for amino acids). As nutrient levels fall, insulin concentration declines and inactivation of mTOR ensues. The insulin pathway is mediated through other enzymes (PI3K/Akt/Ras) or the forkhead O (FOXO) transcriptional factor and inactivation of mTOR leads to enhanced recycling of damaged proteins. Adenosine monophosphate-activated protein kinase (AMPK) is a third possible CR-relevant pathway. The activity of this ATP-producing enzyme is elevated by CR through activation of another kinase, liver kinase B. One additional pathway is that directed by sirtuins, a family of nicotinamide adenine dinucleotide (NAD)-histone deacylases, the activity of which increases with CR. How much each pathway contributes to life extension remains to be determined.

Given the discovery of molecular mechanisms of CR, it is reasonable to look for CR mimetics that would provide all the benefits of CR without the agony of food deprivation. Several have been identified: resveratrol, rampamycin, and metformin. Resveratrol is a polyphenol found in red wine and shown to activate sirtuins in animal models of aging (yeast to rodents) and produce many but not all of the benefits identified with CR. Rapamycin is an immunosuppressant drug that inhibits mTOR. Although it extends the lifespan in mice, it produces unwanted side effects, for example, diabetes. Metformin is currently used in the treatment of type 2 diabetes and is of interest because it activates AMPK.

CR is Analogous to Food Shortage in the "Wild" Organisms throughout the millenniums survived in the presence of environmental stressors such as scarcity of food. It is *hypothesized* that surviving organisms developed a strategy to *more efficiently* metabolize the limited supply of nutrients. Energy for reproduction was diverted into pathways for survival. The sensing of insufficient nutrients initiated the activation of maintenance and protective pathways. *These same survival pathways of metabolic efficiency and cellular protection serve to extend maximal lifespan through CR.*

Caloric Restriction as an Example of Hormesis More than a decade ago, scientists began to consider CR as a phenomenon that elicited a daily mild stress response capable of inducing protective effects. Scientists, such as Masoro (2007) and others, proposed the term *hormesis*. Hormesis is *defined by Webster* as "a theoretical phenomenon of dose–response relationships in which something (as a heavy metal or ionizing radiation) that produces harmful biological effects at moderate to high doses may produce beneficial effects at low doses." CR is deemed a chronic small but beneficial stress that protects animals against other more harmful stresses such as surgery, inflammatory agents, toxic chemicals, and heat stress. In present-day terms, CR is an example of hormesis, as long as it does not morph into malnutrition or starvation. Thus, in the presence of CR, potential life-shortening stresses are overcome due to CR-dependent optimization of repair and maintenance programs, beneficial changes that extend the MLS in animals.

LABORATORY ANIMAL MODELS

Animal Models Are Useful Adjuncts to the Study of the Aging Process

In practical terms, compared to humans, animals are less complex, and have shorter lifespans allowing for repeat studies. Importantly, select animal models display natural survival curves similar to man, in that the mortality rate is higher at older ages. The invertebrate models have been particularly informative. Many genes–proteins relevant to metabolism, stress resistance, and cell death are similar (homologous) to genes–proteins found in man. This conservation allows for cautious application of results from invertebrate studies to man. Additionally, study results from mitotic (dividing) and postmitotic (non dividing) cultured invertebrate cells have helped to identify many cellular activities common to human cells. Finally, the genome (DNA) of the invertebrate models and the mouse are fully delineated and, therefore, may be manipulated (deleted, knockout (KO) or enhanced, knock-in (KI)).

Animal models are appropriate for manipulative studies called *loss of function* (KO) and *gain of function* (KI). KO and KI refer to procedures in which the genome has been selectively altered by the loss (KO) or addition (KI) of a specific gene(s) (Figure 2.1). The expected effect of a KO experiment is the loss of function previously directed by the knocked out gene. Biogerontologists reason that if the KO gene is involved in the acceleration of the aging process, its absence should slow the rate of aging and increase the MLS of the genetically manipulated animal. Alternatively, if the KO gene acts in a way so as to retard aging, its loss would shorten the lifespan of the genetically manipulated animal. However, it is also possible that the KO gene has

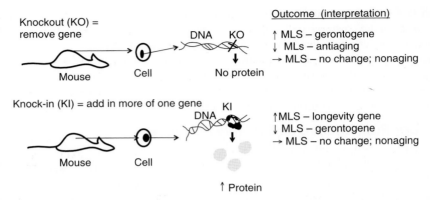

Figure 2.1. Genetic manipulations applied to animal models of aging and possible outcomes on maximal lifespan (MLS).

no influence whatsoever on the aging process and hence MLS of the genetically manipulated animal would be the same as the control (nonmanipulated) animal. The removal of any particular gene is further complicated by the possibility that other genes with similar functions could "take over" for the deleted gene or that the selected KO gene affects important functions unrelated to aging but essential for normal living. Although simple in conception, KO studies are exceedingly difficult to perform. Using this approach, several genes have been identified and labeled *gerontogenes*. *These are genes whose KO confers an increase in longevity*. An example of a gerontogene is Daf-2 (identified in the roundworm) and is determined to be a component of the insulin signaling system.

A second genetic approach is the insertion of a gene (KI or the gain of function). The organism is now fortified with multiple copies of a gene (hence an abundance of protein) directing a certain function. The genetic enhancement creates a functional enhancement. If the KI gene directs an antiaging activity, then mean and MLS should increase. Such genes are called longevity assurance genes. Other outcomes could be a shorter MLS if the KI gene affects proaging activity or no change in MLS if no age-dependent activities are affected. With all experiments using genetic manipulations, the presence or absence of manipulated genes must be confirmed by determination of the presence or absence of the specific gene-related function, for example, the presence or absence of the coded protein.

Yeast: Saccharomyces cerevisiae

Saccharomyces cerevisiae is commonly called Baker's yeast, used as a leavening in baking but not in fermentation of alcohol that requires related yeasts. The entire genome of *S. cerevisiae* has been defined (SGD NIH National Human Genome Research Institute).

Aging in *S. cerevisiae* may be investigated during either the "replicative phase" or the "stationary phase." In the replicative phase, daughter cells form by budding from the mother cell. Aging is defined by the number of buds and rate of their appearance; the senescent phenotype appears as an enlarged cell with numerous bud scars, distorted cell wall, and a slowed budding time. In the stationary phase, aging equates to the amount of time a nonreproducing cell remains viable in culture (essentially its chronological aging). In both phases, *S. cerevisiae* exhibits a species-specific MLS that varies with the yeast strain (subspecies).

The lifespan of yeast (both phases) can be extended by CR. Hundreds of KO and KI manipulations have been conducted on *S. cerevisiae* to identify the genes involved in CR. Two proteins of prominence are Sch9 and RAS; their genetic manipulation suggests they are important in pathways of nutrient control, response to stress, and DNA stability.

The impact of lifespan-regulating genes is evident in the results of a recent experiment. A defective gene that produces a life-shortening condition in man called progeria (see progeria discussion below) was overexpressed (KI) in yeast. The KI-manipulated yeast exhibit signs of accelerated aging compared to nonmanipulated controls. When "progeria" yeast were exposed to CR, the devastating effects of the progeria mutation (life-shortening effect) were prevented. Although not immediately transferable to man, studies of this type create the foundation for future discoveries that have potential to benefit humans.

Roundworm: Caenorhabditis elegans

Caenorhabditis elegans (commonly called *C. elegans*) is a 1 mm long transparent roundworm or nematode. It is free-living and consumes bacteria, for example, *Escherichia coli*. There are two sexes: male and hermaphrodite (containing both sexes and self-fertilizing). *C. elegans* lives 2–3 weeks. The sequence of its genome has been full delineated (*C. elegans* Sequencing Consortium, 1998). All of the cells of *C. elegans* are postmitotic, which means they lack the ability to divide into daughter cells. This model shows classic signs of aging such as loss of muscle mass and abnormal protein metabolism, observations useful in the study of sarcopenia (age-related muscle loss in man).

Under severe environmental conditions, for example, controlled starvation (CR specific to the roundworm), *C. elegans* forms the dauer diapause, a phase where *aging stops*. Organisms continue to exist and return to normal aging when environmental conditions are ameliorated (food is available). Results of studies on this interesting phase have identified genes that modulate aging (gerontogenes, mentioned above). Two important genes that are "turned off" in the dauer diapause are *daf2* and *daf23*. Mutant nematodes that lack these genes (KO) have extended lifespans. The *daf2* gene directs the synthesis of the receptor for insulin/insulin growth factor-1 (IGF-1). Without it, as in the *daf2* KO mutant, no receptors are made and signaling mediated by insulin/IGF-1 is inhibited. Thus, insulin and IGF-1 have no or little effect in the *daf2* KO. Significantly, this leads to an enhancement of the response to stress, beneficial changes in metabolism, depression of growth, and in essence, activation of activities that contribute to a longer lifespan. The *daf2* gene shares a similarity with the insulin/IGF-1 receptor in man. *Daf23* is a protein in the chain of command stimulated by insulin/IGF, so even in the presence of normal IGF receptors, a *daf23* KO (lacking the *daf23* gene) could not respond to IGF and would exhibit an increased response to stress, protective metabolism, reduced growth, and a longer lifespan. Encouraged by these discoveries, work is underway to determine the role of insulin/IGF-1 and the insulin receptor in human aging.

Fruit fly: Drosophila melanogaster

Drosophila melanogaster is a tiny two-winged insect, known as the common fruit fly. At approximately 2 mm in length, it is slightly larger than the *C. elegans* and lives approximately 30 days (if kept at around 84°F). This model has been studied in detail and much is known about its development, reproduction, and aging. The sequence of the genome has been published (Adams et al., 2000).

As with the previous models, *D. melanogaster* has been subjected to a variety of CR protocols and data indicate that mechanisms defined in other models that increase lifespan are also present in *D. melanogaster*. Unfortunately, due to several unresolved technical issues with this model, for example, toxic components in diets of the fruit fly and consequences of mating activity, data from this model regarding CR-related effects have been interpreted cautiously. However, where diet is not relevant, *D. melanogaster* has been of value in studies relating to oxidative stress, organ system aging, and age-related pathologies.

Mouse: Mus musculus

The mouse is a popular laboratory animal model of aging. The mouse is a small mammal, about 3 inches in length with a tail of several inches. Breeding commences at less than 2 months of age. Lifespans average 3–5 years in the laboratory, but are significantly shorter in the wild (~4 months). The mouse genome has been sequenced and is publicly available (www.informatics.jax.org/).

Many studies have documented degenerative changes in several systems of *M. musculus*: skeletal and cardiac muscles, neuroendocrine, and neuronal. CR in the mouse produces an increase in stress resistance and a more efficient cellular metabolism. CR also decreases the incidence of disease, that is, cancers, and increases the lifespan. Some evidence suggests that CR operates by inhibition of the insulin signaling pathway (ISP) in a manner similar to that observed in lower organisms. Not surprisingly, infusion or injection of insulin can reverse the effects of CR. A second pathway, mediated by sirtuin proteins, can be activated by CR and their activation represents a supplemental pathway to enhance stress resistance and optimize nutrient metabolism.

Nonhuman Primate: Macaca mulatta

Macaca mulatta (rhesus monkey) is a nonhuman primate. A study initiated in 1987 placed some rhesus monkeys on CR (30% restriction of calories) and others on an *ad libitum* diet. Data gathered over more than 20 years indicate that CR monkeys are healthier than the non-CR controls. Monkeys on CR exhibit all the favorable changes detailed above relating to blood lipid profiles, cardiovascular and immune function, fasting blood glucose level, oxidative damage, and hormone levels. Additionally, lower levels of inflammatory mediators IL-6 and IL-10 and higher levels of anti-inflammatory interferon-gamma are observed. Reproductive function and locomotor activity remain normal and acoustic responses are enhanced compared to *ad libitum* controls.

The primate CR studies were performed at two sites: NIA and University of Wisconsin (UW) (Wisconsin National Primate Research Center). Data from the NIA showed that CR primates *did not live longer* than controls, that is, mortality rates were similar between CR and control monkeys. This conflicted with *UW findings that showed CR increased survival*. Specifically, only 26% of the CR monkeys died from age-related causes compared to 68% of control monkeys. This statistically significant observation indicated that the death rate throughout in the control group was 2.6 times greater compared to the CR group. Although additional explanations may arise, the current resolution to the discrepancy in survival data from NIA and UW is that *NIA control monkeys were in fact modestly calorically restricted*. Body weight comparisons of NIA with UW control monkeys and NIA controls compared to a database of normal monkeys in captivity of the same gender and age show that NIA controls weighted less. This suggested that the NIA controls were subjected to CR albeit of sufficient degree to modestly extend their lifespans hence negating a statistical difference in mortality rate between them and the CR monkeys.

These results are significant for several reasons. First, the increase in life time survival with CR in the monkey is proof of conservation of CR-mediated mechanisms for life extension. Results of CR-exposed animal models from yeast to rodent and now to monkey suggest that CR is likely to provide benefits in man. Second, CR was initiated in monkeys at age 7–14, unlike rodent studies in which CR is initiated much earlier (after weaning). The increase in survival rate suggests that CR initiated in adulthood is also effective. Third, the explanation for the discordant results generated with the NIA controls implies that even modest CR (less than 30%) has a significant impact on lifespan. Analysis of these crucial studies continues with potentially more insights.

MAN AS MODEL: BALTIMORE LONGITUDINAL STUDY

In 1958 the Baltimore Longitudinal Study on Aging (BLSA) in man was initiated within the gerontology division of the National Heart Institute. It was not until 1974 that the government established the NIA as an entity separate from the National Heart Institute.

The BLSA recruits volunteers to visit NIA every 2 years for a battery of tests (more than 100) and scientists study physiological, biochemical, and other age changes, for example, disease onset and interaction with aging. At present, some 1400 men and women aged 20–90 are enrolled. For more information, visit the Web site: http://www.grc.nia.nih.gov/branches/blsa/blsa.htm.

Biogerontologists of the BLSA have published extensively and have generated a number of significant results (see Table 2.1). The BLSA concluded that aging and disease are distinct processes. Additionally, the BLSA observed that age change in the heart and arteries, for example, cardiac and arterial stiffening, are risk factors for cardiovascular diseases that can be minimized with exercise. Furthermore, medications affecting arterial function, for example, some antihypertensive drugs, could also retard the age-associated decline in arterial function.

As is apparent in Table 2.1, the BLSA evaluates both basic age changes and age-associated pathologies. Ironically, as noted by Hayflick (2007) investigations of age-related pathologies yield no insights into mechanisms of aging; yet studies on aging uncover a plethora of age-related vulnerabilities to disease. Thus it baffles some biogerontologists why research funding for studies of age-related pathologies far exceeds that for research funding to study the basic mechanisms of aging. Additionally, NIA conducts many studies using the cross-sectional study design or a short longitudinal study of two or more cohorts.

TABLE 2.1. Findings from the Baltimore Longitudinal Study of Aging[a]

Heart, arteries
- Cholesterol (low-density lipoprotein (LDL)) is risk factor for cardiovascular disease in men even after age of 75
- Low testosterone predicts arterial artery stiffness
- Moderate alcohol consumption (but not excessive or abstention) associated with minimal arterial stiffness

Cognitive function
- Poor performance on visual recall tests predicts cognitive decline as much as 20 years prior to overt change
- Cognitive aging is complex; vocabulary test scores increase; visual memory declines
- Estrogen replacement therapy preserves cognitive memory in women
- Use of nonsteroidal anti-inflammatory drugs (ibuprofen) reduces risk of Alzheimer's disease
- High testosterone levels are associated with improved cerebral blood flow; may relate to improved mental function in elderly men with higher testosterone level
- Low levels of testosterone and depressive symptoms are predictors of Alzheimer's disease

Personality
- Personality is stable throughout aging
- Adaptation to stress in the elderly is similar or better than in younger individuals
- Personality traits, not situations, determine happiness in aging

Sensory
- Hearing loss at high frequencies (presbycusis) is confirmed
- Hearing loss at low frequencies is also common and accelerates in the eighties
- Decline in taste intensity (especially salt) and perception occurs with age

Diet/metabolism
- Alcohol ingestion: Metabolism of alcohol is not changed; change in body composition is a factor; impact greater on cognitive function and reaction time in elderly
- Body fat relocates from hips/thighs to abdominal area where it poses a risk factor for cardiovascular disease/diabetes; redistribution is prominent in men
- Predictors of diabetes are obesity, accumulation of abdominal fat, and decreased physical fitness

Kidneys
- Slow decline in function from 20 to 80 years is considered a normal age change

Physical exercise
- Reduction in leisure time physical activity in older men is a predictor of increased all-cause mortality

Hormones
- Erythropoietin (hormone produced by kidneys to stimulate red blood cell production in the bone marrow) production increases with age except in those with diabetes/hypertension
- Blood levels of IGF-1 decline with age in men and women

[a] Information found at http://www.blsa.nih.gov/

Another important study relevant to man is the Framingham Heart Study, which began in 1948 with the objective of identifying risk factors for cardiovascular disease. The original cohort (residents of Framingham, MA) included more than 5000 individuals of both sexes, aged 29–62. In 1971 the offspring cohort was recruited consisting of more than 5000 participants of both sexes, aged <10–70 years. Studies with a cohort of the third generation are now in progress (2005) with 4000 participants. Research discoveries (visit http://www.framinghamheartstudy.org) from the Framingham Heart

Study have had a significant impact on reducing cardiovascular disease in our society. Briefly, the Framingham Heart Study identified the following risk factors for cardiovascular disease: cigarette smoking, high cholesterol, high blood pressure, abnormal electrocardiogram, menopause, and psychosocial factors.

Progeroid Syndromes as Premature Aging

Models of accelerated aging in man are represented by the progeroid syndromes of Hutchinson–Gilford syndrome (HG) and Werner's syndrome (Wn). HG and Wn are the infantile and adult forms, respectively, of the genetic disease called progeria. At present, it is unclear whether the accelerated aging displayed by affected individuals is authentic aging that is "sped-up" or just an expression of genetic mutations masking as aging. The *syndromes do not mimic aging in all respects and are termed "segmental."* One of several differences between HG and Wn is the time of onset. Symptoms of HG appears very early in life and death occurs within 20 years. Symptoms of Wn appears somewhat later (adolescence) and affected individuals die at 40–50 years of age. Individuals with HG display skin atrophy and age-related pathologies of hypertension and atherosclerosis. Those with Wn have early onset of cataracts, gray hair, aged skin, joint abnormalities, osteoporosis, thymic atrophy, and age-related pathologies of atherosclerosis, diabetes, and increased susceptibility to some cancers.

HG and Wn are caused by genetic mutations. The genetic defect in HG is in a gene (LMNA) that makes lamin, an essential protein of the cell nucleus. How a lamin deficiency causes the HG syndrome is unknown. The genetic defect *for Wn is a mutation in a WRN gene that produces an enzyme with helicase and endonuclease activity*, activities needed for DNA repair and maintenance. It is postulated that loss of a functional WRN gene allows DNA damage to accumulate. The detrimental effects on cells serve to accelerate development of age-related pathologies.

Evaluation of tissues removed from progeria patients has been helpful in assessing the general role of DNA instability in the aging process, especially as it might relate to abnormal cell division and vulnerability to cardiovascular disease. It is hoped that an understanding of the role played by the helicases and lamins in progeria will shed light on basic mechanisms in aging.

SUMMARY

Biogerontological research faces many hurdles. Humans are heterogeneous in genotype and phenotype; studies with human subjects are expensive; most studies are correlative and do not show cause/effect; animal models of aging are important, but may lack relevance to man.

The most common study designs are the cross-sectional and the longitudinal. The former design infers age changes, is relatively inexpensive, and generally completed within a short period of time. The latter design measures age changes directly, but is expensive and time-intensive.

Animal models of aging are invaluable. Studies with animal models help to unravel the conserved mechanisms that are shared among all organisms and have potential to establish causality to replace statistical correlations.

The BLSA and the Framingham Study are noteworthy longitudinal studies because they have added important insights into aging and cardiovascular disease, respectively. The BLSA defines the aging process as separate from disease pathology; the Framingham Study identifies the risk factors for cardiovascular disease.

CR is an experimentally induced 30% reduction in calorie intake (with adequate essential nutrients) below that required to maintain normal weight. It is analogous to food shortage in the "wild" and is an example of hormesis. Species subjected to long-term CR respond with enhanced physiological health and significant MLS extension. In mammalian models of aging, CR promotes a favorable lipid profile, optimal cardiovascular, immune, and hormonal functions, reduced oxidative damage, delayed onset of age-related pathologies, and decreased mortality rate. CR studies are elucidating new pathways to understand aging.

CRITICAL THINKING

What unique difficulties are encountered in studies of biological aging?

Why are research programs like the BLSA of value?

Are progeroid syndromes convincing models of accelerated aging? What can one learn from these conditions?

Of what value are CR studies? What impact would the discovery of a safe and effective CR mimetic have on our society?

KEY TERMS

Ad libitum term used to describe free access usually to food.

BLSA *B*altimore *L*ongitudinal *S*tudy of *A*ging is an ongoing study funded by National Institute on Aging that obtains physiological and psychological data from volunteers every 2 years and summarizes these findings periodically.

Caenorhabditis elegans scientific name for the roundworm, valuable model of aging.

Caloric restriction (CR) also called dietary restriction (DR) with nutrition. State of reduced consumption of calories (generally 30% decrease compared to normal) for a significant period of time (years).

Cause-and-effect data data relating one variable to another in temporal and related sequence whereby one effect produces another.

Cohort group of similar but not identical individuals, for example, individuals of 20 years of age.

Correlative data data related by statistical or mathematical means; does not demonstrate a cause-and-effect relation.

Cross-sectional study design study design in which a variable is measured in cohorts of different ages. The variable is averaged and the mean is related to changes over time. Age change is inferred.

Dauer diapause a protective phase of the life cycle of the roundworm. Phase is induced by severe environmental conditions. No aging occurs in this phase.

KEY TERMS

Deacylases class of enzymes that hydrolytically (using water) cleaves the functional acyl group. This group is characterized by the formula RCO, where R is variable, C = carbon, and O = oxygen.

Drosophila melanogaster scientific name for the fruit fly, common model of aging.

Enzyme protein with ability to accelerate the rate of a metabolic or biological reaction.

Genotype totality of genes (DNA) in an organism.

Invertebrate organism largest group of organisms; organism without a backbone, for example, yeast, worms, and flies

Kinase general term used to describe enzymes that use ATP or related phosphorylated high-energy compound to facilitate a chemical reaction.

KO/KI animal studies studies in which gene(s) are silenced (knockout (KO)) or additional gene(s) are added (knock-in (KI)) to determine the effect of gene manipulation on function.

Longitudinal study design study design in which a variable is measured repeatedly in the same individual over a set period of time. Age change is measured directly and not inferred.

Mitotic cell cell capable of dividing (proliferating) into two identical daughter cells. Mitotic cells divide repeatedly. Fibroblasts, endothelial cells, and immune cells are examples of mitotic cells.

Mus musculus scientific name for the mouse; common model of aging.

NAD nicotinamide adenine dinucleotide is an important cofactor for enzymatic function.

Phenotype expressed characteristics of an organism, for example, physical characteristics of cells, tissues, and organs; includes all aspects of organism except genotype.

Postmitotic cell cell incapable of dividing; cell is considered terminal and is highly specialized, for example, neurons, heart cells, and skeletal muscle cells.

Progeria a group of diseases that share the characteristics of accelerated aging and early death. Etiology of these syndromes is related to dysfunctional genes.

Saccharomyces cerevisiae scientific name for yeast, a one-celled organism that is a common model of aging.

Sirtuins specific histone deacylase enzymes coded for by the SIRT family of genes; enzymes modify genes to influence expression. Sirtuin activation is one of several proposed mechanisms of CR-induced life extension.

TOR(mTOR) a pivotal enzyme in a signaling cascade involved in nutrient sensing. The name derives from the enzymes sensitivity to rapamycin, an antibiotic and immunosuppressive compound. Rapamycin inhibits TOR, hence target of rapamycin. Inhibition of TOR is considered one of several important pathways for CR-induced life extension.

Transcriptional factors small molecules or enzymes that change DNA structure and as a result turn genes "off or on."

Vertebrate organisms organisms with a backbone or skeletal system, for example, man, mammals, and fish.

BIBLIOGRAPHY

Review

Arking R. 2006. *Biology of Aging: Observations and Principles*, 3rd ed. New York: Oxford University Press.

Fontana L, Partridge L, Longo VD. 2010. Extending healthy life span: from yeast to humans. *Science* **328**(5976):321–326.

Hayflick L. 2007. Entropy explains aging, genetic determinism explains longevity, and undefined terminology explains misunderstanding both. *PLoS. Genet.* **3**(12):e220

Johnson TE. 2008. *Caenorhabditis elegans* 2007: the premier model for the study of aging. *Exp. Gerontol.* **43**(1):1–4.

Masoro EJ. 2007. The role of hormesis in life extension by dietary restriction. *Interdiscip. Top. Gerontol.* **35**:1–17.

McDonald RB, Ramsey JJ. 2010. Honoring Clive McCay and 75 years of calorie restriction research. *J. Nutr.* **140**(7):1205–1210.

Minamino T, Komuro I. 2008. Vascular aging: insights from studies on cellular senescence, stem cell aging and progeroid syndromes. *Nat. Clin. Pract. Cardiovasc. Med.* **5**(10):637–648.

Pereira S, Bourgeois P, Navarro C, Esteves-Vieria V, Cau P, De Sandre-Giovannoli, A, Levy N, 2008. HGPS and related premature aging disorders: from genomic identification to the first therapeutic approaches. *Mech. Ageing Dev.* **129**(7–8):449–459.

Perls T. 1995. The oldest old. *Sci. Am.* **272**(1):70–75.

Piper MD, Partridge L. 2007. Dietary restriction in *Drosophila*: delayed aging or experimental artifact?. *PLoS Genet.* **3**(4):461–466.

Experimental

Adams MD, Celniker SE, Holt RA, Evans CA, Gocayne JD, Amanatides PG, Scherer SE, Li PW, Hoskins RA, Galle RF et al. 2000. The genome sequence of *Drosophila melanogaster*. *Science* **287**(5461):2185–2195.

C. elegans Sequencing Consortium. 1998. Genome sequence of the nematode *C. elegans*: a platform for investigating biology. *Science* **282**(5396):2012–2018.

Colman RJ, Beasley TM, Kemnitz JW, Johnson SC, Weindruch R, Anderson RM. 2014. Caloric restriction reduces age-related and all-cause mortality in rhesus monkeys. *Nat. Commun.* **5**:3557.

Das SK, Gilhooly CH, Golden JK, Pittas A, Fuss PJ, Cheatham RA, Tyler S, McCrory MA, Lichtenstein AH, Dallal GE, Dutta C, Bhapkar M, DeLany JP, Saltzman E, Roberts SB. 2007. Long-term effects of 2 energy-restricted diets differing in glycemic load on dietary adherence, body composition, and metabolism in CALERIE: a one year randomized controlled trial. *Am. J. Clin. Nutr.* **85**(4):1023–1030.

Harrison DE, Strong R, Sharp ZD, Nelson J, Astle C, Flurkey K, Nadon NL, Wilkinson JE, Frenkel K, Carter CS, Pahor M, Javors MA, Fernandez E, Miller RA. 2009. Rapamycin fed late in life extends lifespan in genetically heterogeneous mice. *Nature* **460**(7253):392–395.

Heilbronn LK, de Jonge L, Frisard MI, DeLany JP, Larson-Meyer DE, Rood J, Nguyen T, Martin CK, Volaufova J, Most MM, Greenway FL, Smith SR, Williamson DA, Deutsch WA, Ravussin E, Pennington CALERIE team. 2006. Effect of 6-month calorie restriction on biomarkers of longevity, metabolic adaption, and oxidative stress in overweight individuals: a randomized controlled trial. *JAMA* **295**(13):1539–1548.

Lefevre M, Redman LM, Heilbronn LK, Smith J, Martin CK, Rood JC, Greenway FL, Williamson DA, Smith SR, Ravussin E, Pennington CALERIE team. 2009. Caloric restriction alone and with exercise improves CVD risk in healthy non-obese individuals. *Atherosclerosis* **203**(1):206–213.

Mattison JA, Roth GS, Beasley TM, Tilmont EM, Handy AH, Herbert RL, Longo DL, Allison DB, Young JE, Bryant M, Barnard D, Ward WF, Qi W, Ingram DK, de Cabo R. 2012. Impact of caloric restriction on health and survival in rhesus monkeys from the NIA study. *Nature* **489**(7415):318–321.

McCay CM. 1947. Effect of restricted feeding upon aging and chronic diseases in rats and dogs. *Am. J. Public Health* **37**(5):521–528.

McCay CM, Crowell MF, Maynard LA. 1989. The effect of retarded growth upon the length of lifespan and upon the ultimate body size. 1935. *Nutr.* **5**(3):155–171.

3

EVOLUTIONARY THEORIES OF AGING

HISTORICAL VIEWS AND INSIGHTS

Unsupportable Programmed Aging Is Replaced by Evolutionary Tenets

The eminent scientist August Weismann (1834–1914) proposed the first evolutionary explanation of aging. He suggested that the answer to why organisms age could be attributed to a *programmed phenomenon* that served to reduce functionality and promote increased mortality of older individuals, and thereby facilitate the survival of the younger (reproductively fit) individuals. Thus, mortality of older organisms assured the survival of the species by limiting population size and preventing wasteful consumption of scarce resources by older and generally weaker organisms.

Biogerontologists were uneasy with the idea of programmed aging for several reasons. First, it implied the existence of specific "aging" genes whose expression would aid and/or accelerate biological decline, more or less along the lines of a biological clock that might limit the number of times a cell could divide before dying. It is cautioned at this point not to confuse "aging" genes noted here with gerontogenes introduced in Chapter 1. The latter are genes that drive the mechanisms of metabolism and whose absence, as in genetic deletion experiments in animal models of aging, increases lifespan. Their potentially "deleterious" effects are intertwined with environmental stimuli. They do not constitute deliberate programmed aging.

Human Biological Aging: From Macromolecules to Organ Systems, First Edition. Glenda Bilder.
© 2016 John Wiley & Sons, Inc. Published 2016 by John Wiley & Sons, Inc.

A second reason that generated discomfort with programmed aging was the inability to conceive how a detrimental program could be selected for in the wild. *Aged organisms rarely exist in the wild*. Factors such as predation, infection, harsh weather conditions, and food shortage assure a very short lifespan, early death and no aging. Thus evolutionary development of a program to reduce functionality (aging) is unlikely since its expression would be made impossible in the wild and clearly discordant with the evolution of "beneficial" traits. Since aging represents loss of function, few organisms would live long enough to pass "aging traits" to the next generation *even if* genetically possible. Other than in captivity (the so-called protected environment), aging is *not observed in the wild* and according to evolutionary authorities, Kirkwood and Austad (2000) could not have been selected for or against.

For years aging was considered the "unsolved problem of biology." The significance of aging or why we age remained elusive until the work of the British scientist, JBS Haldane. According to a historical perspective by Rose et al. (2008), Haldane in his studies on Huntington's disease (HD), a progressive degenerative skeletal muscle disorder, suggested that *genes expressed in late life or after reproductive* decline would not experience natural selection pressures and even though detrimental, would remain in the gene pool (reference to the totality of genes in a population of the same species). To Haldane, HD was an example of the *lessening of natural selection pressure*. This seminal concept was applied to senescence in general and further developed by others into the *evolutionary theory of aging, considered the premier theory of aging*.

In addition to JBS Haldane (1892–1964), contributors to the present-day evolutionary theory of aging are distinguished scientists such as Peter Medawar (1915–1987), George C. Williams (1926), William Hamilton (1936–2000), Brian Charlesworth (1945), and Michael R. Rose (1955). These scientists gathered data and developed insights that explained aging within the basic tenets of evolution proposed by Charles Darwin (1859; 1871).

Darwin's Evolutionary Tenets

Natural Selection Favors Survival Traits The theory of evolution proposed by Charles Darwin in his writings (*Origin of the Species by Natural Selection*, 1859; *Descent of Man*, 1871) provides the basis for the present-day theory of evolution that unites a wealth of biological observations. The theory proposes the *slow* modification of a species from a common ancestor. Darwin described this as "descent with modification." Darwin concluded that biological characteristics that sustain the survival of an organism in a *particular environment* would contribute to fitness as defined in terms of reproductive output or number of offspring (also called *fecundity or fertility*) and hence would have a high probability of passage to the next generation. Conversely, it is reasoned that biological characteristics detrimental to survival (fitness) would reduce reproductive success and have a reduced chance of passage to the next generation. This mechanism is called *natural selection* and describes a possible means for the development of an appropriate biological trait in a population in a select environment where the trait has survival value. It is also realized that natural selection or evolution of characteristic traits in a population or group of organisms occurred over an *extremely* long period of time.

Genes and Evolution

Genes (DNA Sequences) Possess the Hereditary Information That Is Passed from Generation to Generation through the Germline (Gametes) Discoveries from the disciplines of molecular biology and genetics provide the molecular mechanisms for evolutionary theory. The generational passage of survival traits is accomplished by the passage of *genes*. In essence, the *gene is the unit of hereditary information that is passed from one generation to the next through the germline* (gametes or in the case of humans, ovum and sperm). In the union of gametes that yields an offspring, each parent donates a *copy* or *allele* of each and every gene. As a result, cells of the offspring (alternatively called *somatic cells*) contain two alleles for each gene, one allele from each parent. Somatic cells *or soma* (all cells in an organism *except* the ovum and sperm) contain at least two alleles per gene (diploid number) and therefore differ from gametes (ovum and sperm) that contain only one allele of each gene (haploid number).

The sum total of genes of an organism is the genotype. As one might expect, the genotype (since it contains all the genes) produces the various behavioral, physiological, and biochemical characteristics of an organism. This composite expression previously introduced in Chapter 1 is designated the *phenotype* (*observable and unobservable traits*). Obviously, genotypes of members of one species (organisms that can mate and produce a fertile offspring) are more similar than genotypes from different species. Yet within a species there is an apparent variation in phenotypes. This is the result of *allelic diversity* in which multiple copies of a particular gene exist within the genotype or gene pool. The variation in human eye color (blue, violet, green, brown, and black) is a relevant example of allelic diversity.

In molecular terms, genes are segments of *deoxyribonucleic acid* (DNA). The *base sequence* of the DNA in each gene provides the information (code) to make one unique protein. Proteins are incredibly important and immensely diverse. Proteins function as enzymes, hormones, structural components, receptors, channels, and mediators that build, maintain, and regulate the organism. In this way the genotype (DNA) determines the phenotype (protein structure and function).

Evolved Traits Arise through Genetic Variations It is now evident that over the early history of a species, the evolved traits that supported high levels of fitness resulted from *variations* in the genes in a particular population (referred to as variations in the gene pool). Data show that genetic variations may arise through several pathways: (i) *genetic mutation*, (ii) *genetic drift*, and (iii) *genetic migration*.

In general, genetic variations occur by chance mutation of DNA. Mutations arise as a consequence of DNA structural errors or mistakes generated during the exceptionally complex process of DNA replication or as an outcome of overt environmental damage, for example, radiation, oxidants, and toxins. Note that only mutations in the DNA of gametes or germline are inherited, and passed to the next generation; mutations in somatic cell DNA are *not* passed to the next generation (Figure 3.1).

Additionally, genetic drift is a factor that produces variations in the genes. As with genetic mutation, genetic drift occurs by chance and refers to the *randomness of*

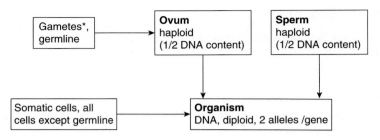

Figure 3.1. Basic concepts: germline and somatic cells with genetic content.

allele selection during chromosomal sorting, a process that occurs during meiosis whereby the *diploid* number of chromosomes is reduced by half, to the *haploid* number in the gametes. This randomness of allele selection is especially significant in populations indiscriminately decreased by environmental catastrophes and is thus considered an important mechanism for genetic variation.

Another source of germline DNA variants arises from genetic migration. This occurs in a population as a result of *migration* of individual organisms into and out of that population. The migratory flux of organisms is facilitated by effective or open means of mobility. Many times, physical barriers, man-made or natural, prevent this.

Germline DNA variants exert one of several terminal effects: (i) neutral, (ii) detrimental, or (iii) beneficial effect. If neutral, variants exist *without effects*. If detrimental, variants are unlikely to contribute to fitness or reproductive success and hence have a *low probability of surviving to the next generation*. However, the genetic variant that is both *beneficial and favored by the environment* has the greatest likelihood of passage to the offspring. In sum, the establishment of survival traits results from a change in the gene pool due to *chance variation in the genes* (Figure 3.2).

Several lines of evidence support biological evolution. There are data from artificial selection experiments in the laboratory, observations on evolution in the wild with environmental constraints, and inferences from fossil history of transitions.

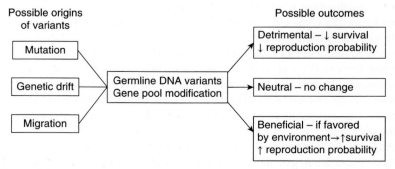

Figure 3.2. Basic concepts: influences and outcomes of gene pool modification.

CONTEMPORARY EVOLUTIONARY THEORIES: DISPOSABLE SOMA THEORY (DST), ANTAGONISTIC PLEIOTROPY THEORY (APT), AND MUTATION ACCUMULATION THEORY (MAT)

Aging Is a Side Effect of Evolution

The three evolutionary theories described below support the view that aging is *not* an adaptive process but, as summarized by Partridge (2010), basically a *side effect of evolution*. Aging is a polygenic (involving many genes) trait; importantly it evolved through a *reduction in selection pressure* following reproduction. The lessening pressure comes about because few organisms reach older age (obvious in those living unprotected in the wild). *Aging as a side effect depends on the beneficial genetic variants that bolster fitness and reproduction.* According to Kirkwood and Austad (2000) and Partridge (2010), the degree to which a species' evolved maintenance and repair systems effectively counteract intrinsic and extrinsic damage and stress during the post reproductive period *directly* influences the lifespan of its members. The conclusion that selection pressure wanes after reproduction has been mathematically supported in several analyses by Hamilton (1966) and Charlesworth and Partridge (1997).

The evolutionary theories of DST, APT, and MAT are not equal with regard to their level of satisfying proof. The most comprehensive and supportable of the three evolutionary theories of aging is the DST. The additional two theories, APT and MAT, at present are modestly convincing. They, however, provide testable biological mechanisms complementary to the DST (Kirkwood and Austad, 2000). APT and MAT are discussed first.

Antagonistic Pleiotropy Theory

Genes that Benefit Fitness in the Young Become Deleterious in the Aged

The antagonistic pleiotropy theory (APT) proposes that *genes that contribute to aging survive into the next generation only if they exert beneficial functions in young organisms*. The genes of the APT are "antagonistic" because although such genes produce beneficial effects in organisms up to and during reproduction, these *same* genes produce negative (antagonistic) effects after the reproductive phase and therefore contribute to senescence. The term pleiotropy emphasizes the multifunctional aspect of these genes, ones that support fitness in the young and others that contribute, albeit indirectly, to deterioration in old age. This concept was developed by Williams (1957) and labeled as antagonistic pleiotropy much later by Rose (1982). It was reasoned that beneficial genes remain in the gene pool even if these same genes subsequently express deleterious effects postreproduction because after reproduction, there is *relatively less or no selection pressure to remove them*. APT is popularly called the "pay later" theory.

The number of examples of potential APT genes or gene programs has steadily increased since the concept was first introduced. A general example is that of the development of elaborate feathers in male birds, for example, peacock. The larger and more colorful the plumage, the more desirable the peacock to the pea hen. Since size and color indicate fitness, reproduction is assured. However, this benefit poses a marked hindrance postreproduction in that larger feathers are heavier and thus dramatically reduce the bird's rate of escape from predators. Another example is apoptosis. Apoptosis is the process that regulates cell death (frequently termed "cell suicide"). Apoptosis is essential in development and growth (remodeling of tissues) and in suppression of tumor growth in the young organism. In the older organism, apoptosis is notably a disposal program of damaged cells with success dependent on an adequate supply of replacement cells, for example, stem cells. However, stem cell replacement markedly declines with age and consequently apoptosis of irreplaceable cells such as the skeletal muscle fibers or cardiac cells leads to reduced tissue size and limited function. Apoptosis benefits young organisms through prevention of cancers and remodeling of tissues but in older organisms apoptosis promotes atrophy and reduced function.

The transcriptional factor, NF-kappa beta (NF-κβ), may be another example in support of APT. NF-κβ is an important proinflammatory inducer that acts at the gene level to initiate a cascade of events culminating in production of cytokines and adhesion molecules that support inflammation. Novel computational methods by Adler et al. (2007, 2008) have identified NF-κβ as a prime initiator of organ dysfunction and the age-associated onset of inflammatory-based diseases such as atherosclerosis and Alzheimer's disease. In contrast to its role in aging, in the young organism, NF-κβ-directed activities are now considered indispensable in fighting infections. This dual function suggests that NF-κβ is an antagonistic pleiotropic factor that is beneficial to the young but detrimental to the older organism.

Mutation Accumulation

Genes Expressed Late in Life Remain in the Gene Pool and May Be Deleterious
Although originated by Haldane in 1932 as discussed above, the mutation accumulation theory (MAT) per se is usually credited to Medawar (1957). Medawar reasoned that since natural selection pressures abate following reproduction, *late-acting alleles originating from a germline mutation will remain in the gene pool with no means of elimination*. Lethal mutations limit the lifespan through mechanisms of gradual deterioration and death.

There are several examples that endorse the MAT. One is the gene mutation that causes HD. This neurodegenerative disease produces a progressive (10–20-year trajectory) deterioration in motor, cognitive, and emotional functions. Classic symptoms are chorea (abnormal involuntary spastic muscle movement), dementia and depression. HD patients possess a mutated HD gene that produces an altered protein. The altered protein through a variety of mechanisms slowly poisons cells of the brain (Huntington's Disease Collaborative Research Group, 1993). Several other neurodegenerative disorders such as spinal and bulbar muscular atrophy, and the spinocerebellar ataxias exhibit mutations similar to the mutated HD gene. Since the onset of symptoms of HD and similar diseases occurs in the fourth to sixth decade of life, its persistence in the gene pool is explained by a reduction in selection pressure at later ages. Thus, according to the MAT, deleterious mutations accumulate passively and cannot be shed since their expression occurs in late life.

One prediction from this theory is that the lifespan of an offspring (progeny) should not be linearly related to the lifespan of the parents. Instead, if late-acting mutations accounted for the age at death, the relation between progeny and parental lifespan would be nonlinear, but with an increasing slope for this relationship in longer lived parents (Ljubuncic and Reznick, 2009). According to Gavrilov and Gavrilova (1998), analysis of the lifespans of European royal and noble families, considered a source of reliable genealogical data, supports this prediction.

Disposable Soma Theory

Evolutionary Life History of a Species Determines Degree of Investment in Germline (Reproductive Success) and in Soma Maintenance (Longevity)

The DST is framed in terms of the germline (gametes) and the soma (all other cells) of an organism. In the wild, the soma is disposable. Clearly, it is all about reproductive success so the soma must be of lesser importance than the germline. If the germline cannot continue, what good is the soma. Therefore, it is reasoned that when extrinsic *mortality is elevated in the wild*, life is short. In these conditions, the force of natural selection favors reproductive achievement (more offspring and assurance that offspring will reproduce). Accordingly, as more resources are devoted to fecundity less are available for general soma maintenance. In species in which nearly all of the resources are devoted to fecundity, there will be few mechanisms for maintenance/repair/replacement, and so environmental and internal damage will go unheeded and accumulate. The unavoidable and eventual outcome is cell/tissue/organ deterioration, for example, rapid aging and death of the organism, albeit for the species, high reproductive success. On the other hand, in an environmental niche that favors a *low extrinsic mortality*, there is less pressure for rapid reproduction, thereby permitting the species to choose resources that favor soma maintenance, repair, and replacement and in effect, a means to counteract intrinsic and extrinsic damages and hence experience a longer lifespan.

In sum, the DST proposes that aging resulted in part from "trade-offs" between reproductive success and longevity during the early life history of a species. In adaptation to the demands of their ecological niche, organisms survived based on pressure to select between resources for storage, growth (and metabolism), and maintenance/repair on the one hand and resources for reproduction on the other (Figure 3.3).

DST Predictions: Relation of Fecundity and Longevity; Relation of Longevity and Maintenance Mechanisms

Several predictions have been derived from the DST. First, the theory predicts that a change in natural selection with a greater or lesser pressure should change both fecundity and longevity as described

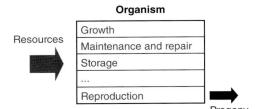

Figure 3.3. Choices that organisms faced according to Darwin's natural selection tenets. (Reproduced with permission from Kirkwood (2005).)(See plate section for color version.)

DST–Extremes force trade-offs	
Hostile environment • Mortality rate high • Animals reproduce in large numbers • High fecundity • Short life span Example: Common meadow mole Life span ~10 months with three litters/every 2 months)	*Friendly environment* • Mortality rate low • No need to expend energy on reproduction • Low fecundity (fewer offspring) • Longer life span Example: Humans Life span >100 years; 1–12 offspring/30 years of fecundity

Figure 3.4. Evolutionary characteristic of the disposable soma theory (DST).

above. Reports show that laboratory manipulations in fruit flies that artificially delay the onset of reproduction simultaneously increased longevity; conversely, laboratory-manipulated fruit flies with increased lifespans also exhibit reduced fecundity. Additional data show that long-term observations on the same species in two different environments reveal that those in a high predation environment demonstrate high fecundity and reduced longevity (Figure 3.4). Interestingly, organisms that have evolved unique structures (e.g., wings in birds, shells in turtles, and highly developed brain in man) to reduce predation thereby increasing survival are relatively long-lived and concomitantly less fertile.

The DST makes a second prediction. It states that organisms with longer lifespans should express an abundance of robust maintenance and repair mechanisms compared to organisms with shorter lifespans. In essence, the extent of maintenance and repair programs should be proportional to the lifespan. Evidence in support of this prediction has focused on the oxidative damage molecules, reactive oxygen species (ROS). ROS are unquestionably injurious molecules that initiate cell and tissue damage and the ability of an organism to prevent ROS-induced damage would be an important indicator of the available maintenance and repair mechanisms. Data show that (i) organisms modified in the laboratory for increased longevity express vigorous maintenance and repair mechanisms that diminish the presence of ROS and (ii) the ROS levels in long-lived rodent species are significantly lower compared to levels found in short-lived rodents.

DST Explains the Lifespan Extension Effects of Caloric Restriction as an Evolutionary Conserved Adaptation to Food Shortage

The effects of caloric restriction (CR) (see Chapter 2) have been analyzed in evolutionary terms and are considered an *adaptation* to food shortage, a common occurrence in the wild. The molecular pathways activated through CR in laboratory animals are those that would support survival during extended periods of starvation in the wild. It is now known that activation of "nutrient sensing" mechanisms during CR maximizes metabolic efficiency, an effect that fosters survival until food becomes available. It is concluded that CR produces an *elevated state of stress-induced resistance without reproduction*. With restoration of sufficient nourishment, reproduction resumes and the CR-activated pathways deactivate. Extensive data in support of this conclusion have been gathered from results of CR studies with yeast, worms, fruit flies, rodents and monkeys emphasizing its evolutionary conservation. The ability to adapt to a changing environment, for example, availability of food is considered a reasonable component of "evolutionary optimization" (Kirkwood, 2005). Many genes currently termed longevity determinant genes are influenced by CR and importantly they are genes involved in one or more aspects of maintenance and repair.

The DST Applies Only to Species that Age and Reproduce Sexually

There are several constraints on the DST. DST applies only to populations that age. When age classes are nonexistent as in some bacteria, organisms show no signs of senescence. Second, the theory relates to organisms that express a distinction between the germline and the soma.

Although controversial, organisms such as the *Hydra* that mostly reproduce by asexual budding, for example, also appear not to age, apparently due to a high content of stem cells. The *Hydra* does not reproduce sexually (it has no germline) but interestingly artificial elevation of expression of a specific longevity gene (FOXO3A) causes the *Hydra* cells to convert to germline-like cells; conversely reduction of gene expression of FOXO3A induces conversion of cells to somatic cells that subsequently senescence. These results suggest a role for this particular gene in stem cell evolutionary history and emphasize the need to unravel the evolutionary significance and relation of germline selection to aging.

SUMMARY

Biogerontologists reject the proposal that aging and death are determined by a genetic program. In terms of Darwinian fitness, it is unrealistic to select for detrimental traits. Furthermore, most animals in the wild never "age" due to predation, disease, and so on, and hence the evolutionary selection of a gene program that produces deterioration and death is unlikely.

Biogerontologists consider that aging evolved through a reduction in selection pressure that occurs following reproduction. Aging is basically a side effect that depends on the beneficial genetic variants that bolster fitness and reproduction. The variants are those involved in DNA repair and soma maintenance that contribute to fitness. The most widely supported explanation is set out in the disposable soma theory that proposes that organisms throughout their life histories made "trade-offs" between reproductive success and longevity and thus allotted resources to reproduction, maintenance, and general housekeeping, according to the environmental pressures at the time. Organisms that invested in high fecundity invested less in maintenance and experience short lifespans. The opposite is true for animals with low fecundity (as man). Thus, investment in maintenance mechanisms in man was significantly high and man has the potential to live to over 100 years.

Predictions from the evolutionary theory of aging are supported by data showing a relation between (i) fecundity and lifespan in laboratory-manipulated organisms, (ii) lifespan length and robustness of maintenance mechanisms, and (iii) fecundity and lifespan in the same species living in different environments.

CRITICAL THINKING

What is meant by natural selection?

Why is it important to understand the origin of an idea?

Which of the three variations on the evolutionary theory of aging is the most convincing? Why?

How do evolutionary biologists currently explain why humans age?

KEY TERMS

Allele a viable DNA sequence (gene or non-gene); individual inherits set of alleles from each parent.

Diploid number of chromosomes in somatic cells, $N = 46$, that is, normal complement in man, one half (haploid) from father and one half (haploid) from mother.

Evolution an explanation for the change in the hereditary characteristics of groups of organisms over the course of generations (Darwin referred to this process as "descent with modification").

Fecundity potential reproductive capacity of an organism or population; fertility.

Gametes generic or category name for the ovum and sperm.

Gene hereditary material passed to offspring. Each gene contains information in the DNA to code for a single protein.

Genetic drift random selection of alleles during meiosis (preparation of haploid number of chromosomes in gametes).

Genetic mutation a random change in the DNA (gene) caused by a replication error or environmental insult; if occurrence is in the gametes, and passed to the offspring, resulting effects may be neutral, beneficial, or detrimental.

Gene pool totality of all the genes of a species or population.

Germline cells with haploid number (half) of chromosomes that combine to form the embryo. In humans, germline cells are the ovum and the sperm.

Haploid number of chromosomes in the gametes (sex cells) that is 23 in humans.

Meiosis cell division of germline cells whereby cells (sex cells, gametes, or also called sperm and ovum) achieve the haploid (reduced) number of chromosomes.

Natural selection greater reproductive success among particular members of a species arising from genetically determined characteristics that confer an advantage in a particular environment.

Pleiotropic effect of a single gene to produce many phenotypic (physically expressed) traits.

Population group of like organisms.

Somatic line all cells in body except ovum and sperm; mutations in somatic cells cannot be passed to next generation.

Species in general, a group of organisms that can potentially breed with each other to produce fertile offspring and cannot breed with the members of other intra-breeding groups.

Variation genetically determined differences in the characteristics of members of the same species.

BIBLIOGRAPHY

Review

Charlesworth B, Partridge L. 1997. Ageing: levelling of the grim reaper. *Curr. Biol.* **7**(7): R440–R442.

Gavrilov LA, Gavrilova NS. 1998. The future of long life. *Science* **281**(5383):1611–1612.

Gould SJ. 1994. Evolution as fact and theory. *Hen's Teeth and Horse's Toes: Further Reflections in Natural History*. New York: W.W. Norton & Company, pp 253–262.

Haldane JBS. 1941. *New Paths in Genetics*. London: Allen & Unwin.

Hamilton WD. 1966. The molding of senescence by natural selection. *J. Theor. Biol.* **12**(1):12–45.

Kirkwood TB. 1977. Evolution of ageing. *Nature* **270**(5635):301–304.

Kirkwood TBL. 2005. Understanding the odd science of aging. *Cell* **120**(4):437–447.

Kirkwood TBL, Austad SN. 2000. Why do we age? *Nature* **408**(6809):233–238.

Kirkwood TB, Holliday R. 1979. The evolution of ageing and longevity. *Proc. R Soc. Lond. B Biol. Sci.* **205**(1161):531–546.

Ljubuncic P, Reznick AZ. 2009. The evolutionary theories of aging revisited-a mini review. *Gerontology* **55**(2):205–216.

Medawar PB. 1957. *Uniqueness of the Individual*. New York: Basic Books.

Nebel A, Bosch TC. 2012. Evolution of human longevity: lessons from *Hydra*. *Aging (Albany NY)* **4**(11):730–731.

Partridge L. 2010. The new biology of ageing. *Philos. Trans. R Soc. Lond. B Biol. Sci.* **365**(1537):147–154.

Roos R AC. 2010. Huntington's disease: a clinical review. *Orphanet. J. Rare Dis.* **5**(1):40–47.

Rose MR. 1982. Antagonistic pleiotropy, dominance, and genetic variation. *Heredity* **48**(1):63–78.

Rose MR, Burke MK, Shahrestani P, Meuller LD. 2008. Evolution of aging since Darwin. *J. Genetics* **87**(4):363–371.

Rose MR, Graves JL. 1989. What evolutionary biology can do for gerontology. *J. Gerontol.* **44**(2):B27–B29.

Walker D. 2006. *Inheritance and Evolution*. Minnesota: Smart Apple Media.

Williams GC. 1957. Pleiotropy, natural selection and the evolution of senescence. *Evolution* **11**(4):398–411.

Experimental

Adler AS, Kawahara T LA, Segal E, Chang HY. 2008. Reversal of aging by NF-κβ blockade. *Cell Cycle* **7**(5):556–559.

Adler AS, Sinha S, Kawahara TLA, Zhang JY, Segal E, Chang HY. 2007. Motif module map reveals enforcement of aging by continual NF-κβ activity. *Gene Dev.* **21**(24): 3244–3257.

Charlesworth B. 2001. Patterns of age-specific means and genetic variances of mortality rates predicted by the mutation-accumulation theory of aging. *J. Theor. Biol.* **210**(1):47–65

Huntington's Disease Collaborative Research Group. 1993. A novel gene containing a trinucleotide repeat that is expanded and unstable on Huntington's disease chromosomes. *Cell* **72**(6):971–983.

SECTION II

BASIC COMPONENTS

INTRODUCTION TO MACROMOLECULES AND CELLS

Loss of Molecular Fidelity Is the Essence of Aging

This overview of biological principles relevant to molecular and cellular aging is presented to help the reader understand aging in these domains. This is important because changes at the macromolecular and cellular levels strongly influence functionality of everything else, tissues, organs, organism. The biological hierarchy is, therefore, incredibly important. Additionally, it is necessary to keep in mind the enduring relation of structure to function and the essential role of electrons in shaping and reshaping cellular components. These principles are emphasized in a definition of aging originated by Hayflick (2004) (and reviewed in Chapter 1) that aging is a loss of molecular fidelity that exceeds repair.

Biological Organization of the Organism Begins with Atoms That Combine to Form Molecules and More Complex Structures: Macromolecules, Cells, Tissues, Organs, and Organ Systems

Biologically important atoms and molecules shape a specific structural hierarchy that defines the basic unit of life, the cell. Cells are organized entities of unique, complex macromolecules created by specific arrangements of molecules originating from a collection of select atoms. Atoms are elements comprised of a nucleus of protons and neutrons surrounded by electrons circulating in a particular space or orbit.

Human Biological Aging: From Macromolecules to Organ Systems, First Edition. Glenda Bilder.
© 2016 John Wiley & Sons, Inc. Published 2016 by John Wiley & Sons, Inc.

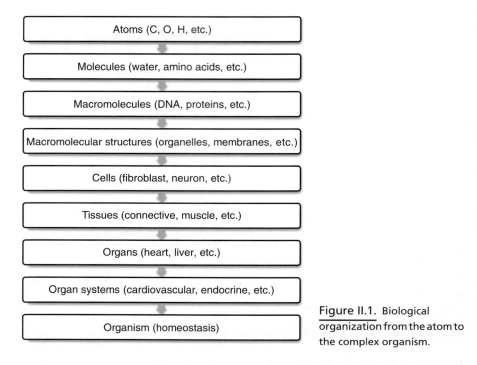

Figure II.1. Biological organization from the atom to the complex organism.

Cells exist individually, for example, amoeba, or as sophisticated arrangements of cooperating cells, for example, muscles. Multicellular organisms survive as an interactive hierarchy of tissues, organs, and systems. In this assembly, cells operate collectively as tissues; tissues unite to form organs; organs interact as organ systems that ensure physiological normalcy called *homeostasis* (Figure II.1).

Biologically Important Atoms

Biological ubiquitous atoms are (i) hydrogen (H), (ii) oxygen (O), (iii) carbon (C), (iv) nitrogen (N) (v) sulfur (S), and (vi) phosphorus (P). Atoms form molecules. The *precise arrangement of the specific atoms* of C, H, N, O, S, and P in each molecule *determines the uniqueness in structure and function* of that molecule.

Atoms of each molecule are *held together by the sharing of electrons of the atoms*. The electron sharing creates a significant attachment between two atoms called the *covalent bond*. Covalent bonds provide the "glue" for molecules and more complex molecular arrangements and at the same time, permit adaptability through bond formation, breakage and reformation. *Unregulated and permanent disruption or irregular formation of these bonds alters structure and function of molecules, a key characteristic of aging.*

Key Molecules Are Amino Acids, Fatty Acids, Sugars, Bases, Water, and Phosphates

Key biological molecules are (i) water, (ii) fatty acids, (iii) sugars, (iv) amino acids, (v) nitrogen bases, and (vi) phosphate.

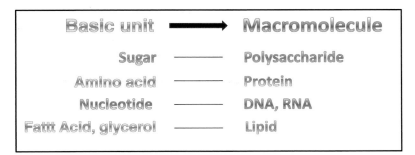

Figure II.2. Building blocks of four basic macromolecules.

Water, composed of oxygen and hydrogen atoms (H_2O), provides nearly three-quarters of the mass of the cell. Most other cellular molecules contain carbon and diverse combinations of aforementioned atoms.

Major Macromolecules Are Proteins, Lipids, Polysaccharides, and Nucleic Acids

Macromolecules formed from molecules are the second most abundant constituents of the cell and include (i) proteins, (ii) polysaccharides, (iii) nucleic acids, and (iv) lipids.

Macromolecules are made from molecular "building blocks" (Figure II.2). In particular, amino acids covalently bond in a linear fashion to form proteins; similarly, nucleotides bond to form strands of nucleic acids; sugars bond to form branched chains of polysaccharides; and fatty acids bond to alcohol to form lipids. Macromolecules combine further to form higher order biological structures, for example, cellular membranes, tubules, protein-producing platforms, and cell attachment structures. Many of these structures use biological forces (hydrogen bonds; van der Waals and electrostatic forces) in addition to the covalent bond to hold them together.

Macromolecules Are Constantly Formed (Biosynthesized) and Broken Down (Degraded)

Macromolecules may be chemically broken apart or degraded into their subunits in a process termed *catabolism*. *Anabolism* is the reverse process in which molecules are biosynthesized or joined together to form new macromolecules. *Catabolism and anabolism are the central components of cell metabolism.* Cell metabolism is a necessary cell function that utilizes oxygen to obtain energy from ingested food to fuel activities of growth, reproduction, and maintenance. This is significant because *cell metabolism generates a plethora of oxidants*, substances that may alter and harm macromolecules in diverse ways (Chapter 4). Influential oxidants are superoxide anion, hydrogen peroxide, and nitric oxide.

The Three-Dimensional Structure of Macromolecules Determines Function; Altered Structure Produces Reduced or Absent Function

Biological compounds occupy space and project a three-dimensional (3D) structure that is unique to each type of molecule and macromolecule. Macromolecular

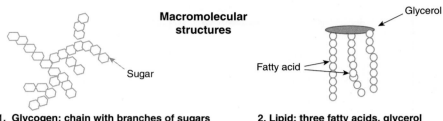
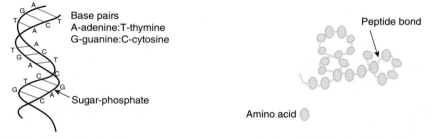

Figure II.3. Representative structures of macromolecules. (See plate section for color version.)

structures are distinct ranging from globular, branched, coiled, flat, chain-like, lattice-like, and/or one of a myriad of other 3D shapes (Figure II.3). *Structure is critical to normal biological function because structure dictates function.* Structure affects the manner in which two or more molecules (macromolecules) interact, a factor that determines the kinds, quantities, and speeds of *all* chemical reactions within an organism. Unwanted changes in structure seriously impact and limit biological function. *Structural changes brought about by the aging process, and in particular those that arise from oxidative stress (Chapter 4), seriously modify macromolecular structure and contribute to cellular dysfunction.*

The Cell Is the Smallest Enclosed Unit of Living Matter

The boundary of a cell is defined by the plasma membrane, an outermost lipid bilayer crowded with an array of integral proteins and glycoproteins (protein with attached sugar(s)). The plasma membrane is complex and possesses domains of lipids and proteins, with and without attachment to internal protein fibers, termed the cyto-skeleton. Many of the proteins in the lipid bilayer are semimobile and act as pores, channels, anchors, and receptors.

Within the cell milieu or cytoplasm, there exist numerous distinct but interdependent compartments called organelles. Predominant cell organelles include (i) the nucleus, (ii) mitochondria, (iii) the endoplasmic reticulum and associated Golgi apparatus, (iv) lysosomes, and (v) peroxisomes. The organelles are illustrated in Figure II.4.

INTRODUCTION TO MACROMOLECULES AND CELLS

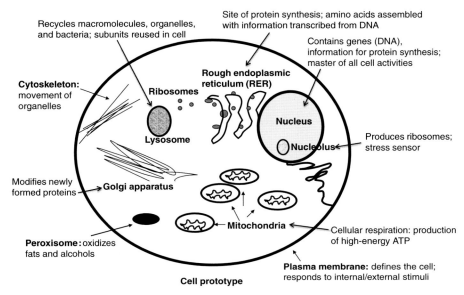

Figure II.4. Cell prototype with organelles and associated functions.

4

AGING OF MACROMOLECULES

INTRODUCTION TO OXIDATIVE STRESS HYPOTHESES

There are two main oxidative stress hypotheses: (i) The free radical theory presently referred to as the oxidative stress hypothesis and (ii) the redox stress hypothesis. Of the two, the former is the oldest, most widely known, most diversely supported (by different models of aging), and also most strongly criticized. The latter is too recent to bear any of these distinctions.

Both *hypotheses underscore the deleterious effects of uncontrolled oxidative chemical reactions (oxidative stress) as the best explanation for the process of aging.* However, they differ significantly with regard to the role assigned to the oxidant, generally termed reactive oxygen species (ROS). For the free radical theory (oxidative stress hypothesis), *ROS are viewed as damaging agents* that destroy the structure of key macromolecules, for example, proteins, lipids, and DNA. Although the plethora of antioxidative mechanisms serve to prevent these attacks, they too eventually become oxidatively damaged and fail. The result is buildup of unrepaired structures that function poorly and gradually limit cell and organ activities. Thus, *ROS cause structural damage that subsequently produces dysfunction.* In contrast, the redox hypothesis proposes that *ROS serve as obligatory second messengers and regulate numerous key functions in cells.* The presence of compounds called redox pairs ensure that as proteins are activated (oxidized) by ROS, redox pairs are nearby to reduce them back to a resting state, ready for another messenger. However, over time the cell becomes more oxidized, a change that overwhelms the redox pairs and retains the signalling proteins in a state of lasting oxidation; hence, cell signaling is sporadically and eventually permanently inhibited. *The redox hypothesis emphasizes protein*

Human Biological Aging: From Macromolecules to Organ Systems, First Edition. Glenda Bilder.
© 2016 John Wiley & Sons, Inc. Published 2016 by John Wiley & Sons, Inc.

dysfunction as the prime initiator of aging due to progressive loss of regenerative capacity from the redox pairs. Structural damage is secondary.

The distinction between the two hypotheses is important because it defines the appropriate intervention to prolong the health span and possibly treat oxidative-dependent diseases. For example, the oxidative stress hypothesis targets optimization of antioxidative enzymes and structural repair systems (and use of antioxidative supplements). This hypothesis measures oxidized macromolecules and relates their levels to the lifespan of various animal models. In contrast, the redox hypothesis emphasizes the importance of redox pairs, for example, glutathione and measures the oxidized state of these pairs. It also looks for protein dysregulation through ROS mediators. Neither hypothesis has been fully explored in mammals, for example, mouse, and none in man. Additional data are essential to favor one over the other.

To understand these hypotheses, it is important to learn about oxidation/reduction reactions and how macromolecules become perturbed over time.

Oxidation/Reduction Principles

Transfer of One or More Electrons between Molecules is Essential for Oxidation and Reduction Reactions

Oxidation and reduction reactions are ubiquitous chemical reactions. The oxidation of iron with the appearance of rust is an observable example. This reaction is characterized by the *transfer of one or more electrons* from one molecule or atom to another molecule or atom, in this case between iron and oxygen. Using traditional terminology, during oxidation, a molecule or an atom loses one or more electrons and is considered oxidized. During the reverse reaction of reduction, a molecule or an atom gains one or more electrons and thus is characterized as reduced (Figure 4.1). *Reduction and oxidation (or **redox**) reactions occur close together in time and space, generally simultaneously.* In the cell, the reaction rate (how fast things change) depends on the affinity (attraction) of the participating atoms for electrons and on the energy change (driving force) that occurs in the redox reaction.

Free Radicals Initiate Damage Because They are Highly Reactive Particles with an Unpaired Electron

Free radicals are by-products of

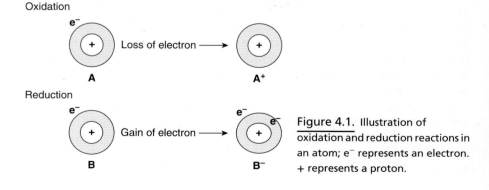

Figure 4.1. Illustration of oxidation and reduction reactions in an atom; e^- represents an electron. + represents a proton.

chemical reactions that occur during physiological and pathological activities (within and outside of cells). Free radicals are *ions, atoms, or molecules that contain an unpaired electron* (either an extra one or a deficiency of one). In this state, free radicals seek to partner with a neighboring atom. This action is indiscriminate and the free radical "attack" (stealing an electron or giving up an electron) has potential to change macromolecular structure and subsequent function. Free radicals *strike adjacent molecules within a narrow radius* (act locally) and often generate a series of intermediate "reactive" compounds that set up a *chain reaction*. Free radicals and chain reactions are *terminated by the activities of antioxidative enzymes or radical scavenging compounds*.

Oxygen is a strong oxidant. It is not a free radical because it has no unpaired electrons, but it has the capacity to withdraw (steal) electrons from nearby molecules or atoms, thereby oxidizing and modifying their atomic structure. However, this seldom occurs due to the position of the free electrons in oxygen and the other unusual characteristics of oxygen. Hence, oxygen is slow acting and requires assistance from enzymes and metals. In cells, and in particular within the subcellular compartments such as the mitochondria (mt) in the presence of a transition metal, for example, iron, oxygen is deliberately, and in a stepwise fashion, reduced to water and carbon dioxide. The reduction of oxygen in a controlled environment (within the cell) generates an abundance of usable energy. However, *during this process, an oxygen radical called superoxide anion is often formed*. Superoxide anion is oxygen that has obtained an additional electron (expressed as $O_2^{-\bullet}$), and although not particularly "reactive" except with iron and copper, it is *abundant and combines with other radicals to produce more harmful products*. Examples of other free radicals or fragments formed as by-products of redox reactions are the hydroxyl radical ($\bullet OH$), the alkoxyl or carbonyl radical ($RO\bullet$) where R represents one or more carbon atoms, and the nitric oxide radical ($NO\bullet$). Their reactivity or propensity to disrupt nearby macromolecules varies from the *most toxic hydroxyl radical to least harmful nitric oxide radical*.

Non-Radical Oxidants are Strong Oxidants with Paired Electrons; They May Act as Signal Molecules and as Mediators of the Oxidative State of the Cell

$O_2^{-\bullet}$ readily reacts with water in the presence of the enzyme superoxide dismutase (SOD) to form *hydrogen peroxide (H_2O_2), a non-radical oxidant* that has no unpaired electrons. H_2O_2 is a potent oxidant and as such may oxidatively (withdrawal electrons) alter structure and function of macromolecules. H_2O_2 is formed enzymatically throughout the cell (mitochondria (mt) and other subcellular compartments) and in the plasma membrane. H_2O_2 *is critically important*, not only for the potential damage it can cause to macromolecules but also because *it readily oxidizes thiol and similar groups (disulfide; sulfur–oxygen combinations) on proteins*, an effect that more recently has categorized it as a *signal molecule*. This is because *protein thiols and variations thereof act as a protein "switch" to regulate many physiological, membrane, and biochemical functions* that include metabolism, gene expression, and immune response. Maintenance of H_2O_2 at homeostatic (normal) levels is a good thing for optimal cellular function. However, a persistent elevation of this oxidant (and others) will produce a steady and persistent oxidation of redox-sensitive protein thiols that subsequently disrupts normal cellular activities. Some important free radicals and non-radical oxidants are given in Table 4.1.

TABLE 4.1. Components of oxidative stress theories: radicals, oxidants, antioxidants, redox pairs, biomarkers of oxidation

Species	Symbol/abbreviation	Comments
Radical		
Superoxide	$O_2^{-}\bullet$	Abundant production during oxidation of nutrients in mitochondria
Hydroxyl	$\bullet OH$	Potent free radical
Nitric oxide	$NO\bullet$	Weak free radical
Non Radical		
Hydrogen peroxide	H_2O_2	Abundant oxidant; formed through SOD action on superoxide
Peroxynitrite	$NO_3\bullet$	Oxidant resulting from combination of superoxide and nitric oxide
Lipid hydroperoxide	Lipid-OOH	Oxidation of carbons of double bond (PUFA)
Antioxidant Enzymes		
Superoxide dismutase	SOD	Converts superoxide to hydrogen peroxide and oxygen
Catalase	CAT	Converts hydrogen peroxide into oxygen and water
Glutathione peroxidase	GPx	Same action as catalase; also converts lipid hydrogen peroxides to alcohols
Redox couples		
Glutathione	GSSG/2GSH	Keeps thiols in reduced state; prevents cross linkage
Nicotinamide adenine dinucleotide; Nicotinamide adenine dinucleotide phosphate	$NAD^+/NADH$; $NADP^+/NADPH$	Contributes to reduced cell environment
Thioredoxin	$TrxSS/Trx(SH)^2$	Keeps thiols in reduced state
Ascorbic Acid	L-ascorbate/dihydroascorbic acid (vitamin C)	Scavenger of free radicals
Markers of Oxidation		
Ethane, pentane	C_5H_{12}	Volatile compounds (measured in exhaled air); Markers of PUFA peroxidation
Pentosidine	Glycation of protein; AGE	Fluorescent product of sugar reacting with lysine/arginine of protein
8-oxodG	Oxidation of guanine base of DNA	Assayed in urine/serum/cells for evidence of oxidative stress

TABLE 4.1. (Continued)

Species	Symbol/abbreviation	Comments
Lipofuscin	Oxidized lipid and protein degradation products	"age pigment"; accumulates within lysosomes of cell; histologic fluorescent detection

OXIDATIVE STRESS

Oxidative Stress Represents the Measureable Increase in Radical and Non-Radical Oxidants in an Organism The term reactive oxygen species refers to *both free radical and non-radical oxidants*. Oxidation reactions also occur in the presence of oxidants that contain nitrogen (reactive nitrogen species (RNS), for example, nitric oxide) and carbon (reactive carbonyl compounds (RCCs), for example, aldehydes). All three species of oxidants contribute to the phenomenon of oxidative stress.

Oxidative stress is an increase in measureable oxidant activity in an organism due to generation of ROS, RNS, and/or RCC and encompasses transient and chronic oxidative events in cells. With current methodologies, oxidative stress indirectly predicts the "health" of the cell or an organ. Oxidative stress represents the imbalance between oxidant production and "requisite" suppression by antioxidants, redox pairs, and repair mechanisms. Clearly, as oxidative stress increases, the potential for macromolecular damage also rises and, as more recently recognized, may additionally perturb cell signaling, both of which *if not repaired or reversed*, may lead to abnormal changes indicative of aging.

The extent of oxidative stress is estimated from (i) cellular or extracellular concentration of oxidized macromolecules or (ii) calculation of the overall oxidative state of a cell. The former quantifies oxidized proteins, for example, pentosidine, oxidized DNA, for example, 8-oxo-2'-deoxyguanosine, and/or oxidized lipids, for example, hydroxynonenal (see Table 4.1). Measurement of redox pairs (generally, the amount of reduced glutathione) is needed to calculate the redox potential or the oxidation state of a cell. Since *oxidative stress increases with age* as evidenced by its detection in isolated tissues and cells from aged organisms, including man, it is important to know how this comes about and how it might be constrained.

Sources of Oxidative Stress

Oxidative Stress Arises from External and Internal Sources under Controlled and Uncontrolled Conditions ROS (and RNS and RCC) develop in two ways. They are produced as (i) unwanted by-products of various metabolic reactions, either enzymatic or nonenzymatic (spontaneous), and as (ii) important cell signaling mediators under normal conditions.

Multiple processes located in membranes or subcellular compartments as well as outside the cell potentially contribute to oxidative stress. Inspiratory oxygen (oxygen in the air) as described above is the main source of cellular ROS. One to three percent of inspired air gives rise to superoxide anion, a ROS that is generated as a by-product of cellular respiration in the mt of the cell.

Oxidants generated by immune cells serve to destroy invading microorganisms. On encountering bacteria, immune cells discharge oxygen that engenders large quantities of ROS. Phagocytic (debris engulfing) cells called macrophages generate superoxide that is subsequently converted to more reactive oxidants by enzymes found in nearby immune cells (neutrophils and eosinophils). Immune cell-generated oxidants eliminate a range of microorganisms, but if unrestrained, these same products have the propensity to induce inflammatory tissue destruction.

Several organs, the liver and kidney, generate a variety of oxidants in the normal course of the metabolism of drugs and other foreign substances. The responsible enzymes are the cytochrome P450 reductases (CYP enzymes). Activity of CYPs decline after the age of 70, a change that impacts the selection of drug dosage in the elder (see Chapter 11). However, the contribution of the P450 enzymes to oxidative events and aging remains to be defined.

ROS may be produced in a regulated fashion in response to physiological stimuli. *ROS then serve as "second messengers" and transmit cellular signals to nearby molecules and macromolecules to perform cellular activities.* Hydrogen peroxide stimulates cell growth and differentiation (changes that lead to a specific cell phenotype), assists with cell division, and influences gene expression. It is also a potential inducer of the inflammatory mediator, NF-κβ, whose activation sustains the inflammatory state. Another important second messenger is the RNS, nitric oxide. Nitric oxide regulates the formation (biogenesis) of new mitochondria and also the rate of mitochondrial energy production. Therefore, *these oxidants act as important signal molecules as well as potentially harmful agents.*

Outside the cell, free radicals and non-radical oxidants are readily generated by UV light, X-rays, excessive heat, pollution (ozone), and toxins, for example, asbestos and cigarette smoke. Many oxidants, including superoxide, have been identified in these extracellular processes, although the formation of most oxidants is complex and poorly characterized (Figure 4.2).

Targets of Oxidative Stress: Nucleic Acids, Proteins, and Lipids

Nucleic Acids as Oxidative Targets
There are two types of nucleic acids: DNA (deoxyribonucleic acid) and RNA (ribonucleic acid). DNA contains the "genetic blueprint" for each cell. It is located primarily in the nucleus of the cell, although a small amount resides in the mt. DNA contains the information to produce all of the proteins needed by the organism for survival, growth, and reproduction. During the synthesis of each protein, *each gene (linear sequence of bases) is transcribed from DNA to RNA. RNA translates the information to correctly biosynthesize one protein.* Once formed, proteins function as *enzymes* (facilitators of metabolism) and thus are instrumental in the formation (and breakdown) of other macromolecules, for example, lipids and polysaccharides. In sum, *genes (DNA sequences) relay their information (via RNA) such that proteins are made.* Proteins

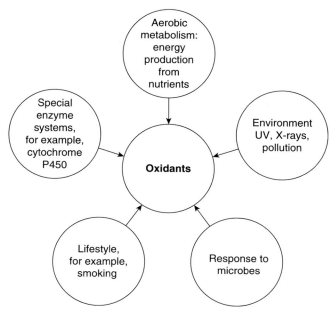

Figure 4.2. Internal and external sources of oxidants.

then function as enzymes, hormones, receptors, and structural components and hence regulate metabolism and cell activities and influence all other cellular functions. *Disruption of the base sequence or base pairing by oxidative events has devastating consequences* that are discussed below.

Box 4.1. More About Nucleic Acids

The *nucleotide* is the basic unit of nucleic acids. Nucleotides are comprised of three smaller units: a nitrogen (N)-containing base (compounds with a double ring), a 5-carbon sugar, and a phosphate group. Nucleotides assembled for DNA contain one of four specific bases: adenine (A), guanine (G), thymine (T), and cytosine (C), and a deoxyribose sugar (lacking an oxygen molecule). Nucleotides assembled for RNA contain the same N-containing bases as in DNA *except* that the base, thymine, is replaced with the N-containing base uracil (U) and that the deoxyribose sugar is replaced with a ribose sugar that has a full complement of oxygen molecules. Nucleotides are *assembled linearly into strands* or polymers: two for DNA; one for RNAs.

To *transcribe a gene* (to achieve protein production), the tightly coiled DNA (chromosomes) is *unwound and straightened out.* The DNA strand reveals a *linear sequence of bases* (the code) that *dictates the final amino acid (AA) sequence of each protein.* In particular, the base sequence of an active gene is *transcribed* into RNA by the *Rule of Complementary Base Pairing* (G pairs with C; T pairs with A). Similarly, the RNAs translate (assemble) the protein using the Rule of Complementary Base Pairing. Messenger RNA contains the transcribed gene that with the help of transfer RNA and ribosomal RNA translate the protein.

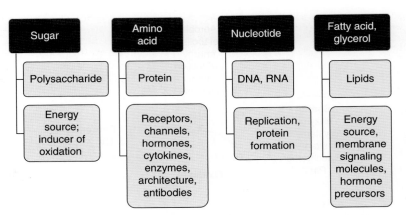

Figure 4.3. Summary of macromolecules: building blocks and functions.

DNA is also involved in cell division and its complex replication is a source of errors that can contribute to cellular dysfunction. This is addressed in Chapter 5. See Figure 4.3 for summary of nucleic acid function.

Oxidation of Nucleic Acids Cause Major Detrimental Effects on Gene Expression and Cell Division

DNA may be damaged by environmental insults (UV radiation, X-rays, and genotoxic agents), by inherent instability of DNA in the form of replication errors and spontaneous chemical disintegrations during cell division, and by oxidative attacks from ROS, RNS, and RCC. The *guanine* base of DNA is especially susceptible to damage and may be oxidized to 8-oxo-7,8-dihydro-2′deoxyguanosine (8-oxodG). This means that at a specific location in this base (carbon 8 on the small ring of this two-ring compound), there is now an oxygen in place of a hydrogen. This relatively minor change leads to significant outcomes such as mutations or cell death.

Oxidative damage alters DNA structure in several different ways: (i) Breaks one or both strands of DNA, (ii) cross-links strands, (iii) modifies the bases (as described above), and/or (iv) modifies the nucleotide prior to incorporation into DNA.

The consequences of DNA damage are *variable but generally serious. The extent of DNA damage determines the response of the cell*. For example, if DNA damage is severe, factors are released that immediately start the cell on its way to cell disappearance, a process called apoptosis. If DNA damage is less severe, DNA repair may be possible. However, if no repair occurs, DNA damage remains to possibly contribute to cancer formation, gradual cell instability, and/or age changes.

To prevent a negative outcome, cells utilize an extensive repertoire of mechanisms that, according to Garinis (2008), continually "scan, detect and repair DNA injuries." The nucleotide excision repair (NER) and base excision repair (BER) systems are examples of two repair mechanisms that focus on DNA (discussed below as oxidative stress countermeasures). Data from studies with normal and transgenic mouse models (expressing inserted genes from another species) and human leukocytes show that *DNA base damage increases with age simultaneously with an age-related decline in BER* in various tissues and cells. Furthermore, age-related BER decline is reversed with a caloric restriction (CR) protocol in the mouse model. Interestingly, individuals with *accelerated aging syndromes*, for example, progeroid Werner's or Cockayne syndrome, *have deficient DNA repair systems* (Chapter 2). New transgenic models with partial deletion of BER are in progress and could define the role of DNA repair mechanisms as major players in the determination of longevity.

RNAs are also potential candidates for oxidation. There is evidence to show that in human cells, oxidation of the messenger RNA, the RNA carrying the code for the amino acid sequence, results in production of an aberrant protein. Additionally, in human neurons (autopsy samples), oxidized RNA (detection of 8-hydroxyguanosine) is elevated in the elderly and is markedly increased in patients diagnosed with cognitive impairment or Alzheimer's disease.

Oxidative Target: Proteins

Protein Activity is Diverse, Essential to Normal Cell Function, and Tightly Regulated Proteins perform *many different* functions. They act as hormones to modulate growth, metabolism, and reproduction; as cytokines (small molecules) to influence immune and inflammatory processes; as enzymes to facilitate anabolic or catabolic reactions; as antibodies to participate in immune reactions against pathogens and tumors; as structural components to provide strength, elasticity, and support; as receptors with to regulate signaling (direction-providing) pathways in cells; as ion channels to regulate nerve, muscle, and secretory activity; and as components of the basic motor apparatus of muscles or motile cells. See Figure 4.3 for summary of protein function.

Table 4.2(a) lists some proteins with known susceptibility to oxidation. Note that as protein structure changes with oxidation, so does function, for example, decline in connective tissue strength and elasticity and altered metabolism.

Box 4.2. More About Amino Acids

Twenty chemically different amino acids are aligned linearly in different arrays to create the plethora of diverse proteins for the organism. The atomic composition of each AA dictates its biological essence, including three-dimensional (3D) structure, reactivity (interaction with and ability to bond to other biological molecules), and physical/chemical characteristics such as solubility in body fluids, for example, water and fats. Many AAs are biosynthesized in the cells of the body and are called *nonessential* AAs. Others must be obtained from the diet and are called *essential* AAs.

Amino acids are biochemically joined in a linear fashion to form peptides (2–50 AAs) or proteins (typically hundreds of amino acids). Peptides and proteins are biosynthesized on complex macromolecular structures called ribosomes in the process of translation, noted above. Once formed, proteins fold into *a 3D* structure dictated by the lowest energy requirement of the linear sequence of its AAs and the formation of the bond attractions among the various AAs.

TABLE 4.2. Proteins Affected by Oxidative Stress

a. Damage through oxidative stress

Collagen

- Most abundant protein of connective tissue; provides platform for cells, strength and structure to tissues
- Target of oxidative stress and subsequent cross-linkage causes loss of strength and support

Elastin

- Structural proteins with "elastic" properties
- Susceptible to cross-linkage; biosynthesis declines; give rise to wrinkles and sags

Hemoglobin

- Oxygen carrier in the red blood cell (RBC)
- Glycation target, accelerated in diabetes, changes structure and oxygen affinity

Apoprotein B

- Protein found in low-density lipoprotein (LDL); molecule that transports fats to cells
- Oxidized lipids in LDL generate reactive particles (RCCs) that attack and alter apoprotein B — essential first step in atherosclerosis

Aconitase

- Widely distributed enzyme involved in nutrient and iron metabolism
- Enzyme activity inhibited by ROS

Thrombomodulin

- Acts in concert with other proteins to promote blood clotting
- Oxidation of methionine reduces coagulation in cell cultures

Cytochrome c oxidase

- Enzyme in electron transport chain of mt
- Structural change with oxidation

Lysozyme

- Mucosal secretory antibacterial enzyme — non specific immune response
- Oxidation reduces enzyme activity

b. Altered signaling through oxidative stress

15-lipoxygenase

- Generates bioactive lipids
- Function inactivated with oxidation

K channels

- Voltage-gated channelthat opens/closes in response to voltage change
- Oxidation slows inactivation rate; Msr increases inactivation rate

Phosphatase PTP1B

- Regenerates the insulin receptor
- Oxidation prevents insulin signalling (utilization of sugars, AAs)

Insulin receptor

- Stimulation by insulin enhances glucose and AAs uptake in muscle and liver
- Redox regulated cysteines influence function

N-Methyl-D-aspartate brain receptor

- Responds to the neurotransmitter, glutamate; postulated involvement in memory
- Redox sensitive

Protein Glycation Produces Cross-Linkage; Oxidation Disrupts Enzyme Activity A reliable indication of protein oxidation is the detection of pentosidine. Proteins that are spontaneously oxidized by sugars usually a ribose undergo a glycoxidation or glycation. This results in a covalent attachment of a portion of the sugar to lysine or arginine to form an adduct termed an *advanced glycation end product (AGE)* that may additionally link with other proteins. AGEs are detected by measurement of the protein–ribose fragment with classical biochemical methods. Glycation is enhanced in states of persistent hyperglycemia evident in poorly controlled diabetes. In this disease, the hemoglobin protein is one target of glycation and its oxidation is monitored (clinical HbA1c assay) to assess the state of glucose control.

Over the long term, *AGEs promote an unwanted joining or cross-linkage between proteins and in this way permanently change protein structure* and diminish normal function. AGEs found on collagen in connective tissue contribute to *stiffness and reduced compliance* of organs such as in the heart and blood vessels. Furthermore, the collagen scaffolding (attachment and resting place) for cells that facilitate cell–cell interactions and responses to external stimuli are reduced in the presence of abnormally cross-linked collagen. Since AGE-containing collagens are resistant to normal degradation, they accumulate and eventually dilute the pool of normal collagen, further diminishing collagen-dependent functions. AGEs may also be consumed in the diet, for example in foods exposed to high heat and once absorbed, have potential to induce inflammation.

Oxidants not only generate protein cross-linkages but also prevent enzyme functions. An example of ROS inactivation of protein function is that of *oxidation of aconitase*, a cytoplasmic and mitochondrial enzyme with a dual function: converts citric acid to isocitrate in the metabolism of nutrients and acts to optimize iron levels in the body. Aconitase appears to be reversibly inhibited by superoxide anion. As studied in bacterial and human cells, the antioxidation enzyme, superoxide dismutase, reactivates aconitase and restores function. However, continuous oxidation permanently inhibits aconitase function.

The AA *methionine in proteins is a prime target of oxidative attack because it contains a sulfur-hydrogen (thiol) group*. The methionine sulfoxide reductase (Msr) system (discussed below) is devoted to preventing its oxidation. However, methionine sulfoxidation has been documented to increase with age in man. Why the Msr fails or why these oxidized proteins are resistant to degradation (and so accumulate) is unknown.

Direct Damage or Loss of Cell Signaling Deprives the Cell of Important Protein-Dependent Activities Clearly, there is evidence that oxidative modification of proteins changes structure and function. However, *oxidation of cysteine thiols may not necessarily lead to inactivity* and loss of function. Recent evidence suggests that redox reactions regulate (activate or depress) specific protein functions through a thiol "switch" *as long as the overall state of oxidation in the cell is within the homeostatic range*. However, continuous oxidative stress as in aging interrupts this regulation and leads to irreversible bonding of proteins with glutathione and other redox compounds (see below). Cell signaling fails as a consequence. Examples of proteins regulated through the redox thiols are given in Table 4.2(b).

Lipids as Oxidative Targets Fatty acids (FAs) are vital subunits of most cellular lipids. FAs are long chains of carbon (C) and hydrogen (H) molecules with carboxylic acid (COOH) at one end. Unlike AAs and nucleotides, *FAs do not contain nitrogen*. An important characteristic of FAs is the degree of *saturation*. Saturation is a measure of the extent to which *H atoms are bound to C atoms* in the fatty "acyl" chain. *The greater the number of H atoms, the greater will be the degree of saturation*. Fatty acids with a lesser number of H atoms per C atom are less saturated or unsaturated and possess a chemical arrangement marked by a "double bond," a feature that limits carbon–carbon bond rotation and introduces a "kink" into the chain at the location of the double bond(s). Several examples of unsaturated fatty acids, also termed monounsaturated (containing one double bond), are oleic acid (number of carbons:number of double bonds, 18:1)), or polyunsaturated (containing at least two double bonds), are linoleic acid (18:2) and alpha-linolenic acid (18:3). The latter two are *essential* polyunsaturated fatty acids (PUFAs) that must be obtained through the diet. Linoleic and alpha-linolenic acids are known as omega-6 and omega-3 fatty acids, respectively.

> **Box 4.3. More About Fatty Acids and Lipids**
> In the formation of a lipid, three fatty acids are combined with glycerol (3-carbon alcohol) to form a triglyceride. *Stored triglycerides represent an energy reserve* that is accessed by specific enzymes that cleave fatty acids from their storage sites. In addition, the accumulation of triglycerides as body fat effectively insulates the body from heat loss.
>
> There exist two other principal lipids: phospholipids and cholesterol. Phospholipids are of substantial importance because they are major components of all cellular membranes. The phospholipids differ in structure from triglycerides in that *only two fatty acid chains are attached to the glycerol molecule* and in place of the third fatty acid, there is a single phosphate molecule and a small organic molecule, such as choline. This arrangement allows for water solubility (phosphate and carbon end) at one end and fat solubility at the other end (two fatty acids) that facilitates the *assemble of phospholipids into lipid films or membranes*.
>
> Cholesterol and similar lipids are sterols that share some of the same physical and chemical characteristics of glycerides. However, they are distinct in that sterols are structures of four large carbon rings with a hydroxyl group (−OH) and an acyl (hydrocarbon) side chain. This contrasts with fatty acids that are long chains of hydrogenated carbons.

Three fatty acids combine with an alcohol, glycerol, to form a triglyceride, a well-known source of storage energy. *Phospholipids differ in that they contain only two fatty acids and in place of the third fatty acid, a water-soluble carbon phosphate compound* (see Box 4.3 for further description). Because of the ubiquitous presence of phospholipids in all cell membranes, their role in aging is of interest. See Figure 4.3 for summary of lipid function.

The *degree to which phospholipids pack together in the membrane influences membrane characteristics* such as "fluidity" and "leakiness." A "fluid" membrane allows integral components (pores, ion channels, and receptors) to move freely within the membrane. A reduction in fluidity is thought to adversely affect cell function by slowing or preventing necessary activities. However, if the membrane composition permits *too much movement, a greater metabolic cost is exacted to maintain membrane gradients*. This is thought to be significant and is based on the observations that the *composition of the membrane acyl chains correlates with longevity* such that the higher the basal metabolic rate (dependent in part on membrane leakiness), the shorter the lifespan.

Spontaneous Oxidation of Membrane Unsaturated Fatty Acids Produces a Variety of Toxic Compounds; Enzymatically Controlled Oxidations Yield Important Signaling Molecules Oxidation of phospholipids is complex. Reactive intermediaries may be generated enzymatically or nonenzymatically and, more importantly, some intermediaries are destructive and others participate in cellular activities. Identification of the multitude of oxidized phospholipids is a rapidly developing area of study aided with mass spectrophotometry, an analytical biochemical methodology.

Four classes of oxidized phospholipids have been identified: mildly oxygenated; oxidatively truncated; those with cyclized acyl chains; and oxidatively N-modified phospholipids. It is mostly the nonenzymatic (spontaneous) pathway of lipid oxidation or peroxidation that are considered important in aging. *Mono- and polyunsaturated fatty acids are targets of oxidation* because they have *one or more double bonds that are susceptible to ROS and RNS attacks*. These attacks generate an abundance of reactive products ranging from lipid peroxides to hydroxyl radicals, hydrocarbon radicals, and aldehydes. Two products in particular, hydroxynonenal (HNE) and hydroxyhexanol (HHE), are cytotoxic. Other oxidation products induce cell death, damage mitochondria and other subcellular structures, and initiate mutations in genes associated with cancer formation. *Reactive intermediaries can migrate to distant sites* with potential to injure an array of other lipid molecules by the formation of *advanced lipid end products (ALEs)*. ALEs, similar to AGEs, produce cross-linkages between proteins or DNA. On the other hand, many oxidized phospholipids, for example, those generated from arachidonic acid by the enzyme, 15-lipoxgenase, participate in cell signaling, platelet aggregation, and regulation of immune responses, to name but a few. Nevertheless, some metabolites in this pathway are also harmful proinflammatory mediators.

Saccharides as Oxidative Initiators

The sugars (disaccharides) and carbohydrates that humans consume are catabolized (degraded) into simple sugars and used immediately as energy or stored as glycogen

> **Box 4.4. More About Sugars and Polysaccharides**
> Simple sugars are molecules comprised of carbon, hydrogen, and oxygen atoms that form small open or closed rings of 3–8 carbons. Three biologically important sugars, also called monosaccharides, are glucose, fructose, and galactose. Within cells of the liver and skeletal muscle, glucose molecules are covalently linked to form glycogen (polysaccharide), the storage form of glucose. As energy is needed, glucose is released from the glycogen by enzymatic cleavage. Organisms obtain energy from various *dietary forms of sugars*, which are often present as disaccharides (combination of two sugars, for example, sucrose or table sugar) or carbohydrates, for example, starch.

in the liver and muscles. Sugars also function as unique recognition sites when covalently bound to proteins (glycoproteins) or lipids (glycolipids). These complex macromolecules are generally found on the plasma membrane.

Elevated Levels of Sugars Pose Serious Oxidation Threat Unlike the other major macromolecules, *polysaccharides are not targets of oxidative attack.* Instead, they (as sugars) do the "attacking." It is the monosaccharides at high concentrations that initiate oxidative changes by generating AGEs with proteins such as hemoglobin, collagen, and others. The presence of AGEs eventually leads to cross-linkage.

Countermeasures

Maintenance Mechanisms that Suppress Oxidative Stress: Antioxidant Enzymes; Redox Pairs; NER/BER System; Msr System; Cell Organelles Cellular protective responses to the onslaught of oxidants are abundant and include (i) antioxidation enzymes: superoxide dismutase, catalase, glutathione peroxidase; (ii) redox pairs of glutathione (GSH/GSSG), thioredoxin (TrxSS/Trx $(SH)^2$), nicotinamide dinucleotide phosphate (NADPH/NADP$^+$), nicotinamide dinucleotide (NAD/NAD$^+$), alpha-tocopherol (vitamin E) and ascorbic acid (vitamin C), and their associated regenerating (reducing) enzyme systems (peroxidases/reductases); (iii) methionine sulfoxide reductase system (Msr); (iv) DNA repair enzymes, base excision repair (BER), and nucleotide excision repair (NER); and (v) coordinated cell pathways devoted to repair of damage, for example, the lysosomal removal of oxidized molecules (Figure 4.4).

In general, *oxidants in the cell are converted to less reactive compounds* thanks to the above mechanisms and if possible damaged macromolecules are repaired or eliminated through cellular degradation pathways (Chapter 5).

SOD/Catalase/Glutathione Peroxidase are Antioxidant Enzymes that Convert Oxidants to Less Reactive Species SOD is an important antioxidant enzyme because of its ability to convert the free radical, superoxide, to the non-radical H_2O_2. In addition to its location in the mt, SOD subtypes exist in the cell cytoplasm and also outside the cell. Other key antioxidant enzymes are catalase and glutathione peroxidase that enzymatically convert H_2O_2 to water. Catalase is located in the mt and elsewhere (cell organelles, for example, peroxisomes). Glutathione peroxidase, an intra- and extracellular enzyme, acts to reduce oxidized lipids (peroxides) as well as H_2O_2.

Glutathione and Thioredoxin are Redox Pairs that Shuttle Two Electrons to Prevent Persistent Oxidation of Thiol and Similar Groups Two redox pairs of major importance are glutathione and thioredoxin. *Redox pairs exist in two forms, reduced and oxidized, and serve to keep the sulfhydryl groups of proteins in a reduced state.* Oxidized glutathione is regenerated by the enzyme, glutathione reductase.

Glutathione is the most influential redox pair in the cell and thus *determines in large part the oxidative state (also termed redox potential) of the cell.* Its abundance and high capacity to share electrons (in this case, two electrons) ensure its prominence.

Thioredoxin is distributed throughout the cell and regulates redox changes generated by ROS and RNS. Thioredoxin is also tasked with the preservation of a *reduced* cellular environment and additionally plays a role in the prevention of apoptosis. Similar to glutathione, thioredoxin must be regenerated by a reductase enzyme, in this case thioredoxin reductase.

Other significant redox pairs are alpha-tocopherol (vitamin E) and ascorbic acid (vitamin C). The former scavenges radicals (combines with radicals to reduce damage) among the membrane lipids; the latter distributes in the cytoplasm and acts there. Both may be regenerated with glutathione and thioredoxin pairs.

Msr System is a Selective System that Prevents Oxidation of Thiol Groups on the Amino Acid Methionine Whereas glutathione keeps the thiol groups of the amino acid cysteine in the "healthier" state of reduction, a special system additionally serves to reverse the oxidation of the ($-SH$) groups of methionine. This was mentioned earlier in relation to structural damage produced by oxidative stress on methionine and the use of this system to reverse it. This special mechanism is actuated by methionine sulfoxide reductases, collectively called methionine sulfoxide reducing (Msr) system. The Msr system, in reducing methionine, becomes oxidized itself and requires reduction and regeneration by "helper" reducing cofactor compounds such as thionien and selenium.

NER and BER Systems Protect Nucleic Acids from Oxidation DNA damage occurs at the rate of 100,000 oxidative events per day. As noted above, the BER and NER are two important repair mechanisms that fix oxidative damage in DNA. BER and NER are enzyme systems located in the nucleus and mt. BER restores nucleotide bases damaged by internal oxidants, toxins, and irradiation, while NER acts to repair distortions in the DNA helical structure.

The Composition of the Macromolecule is Important In addition to these varied systems, macromolecules of long-lived species appear to have an "innate" higher resistance to oxidative damage. This means that the macromolecules of long-lived species are comprised of molecules that are less susceptible to oxidation. For example, proteins of long-lived species, such as proteins of humans, have a smaller number of oxidative sensitive amino acids (e.g., methionine).

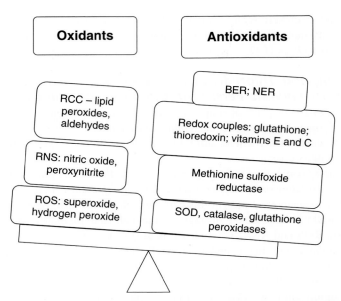

Figure 4.4. Balance between generation of oxidative stress and its prevention.

Similarly, membranes contain phospholipids with variable amounts of saturated/unsaturated fatty acids. The importance of membrane composition is emphasized by the *membrane pacemaker hypothesis by* Hulbert et al. (2007) *that correlated the degree of PUFA in membranes from different species with maximal lifespan* (MLS). The *membranes of short-lived species contain more PUFA* compared to membranes of long-lived species. It is postulated that the susceptibility of PUFA to oxidation contributes to the shorter lifespan. Additionally, *membranes with an abundance of PUFA would be less compact and might require more energy for maintenance.*

STRENGTHS AND WEAKNESSES OF OXIDATIVE STRESS THEORIES

Oxidative Stress Hypothesis

The supportive evidence for the oxidative stress hypothesis is mostly correlative, although in some models of aging, mainly fruit fly, there exist some causal data. The *oxidative stress hypothesis predicts a negative correlation between production of ROS and species-specific maximal lifespan*. For example, it is predicted that species that generate large quantities of ROS will have short MLS and species that generate small amounts of ROS will exhibit long MLS. This is sustained with data from fruit flies (*within species analysis*) that show long-lived flies produce less ROS (superoxide and H_2O_2), exhibit higher levels of antioxidative enzymes (SOD, catalase, and glutathione peroxidase), and experience less protein and DNA oxidative damage compared to short-lived flies. Additionally, among *different species*, there exist the predicted inverse relation between the production of superoxide and H_2O_2 and MLS. In another example, the rat and pigeon are similar in size (mass), yet the MLS of the rat is approximately 4 years compared to the 30-year MLS for the pigeon and in this comparison, ROS production is lower and amount of antioxidant enzymes is higher in the longer lived species, the pigeon. On the other hand, ROS may account for only part of the lifespan difference. For example, others have reported differences in membrane fatty acid composition (rat exhibiting more PUFA compared to the pigeon) that would influence not only the amount of lipid peroxidation but also the basal metabolic rate.

The oxidative stress hypothesis also predicts that caloric restriction (CR), a highly reproducible MLS extending experimental procedure, should decrease ROS production, decrease accumulation of oxidized macromolecules, and fortify the organism's response to elevated levels of oxidative stress. MLS is extended in all animal models of aging subjected to CR (discussed in Chapter 2) and similarly, ROS production and macromolecular damage are less in CR animals compared to controls with free access to food.

The *oxidative stress hypothesis predicts that genetic manipulation of antioxidant enzymes should determine the MLS*. In particular, knockout (KO) (deletion) of an antioxidative enzyme(s) should shorten MLS; knock-in (KI) (addition) of an antioxidative enzyme(s) should lengthen MLS. *Results have been mixed*. In fruit flies, KO of copper (Cu) and zinc(Zn)-SOD1 (located within the cell and between the mt membranes) and partial KO of mt matrix manganese (Mn)-SOD2 (full KO is lethal) shorten MLS as predicted. KI flies with extra copies of genes for SOD1 or SOD2 (hence producing more SOD1 and SOD2 and thereby providing more antioxidative protection) exhibit an increase in mean lifespan but not MLS. However, *flies expressing higher levels of both SOD1 and SOD2 live longer than controls and flies fortified with extra CuZnSOD1 and catalase exhibit a 30% increase in MLS, show less accumulation of oxidized DNA and protein, and are more resistant to external oxidative threats*. Additionally, overexpression of MsrA (one form of Msr) (thereby reversing oxidation of methionine thiol groups) extends MLS in flies and yeast. In contrast, genetic manipulation of antioxidant enzymes in the roundworm, another invertebrate model of aging, has generally been without effect on MLS. It has been noted that evaluation of the oxidative stress hypothesis in the roundworm may be inappropriate since oxygen is of lesser importance to this anaerobic organism.

Results with *vertebrates have been disappointing* and in fact have encouraged a search for a more meaningful interpretation of the oxidative stress hypothesis (see redox stress hypothesis below). CuZnSOD1 KO in mice exhibits a 30% decrease in mean and MLS, increased sensitivity to oxidative stress, evidence of increased DNA mutations, and low fertility in females; mice with partial KO for MnSOD2 show increased mt DNA oxidation but no shortening of MLS. KO of glutathione peroxidase 1 (Gpx1), one member in this large enzyme family, accelerates the onset of cataracts and enhances sensitivity to the chemical oxidant and herbicide, paraquat, but determination of MLS was not done with this manipulation. Furthermore, *overexpression of CuZnSOD1 in mice did not increase MLS*. One study in particular (Perez et al., 2009) performed an extensive analysis that included 18 genetic manipulations in mice ranging from (i) full KO of CuZnSOD1 and glutathione peroxidase 1 (Gpx1) to (ii) partial KO of Gpx4, MsrA, thioredoxin, and MnSOD2 and (iii) KI of SOD1, SOD2, catalase, and Gpx4. Among these, *only one manipulation, SOD1 KO, influenced MLS in mice*. In summary, *manipulation of antioxidant enzymes only minimally affects aging in mice* suggesting to some that there is either considerable redundancy in antioxidative mechanisms such that deletion or enhancement of one or two enzymes would necessarily be without effect and would only modestly contribute to aging or that other pathways are more critical for aging.

Data on humans are modest. It has been shown that (i) human DNA is sensitive to oxidative stress (measurable DNA damage occurs in the presence of oxidative stress), (ii) human cells in culture senesce during conditions that generate free radicals, for example, high oxygen tension, presence of drugs such as paraquat, (iii) human mt DNA from cultured cells and tissue samples accumulate damage under conditions that increase ROS production, and (iv) mt DNA damage assessed in tissues from centenarians is low, *suggesting* that ROS production is also low in these individuals and this low level may contribute to longevity.

Efforts to obtain *supportive evidence for the oxidative stress hypothesis from results of clinical trials have been disappointing.* Results of over 17 randomized, placebo-controlled clinical trials (meaning that each study was conducted in a manner consistent with good scientific practice) that administered vitamin E and C supplements to reduce select diseases (considered dependent in part on excessive oxidative stress) showed *no effect on these age-related pathologies.* Whereas these failed trials suggest to some that at least in man, the oxidative stress hypothesis is irrelevant, it is noted that all of the antioxidant trials with one exception studied the effect of antioxidant supplementation on disease prevention/progression. Since biogerontologists make a distinction between disease and aging, failure to show a benefit of antioxidant treatment in various diseases says little about oxidative stress hypothesis with regard to its role in aging. Furthermore, it has been argued that the dose and formulation of the antioxidant, the patient population, and the duration of treatment all need further evaluation. Results of several small trials that assess markers of oxidative stress show positive effects with chronic administration of vitamin E. Measurement of functional changes would be of value.

Redox Stress Hypothesis

Unlike the oxidant stress hypothesis that postulates a major role for oxidized macromolecules that are structurally altered and become functionally deficient, the *redox stress hypothesis shifts the emphasis away from the oxidant as a structural damage initiator to one of cell regulators (second messenger)* and states that oxidants such as H_2O_2 in conjunction with the redox systems of the cell act to modulate protein activity by reacting with sensitive cysteine thiol (also methionine thiols) groups of redox-sensitive proteins. Some important redox systems already noted are glutathione (reduced/oxidized) (reduced/oxidized), NADPH/NADP$^+$, thioredoxin (reduced/oxidized), and glutaredoxin (reduced/oxidized) and their associated enzymes. *Collectively, these compounds determine the oxidative state of the cell known as the redox potential.* How readily redox couples lose and gain electrons and their concentrations in the cell determine their relative importance in maintaining the balance between oxidative and reductive events.

The most important redox couple is glutathione and *a redox potential for the cell may be calculated from the ratio of the reduced glutathione to the oxidized glutathione using the classic electrochemical Nernst equation*, a formula employed for calculation of voltage across an electric cell or in physiology for the potential generated by ions distributed across a membrane.

In support of the redox stress hypothesis, it is argued that if oxidative damage was the main cause of aging, then overexpression of antioxidative enzymes would consistently increase MLS in all species but as noted above, there is little evidence for this except in the fruit fly. Whereas there is agreement that macromolecules become oxidized, for example, measurement of AGEs, 8oxodG, and lipid peroxides, and hence structurally altered, the dispute centers on *whether this damage is sufficient in quantity to produce cellular dysfunction*, hence aging. Supporters of the redox stress hypothesis indicate there are no studies that have actually measured this. The possibility remains that accumulated oxidatively modified macromolecules exist in amounts *too small* (compared to unoxidized macromolecules) *to explain the plethora of changes indicative of senescence*. Furthermore, results of genetic manipulations in mice showed that mice with reduced levels of MnSOD2 (due to partial MnSOD2 KO) experienced nuclear and mt DNA damage (as measured by 8-oxodG) to an extent far greater (some 15–60% at 26 months of age) than in normal mice and yet normal and partial KO mice have the same mean and MLS (i.e., in this study excessive oxidative DNA damage did not shorten the lifespan).

Thus, according to the redox stress hypothesis, *the redox state of protein thiols (either oxidized or reduced) is a more reliable indicator of oxidative stress* and hence functional loss. The reason for this is that modulation of cysteinyl thiols is the mechanism that influences the behavior of enzymes, gene expression, proliferation, differentiation, apoptosis, and general metabolism. Specific cysteine sulfur groups on proteins act as a "switch" to promote or inhibit protein activity. As ROS are generated, they are tightly controlled within the redox systems so that homeostasis is achieved. However, as levels of ROS gradually increase with age (as evidenced by the gradual oxidation of the cell), redox homeostasis begins to waver. In this case, protein thiols convert to mixed disulfides or exist bound to glutathione (glutathiolation) and proteins are partially or permanently inactivated.

Data in support of the redox stress hypothesis point to (i) comparison among species that show a correlation of high levels of mt H_2O_2, oxidized glutathione, and protein-mixed disulfides with a short MLS, (ii) redox potential measurements taken throughout the lifespan of the cultured cell that show a gradual shift to a more oxidized state, (iii) mice subjected to CR (40% from 4 to 22 months) that exhibit less oxidized glutathione and less evidence of binding of oxidized glutathione to proteins compared to mice given free access to food, and (iv) overexpression in the fruit fly of the two essential enzymes that maintain the redox system by synthesizing glutathione (glutamate cysteine ligase) and generating NADPH (glucose-6-phosphate dehydrogenase) that produces a 40–50% increase in MLS without a loss of fitness, an effect generally lacking with KO/KI of antioxidative enzymes.

The redox stress hypothesis attributes the failure of clinical trials with antioxidants trials to the faulty premise that free radicals cause oxidative-dependent diseases (and not to the arguments given above). The redox stress hypothesis proposes that future clinical trials should test compounds that reduce non-radical oxidant levels, for example, H_2O_2, rather than free radical such as superoxide anion.

The redox stress hypothesis is a testable hypothesis. However, because of its fairly recent formulation, it remains to be determined whether future experiments will

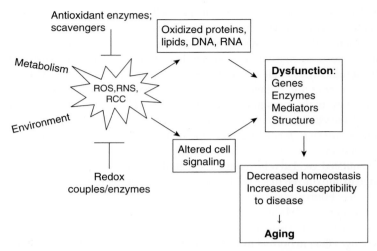

Figure 4.5. Role of oxidative stress in aging. ROS: reactive oxygen species; RNS: reactive nitrogen species; RCC: reactive carbonyl compounds.

provide causal data in mammalian models of aging and whether modulation of the redox potential in clinical trial will retard measureable aspects of aging (Figure 4.5).

SUMMARY

The basic biological macromolecules are polysaccharides, proteins, lipids, and nucleic acids. Each is comprised of essential building blocks: sugars in polysaccharides, amino acids in proteins, nucleotides in nucleic acids, and fatty acids/glycerol in lipids.

Macromolecules exhibit diverse functions. Polysaccharides (glycogen) and lipids (fat) provide a ready source of energy. Lipids are inflammatory mediators and signaling molecules and phospholipids are the chief components of all membranes. DNA contains the genetic information of the cell and relays instructions through RNA in the process of transcription. This information is subsequently translated into proteins. Proteins perform a myriad of functions ranging from regulatory to architectural.

The oxidative stress hypothesis explains age changes as an accumulation of oxidized macromolecules whose functions are compromised due to structural damage. Oxidants, the source of oxidative stress, are plentiful and generated daily from normal metabolic activities and from environmental insults. ROS (free radicals, for example, superoxide and non-radicals, for example, hydrogen peroxide) exert random oxidative attacks on macromolecules that oxidize the thiols of proteins, the guanine of DNA, and the carbons of lipids. Adducts such as AGEs and ALEs are frequently formed that facilitate cross-linkage of proteins or enhance membrane damage. Mechanisms that thwart oxidative damage are abundant and include antioxidant enzymes SOD, catalase, glutathione peroxidases/reductases, and Msr; antioxidant redox pairs such as glutathione, thioredoxin, vitamins C and E; and DNA repair systems such as the BER/NER. However, according to the oxidative stress hypothesis, oxidative damage eventually exceeds repair and aging ensues as accumulated damaged lipids, proteins, and nucleic acids fail to maintain normal function.

The redox stress hypothesis is an alternative assessment of the role of oxidative stress in aging. The redox stress hypothesis proposes that oxidants serve important functions in modulating the functional cysteinyl thiol groups on redox-sensitive regulatory proteins. Thus, redox pairs such as glutathione and thioredoxin maintain the redox state of the cell, thereby ensuring the requisite reduced state for protein thiols and normal cell signalling. Increased oxidative stress perturbs the redox pairs and prevents essential regeneration of protein thiols. Thus, inhibition of homeostatic control mechanisms within the cell contributes to the aging phenotype.

CRITICAL THINKING

In biological systems, what is the relation between molecular/macromolecular structure and function?

What happens to the structure–function relation with age and what are some specific consequences?

Which of the two oxidative stress theories is more compelling? Why?

Would you recommend a daily intake of over-the-counter antioxidative supplements to retard age changes? Explain?

In your opinion, which of the macromolecules is the most important? Why? What change might occur with age to your favorite macromolecule?

KEY TERMS

AGEs advanced glycation end products. These are fragments called adducts that are bound to the amino acids, lysine or arginine, as a result of spontaneous oxidation of sugar. AGEs are generated in foods during excessive heat, for example, caramelization. The adducts are highly reactive and have the potential to alter protein structure and induce inflammation.

Alpha-tocopherol one of the many forms of vitamin E. This form is the main one found in blood and tissues in man. Recommended daily allowance (RDA) is 15 mg.

Ascorbic acid vitamin C; must be obtained in diet since humans cannot synthesize this molecule. It acts as an antioxidant and cofactor from several enzymatic reactions. The RDA is 90 mg with an upper limit of 2000 mg.

Chromosomes name given to the tightly wound DNA strands. In humans there are 23 pairs of chromosomes containing DNA (genes) inherited from an individual's parents. Chromosomes are surrounded with proteins called histones.

DNA deoxyribonucleic acid; the macromolecule containing all the information (cell blueprint) necessary to make proteins (information is first transcribed to the RNA; RNA then translates the information into a specific protein).

Free radical fragment or molecule that contains an unpaired electron in the outer shell, resulting in a high level of reactivity. A free radical combines with other molecules and generally disrupts the electron nature of its victim. If not reversed, interaction causes oxidative changes and possible structural/functional damage.

Glutathione a small protein consisting of three amino acids that is found in high concentrations in the cell. It exists in both the oxidized and reduced forms (redox pair) and is an effective antioxidant.

Glycation the process whereby reactive carbon molecules (as in sugar) react oxidatively with proteins, resulting in unwanted adducts (fragmented carbon molecules attached to the protein).

Homeostasis the set point of normal physiological function in an organism. Homeostatic mechanisms minimize negative effects of stress and return the system to normal.

Lipids a diverse group of macromolecules that include the glycerides (mono-, di-, tri-), phospholipids, sterols, for example, cholesterol, and waxes. Lipids act as membrane components, inflammatory mediators, hormones, and cell signaling agents.

Nucleic acids macromolecules that contain the genetic information that directs cell activity. Deoxyribonucleic acid (DNA) contains inherited genes and transcribes this information to ribonucleic acid (RNA). The latter uses the information to make proteins (translation).

Oxidation/reduction reaction also called redox reaction; type of chemical reaction in which electrons are shuttled between molecules.

Oxidative stress hypothesis originally called the free radical theory of aging. It is a long-standing theory to explain deterioration of cell function that leads to aging. It states that age changes result from unrepaired oxidant damage to key macromolecules in the cell. Introduced by D. Harmin in the 1950s. Support is predominately correlative in nature.

Pentosidine an oxidant generated in glycation reactions and frequently measured to assess the extent of oxidative stress in an organism.

Polysaccharides long chains of sugar molecules that are stored in the liver as glycogen or consumed as carbohydrates.

Polyunsaturated fatty acids (PUFA) long hydrocarbon chains with more than one double bond. The bond creates "kinks" in the structure as well as a susceptibility to peroxidation.

Proteins chains comprised of combinations of any of the 20 amino acids. The sequence and chemical structure of the individual amino acids contributes to the final 3D structure of the protein.

Redox stress hypothesis recently proposed hypothesis to explain age changes at the macromolecular level. It suggests that non-radical oxidants are important cell messengers. Oxidative stress slowly changes the redox potential of the cell, thereby disrupting the normal function of non-radical oxidant signaling molecules and creating oxidized protein thiols. Loss of cell control mechanisms leads to cell dysfunction and deterioration.

RNA ribonucleic acid; the macromolecule that contains information to make a protein. It is called messenger RNA and requires help from two other RNA types, the transfer RNA and the ribosomal RNA. The formation of each RNA depends on the information in the gene (specific DNA sequence).

ROS acronym for *r*eactive *o*xygen *s*pecies. An example is superoxide anion, but includes all oxidants, radical or not.

Saturated fatty acids long chains of carbons to which are attached the greatest number of hydrogen molecules. Saturated fatty acids are flexible, pack tightly in the membrane, and are resistant to peroxidation. An example is stearic acid that can be found in animal fat.

Superoxide anion oxygen that is highly reactive because it has an extra unpaired electron. It is a free radical anion and is generated in several locations in the cell, most notably in the mt during cell respiration.

Thiol compound containing a sulfur-hydrogen (SH) group attached to a carbon; found on amino acids, cysteine and methionine.

Transcription the process whereby DNA relays information of a gene by the formation of RNA. This works through complementary base pairing in which each base pairs with a partner, for example, guanine–cytosine, thymine–adenine.

Transgenic describes an organism into which a new gene has been added.

Translation the process whereby RNA (containing information for the amino acid sequence of a protein) relays this information and enables the biosynthesis of a protein. The process requires cooperation among messenger RNA, transfer RNA, and ribosomal RNA. Complementary base pairing is required in this process.

BIBLIOGRAPHY

Review

Alberts B, Johnson A, Lewis J, Raff M, Roberts K, Walter P. 2002. *Molecular Biology of the Cell*, 4th edition. New York: Garland Science.

Avery NC, Bailey AJ. 2006. The effects of the Maillard reaction on the physical properties and cell interactions of collagen. *Pathol. Biol.* **54**(7):386–395.

Baraibar MA, Liu L, Ahmed EK, Friguet B. 2012. Protein oxidative damage at the crossroads of cellular senescence, aging and age-related diseases. *Oxid. Med. Cell Longev.* **2012**, 919832.

Davies SS, Guo L. 2014. Lipid peroxidation generates biologically active phospholipids including oxidatively *N*-modified phospholipids. *Chem. Phys. Lipids* **181C**:1–33.

Garinis GA. 2008. Nucleotide excision repair deficiencies and the somatotropic axis in aging. *Hormones (Athens)* **7**(1):9–16.

Harman D. 1956. Aging: a theory based on free radical and radiation chemistry. *J. Gerontol.* **11**(3):298–300.

Harman D. 1981. The aging process. *Proc. Natl. Acad. Sci. USA* **78**(11):7124–7128.

Hulbert AJ, Pamplona R, Buffenstein R, Buttemer WA. 2007. Life and death: metabolic rate, membrane composition, and life span of animals. *Physiol. Rev.* **87**(4):1175–1213.

Jacob KD, Noren Hooten N, Trzeciak AR, Evans MK. 2013. Markers of oxidant stress that are clinically relevant in aging and age-related disease. *Mech. Ageing Dev.* **134**(3-4):139–157.

Jones Dean P. 2008. Radical-free biology of oxidative stress. *Am. J. Physiol. Cell Physiol.* **295**(4):C849–C868.

Lee S, Kim SM, Lee RT. 2013. Thioredoxin and thioredoxin target proteins: from molecular mechanisms to functional significance. *Antioxid. Redox Signal.* **18**(10):1165–1207.

Nance MA, Berry SA. 1992. Cockayne syndrome: review of 140 cases. *Am. J. Med. Genet.* **42**(1):68–84.

Pamplona R, Barja G. 2007. Highly resistant macromolecular components and low rate of generation of endogenous damage: two key traits of longevity. *Aging Res. Rev.* **6**(3):189–210.

Sohal R, Orr WC. 2012. The redox stress hypothesis of aging. *Free Radic. Biol. Med.* **52**(3):539–555.

Steinhubl SR. 2008. Why have antioxidants failed in clinical trials? *Am. J. Cardiol.* **101**(10A):14D–19D.

Xu G, Herzig M, Rotrekl V, Walter CA. 2008. Base excision repair, aging and health span. *Mech. Ageing Dev.* **129**(7–8):366–382.

Experimental

Commoner B, Ternberg JL. 1961. Free radicals in surviving tissues. *Proc. Natl. Acad. Sci. USA* **47**:1374–1384.

Fan X, Zhang J, Theves M, Strauch C, Nemet I, Liu X, Qian J, Giblin FJ, Monnier VM. 2009. Mechanism of lysine oxidation in human lens crystallins during aging and in diabetes. *J. Biol. Chem.* **284**(50):34618–34627.

Gardner PR. 1997. Superoxide-driven aconitase Fe-S center cycling. *Biosci. Rep.* **17**(1):33–42.

Orr WC, Radyuk SN, Prabhudesai L, Toroser D, Benes JJ, Luchak JM, Mockett RJ, Rebrin I, Hubbard JG, Sohal RS. 2005. Overexpression of glutamate–cysteine ligase extends life span in *Drosophila melanogaster*. *J. Biol. Chem.* **280**(45):37331–37338.

Paredi P, Kharitonov SA, Leak D, Ward S, Cramer D, Barnes PJ. 2000. Exhaled ethane, a marker of lipid peroxidation, is elevated in chronic obstructive pulmonary disease. *Am. J. Respir. Crit. Care Med.* **162**(2 Part 1):369–373.

Pérez VI, Bokov A, Van Remmen H, Mele J, Ran Q, Ikeno Y, Richardson A. 2009. Is the oxidative stress theory of aging dead? *Biochim. Biophys. Acta* **1790**(10):1005–1014.

Schriner SE, Linford NJ, Martin GM, Treuting P, Ogburn CE, Emond M, Coskun PE, Ladiges W, Wolf N, Van Remmen H, Wallace DC, Rabinovitch PS. 2005. Extension of murine life span by overexpression of catalase targeted to mitochondria. *Science* **308**(5730):1909–1911.

Sun J, Folk D, Bradley TJ, Tower J. 2002. Induced overexpression of mitochondrial Mn-superoxide dismutase extends the life span of adult *Drosophila melanogaster*. *Genetics* **161**(2):661–672.

Sun J, Molitor J, Tower J. 2004. Effects of simultaneous over-expression of Cu/ZnSOD and MnSOD on *Drosophila melanogaster* life span. *Mech. Ageing Dev.* **125**(5):341–349.

Tanaka M, Chock PB, Stadtman ER. 2007. Oxidized messenger RNA induces translation errors. *Proc. Natl. Acad. Sci. USA* **104**(1):66–71.

5

AGING OF CELLS

ROLE OF ORGANELLE

Organelles Separately and Together Maintain the Cell

Organelles are dynamic substructures of the cell. They increase and decrease in number and size and perform diverse functions. Organelle function is modified by many factors, including nutrients, pro- and anti-inflammatory mediators, hormones, reactive oxygen species (ROS), protein complexes, and ions. Organelles maintain characteristic pursuits that can be adapted to foster organelle intercommunication (signal crosstalk). For example, there are signal relays between the mitochondria (mt) and the endoplasmic reticulum (ER), the nucleus and lysosomes, and additionally between lysosomes and mt. This *intercommunication facilitates homeostasis, the physiological normal condition.* Thus, the independent functional contributions of organelles are coordinated and collectively realized in major cellular activities of metabolism, secretion, division, apoptosis, differentiation, and growth. Many of these activities are discussed below.

Organelles implicated thus far in aging are the nucleus, mt, peroxisomes, and lysosomes. The bulk of data to date describe structural and functional changes in organelles of animal models of aging but where possible human data is noted. There exist several hypotheses on the mechanism of age-related changes in organelles, for example, mitochondrial free radical theory of aging and the mitochondrial–lysosomal axis theory of aging. However, these theories are presently under revision. There are only two interventions, caloric restriction and exercise (relevant to skeletal and cardiac muscle, possibly cognition) that have been shown to minimize aging in organelles and they will be discussed in separate chapters (Chapters 7, 9, and 12).

Human Biological Aging: From Macromolecules to Organ Systems, First Edition. Glenda Bilder.
© 2016 John Wiley & Sons, Inc. Published 2016 by John Wiley & Sons, Inc.

Other organelles of the cell such as the ER and the Golgi apparatus most certainly contribute to cellular aging, but their role in this process is at present poorly understood.

HOW ORGANELLES AGE

Mt Contain Their Own DNA Mt are double membrane organelles. They have a distinct convoluted continuous inner membrane surrounded by a smooth outer membrane. Unlike all other organelles, *mt contain their own DNA* that code for RNA and a limited number of proteins.

Mt Biosynthesize ATP to Power the Cell The designation "power houses of the cell" is applied to mt in acknowledgement of their role as *producers of ATP* (adenosine triphosphate), a molecule that supplies the energy for virtually all cellular activities. ATP synthesis occurs through *oxidative phosphorylation*, a process in which oxygen and metabolites from sugars, fats, and proteins are converted to ATP, carbon dioxide, and water. Through a series of redox reactions enabled by enzyme complexes, designated I–V on the inner membrane of the mitochondria, electrons from metabolites are shuttled to oxygen via complexes I–IV (also called the electron transport chain (ETC)). Protons from the same metabolites are funneled into an electrochemical gradient that drives the formation of ATP in complex V. As illustrated in Figure 5.1, oxidative phosphorylation *is responsible for the generation of ROS in this organelle*. ROS formation is known to occur at numerous sites along the ETC.

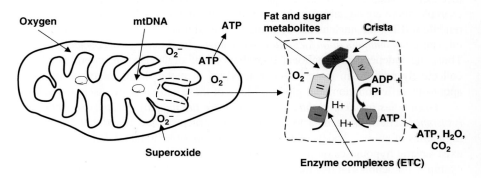

Enzyme complexes facilitate use of oxygen and electrons/protons from nutrients to generate ATP, carbon dioxide, and water

Figure 5.1. Mitochondrion prototype. Components are mitochondrial DNA (mtDNA), electron transport chain (ETC) with enzyme complexes, free radicals (superoxide, O_2^-), ATP production (carbon dioxide, CO_2 and water, H_2O), and hydrogen ion gradient (H^+). (See plate section for color version.)

Mt Determine the Fate of the Cell *Mt have the capacity to terminate the existence of a cell.* Under certain conditions, generally associated with severe stress, a pathway that leads to *cell suicide, termed apoptosis,* is initiated by this organelle. Activation of apoptosis begins with stoppage of oxidative phosphorylation, disruptions in mt membranes, and release of cytochrome C, an mt protein that signals continuation of the apoptotic process. As unfortunate as cell loss seems, avoidance of apoptosis retains the cell in a state of reduced capacity that may initiate or promote the progression of diseases such as atherosclerosis and Parkinson's disease. However, *both cell disappearance by apoptosis and persistence of dysfunctional cells contribute to aging.*

Mt Regulate the Level of ROS Mt exert tight control over the concentration of ROS and maintain low levels through (i) regulation of the activity of antioxidative enzymes, (ii) the efficiency of the ETC (redox state and electrochemical gradients), (iii) the location of ROS generation within the mitochondria, and (iv) the number of mt in the cell. A wealth of data (mostly from animal models of aging with one exception of cultured human cells) suggest that *mt ROS (in particular hydrogen peroxide) play a vital role* in response of the cell to stresses that include hypoxia (oxygen shortage), pathogens, starvation, and growth factor stimulation. Additionally, cell culture results show that without ROS, human stem cells could not differentiate into mature adipocytes. ROS appear indispensible to cell survival.

Sena and Chandel (2012) hypothesize that in the presence of a mild stress, the *increase in mt ROS actually mediates adaptation to that stress.* This adaptation is termed *hormesis* (introduced in Chapter 2) and describes a protective response that fortifies the cell against more severe stresses. In contrast, fixed low levels of ROS *maintain* homeostasis, for example, metabolic stability. In situations of extreme stress, an *excessive increase in mt ROS produces oxidative damage, senescence, or death.* Figure 5.2 summarizes the current view on the different roles played by mt ROS. Attainment of the appropriate cellular level of ROS is critical to the health of the cell.

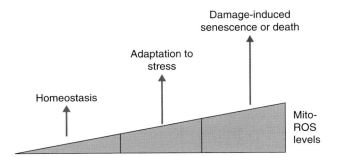

Figure 5.2. The effect of increasing levels of mitochondrial reactive oxygen species (mito-ROS) on the physiology of the cell. (Reprinted with permission from Sena and Chandel (2012).)

Mt Dysfunction Is Present in Tissues from Aged Humans
Mt become dysfunctional with age. The majority of studies and certainly the most extensive ones have evaluated mt structure/function in biopsy samples from human skeletal muscle. Others have assessed postmortem samples from brain and liver. *Mt from aged humans exhibit a reduced abundance of mtDNA and mtRNA transcripts, decreased protein content and a decline in key mt enzyme activity, less oxidative phosphorylation per mt, more extensive damage of mtDNA damage (8-oxodG) and membrane lipids (peroxides), decreased number of mt per cell, and evidence of an increase in mtDNA mutations (live*r). Whereas biopsy samples in these studies were obtained from "healthy" volunteers, confounding effects from uncontrolled factors such as hormones, diet, and physical exercise may bias the results and must be considered.

The mt Free Radical Theory of Aging Proposes mt ROS as a Cause of Aging
Biogerontologists early on identified mt as the main source of cellular oxidants arising as byproducts of cellular respiration. Harman, author of the free radical theory of aging (Chapter 4) revised his theory to align with this observation and proposed the mt free radical theory of aging (MFRTA). According to MFRTA, it is mt ROS that oxidatively damage the poorly protected mtDNA. Unrepaired DNA damage yields mutated mtDNA that fail to produce the requisite proteins for the ETC and oxidative phosphorylation, hence ATP production eventually declines. Thus, MFRTA predicts that mtDNA mutations *increase* with age and *are responsible* for the age-associated functional decline in this organelle. This results in generation of higher and higher levels of ROS and initiation of a "vicious" cycle of deterioration. As additional mt become involved, the pathways to senescence or apoptosis are activated.

MFRTA is supported by several lines of evidence.

First, overexpression of catalase in the mt of mice reduces mt ROS and extends the maximal lifespan (MLS) of the manipulated mice. Second, artificially induced elevations in mtDNA mutations in mice (model called the mutator mouse) induce an accelerated aging syndrome. Third, reduction of activity of an enzyme (by genetic mutation) that normally produces mt hydrogen peroxide decreases mt ROS and extends the MLS. However, other data fail to support MFRTA. Genetic manipulations in yeast, roundworm, and also in mice that *increase* mt ROS paradoxically *extend the MLS*. Findings from cell cultures (primary fibroblasts) from 13 primates, including man show lifespan is *not related to mt ROS production per se* but to the *extent of cellular resistance to stress* (in this study the stress of hyperglycemia). Furthermore, caloric restriction (CR) in animals extends the lifespan even in the presence of high levels of ROS possible through a mechanism of mt-nucleus crosstalk.

Where does that leave MFRTA? As illustrated in Figure 5.2, since ROS exert both beneficial and detrimental effects, the MRFTA has become less persuasive. Additional studies to understand the plethora of potential mt functions and actions with other organelles are desirable.

Mt are Mobile with Changing Morphology
The mass and number of mt in a cell depend on the cell type, cell function, and mt biogenesis capacity. The

number of mt ranges from a few to hundreds. The process of *biogenesis in which mt originate de novo (from basic building blocks) is a prime regulator of mt number*. It is influenced by factors such as exercise, temperature (cold), and hormones. *Mt number is also influenced by autophagy* (a type of recycling). Two other processes of *fission* (division into daughter mt) and *fusion* (merging with another mt) contribute to a *dynamic system of changing mt morphology* that optimizes cell function. The *imbalance between fusion and fission as well as reduced biogenesis and autophagy are considered initiators of cell dysfunction*. All of these processes are tightly regulated by cascades of proteins of which only a few, for example, PGC-1a (peroxisome proliferator-activated receptor gamma coactivator-1 alpha), have been identified.

Altered mt Dynamics as Effecter of Cell Aging

The idea that not only is the *turnover of mt* (the number of new mt generated through biogenesis and the number that disappear through recycling) but also the *morphology of the mt within the cell determined by fusion and fission are all important drivers of mt deterioration and cell death* is new. The benefits of fission and fusion are not as obvious as biogenesis and autophagy. *Fissions is a type of renewal* mechanism in which the unequal division of mt allows the elimination of defective mtDNA and other unwanted components in a daughter mt destined for recycling. *Fusion, on the other hand, creates sharing of resources*. It is postulated that if fusion increases more than fission, there is a preponderance of elongated inefficient mt; if fission exceeds fusion, cells accumulate fragmented mt that tax lysosomal-mediated recycling. The importance of normal mt dynamics is evident in several human genetic diseases, for example, Charcot-Marie-Tooth Type 2A and autosomal dominant optic atrophy in which proteins required for normal fission and fusion are defective, and this deficit produces altered mt morphology, evidence of increased oxidative stress, abnormalities in mtDNA, cell apoptosis, tissue damage, and neurodegeneration. Immerging evidence shows that in *skeletal muscles of the elderly, mt biogenesis and some mt regulatory proteins are decreased*. In mouse models these deficiencies result in muscle atrophy. The role of impaired mt biogenesis and dynamics is of considerable importance and needs further investigation.

Box 5.1. More about Mitochondria

In addition to energy production (ATP) and cell survival responsibilities, mt biosynthesize heme (chemical group associated with many redox or gas transport proteins, for example, hemoglobin), steroids, sugars, and ketones, for example, acetoacetate and acetone, and produce iron–sulfur clusters for exportation. Mt are also an important storage site of calcium. Calcium signaling is the essential signal that matches ATP production with the energy needs of the cells.

It is not unreasonable to expect that as mt function declines with age, the biosynthesis of these compounds would be reduced. Age-related changes in man have not been defined. It is known that the presence of iron–sulfur clusters within the mt contributes to recycling.

Lysosomes Recycle Defective Substructures, Oxidized Macromolecules, and Other Cell Components

The lysosome is a double-membrane vesicle-like subcellular structure filled with an array of acid-loving hydrolytic enzymes with catabolic capacity. The main role of the lysosome is that of autophagy (self-eating). *Autophagy is basically a recycling process.* Substances degraded in lysosomes include abnormal organelles, bacteria, oxidized macromolecules, aggregated proteins, and inflammatory mediators. Once these items are broken down, the subunits or "digested" material exit the lysosome for use as basic anabolic resources. As a result of autophagy, genome stability is preserved, oncogenic proteins (cancer inducers) are eliminated, and mitochondrial turnover is maintained. Adverse interference with autophagy stresses the cell, prevents adaptation, and permits transformation (to a cancerous state) or initiates cell death.

Autophagy Occurs by Three Different Pathways

There are at least *three pathways* available for lysosomal autophagy: *macroautophagy*, *microautophagy*, and the more recently discovered *chaperone-mediated autophagy* (see Figure 5.3). During macroautophagy, internal macromolecular structures, generally bacteria and damaged organelles targeted for destruction, are enclosed in a membrane (autophagosome) that fuses with the lysosomal membrane. Once inside the lysosome, the imports are degraded and the resulting units, primarily amino acids, are transported into the cytoplasm for reuse. In the course of mt fission, the portion of the mt filled with

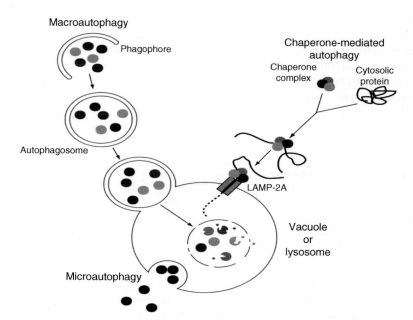

Figure 5.3. Illustration of the three pathways of autophagy in the cell. (Reprinted with permission from Lynch-Day and Klionsky (2010).) (See plate section for color version.)

damaged macromolecules is recycled by this route. This select process, termed *mitophagy,* serves as an indispensible regulator of mt homeostasis.

Microautophagy differs from macroautophagy in that *smaller amounts of cytosolic material are imported by infolding of the lysosome membrane* to create small enclosures or vesicles. The contents of newly formed vesicles are subsequently degraded. The *third pathway, chaperone-mediated autophagy,* allows for selective uptake of cytosolic proteins via a specific protein escort called a chaperone. This pathway requires the lysosomal receptor, lysosomal-associated membrane protein (LAMP-2A), and a recognition site on the protein destined for degradation. As with other entrance routes, physical degradation is achieved through the activity of numerous degradative enzymes and the acidic environment of the lysosome.

Autophagy Declines with Age Possibly Due to Loss of the Receptor–Transporter (LAMP-2A) and/or Defective mt

Several observations suggest that lysosomal function declines with age. First, *toxic products such as lipofuscin accumulate in cells with age implying a possible reduction in autophagy.* Second, results of genetic studies in roundworms, fruit flies, and rodents show that *inhibition of autophagy leads to a shortened lifespan,* and third, CR and CR mimetics enhance autophagy and extend the lifespan.

Lipofuscin is a brown-yellow fluorescent material. Its slow accumulation is taken as a *marker of inadequate degradation of cell components.* Lipofuscin is a mixture of poorly degraded oxidized proteins, lipids, polysaccharides, trace amounts of metal ions, and cell components. Cell and tissue studies suggest that once formed, lipofuscin *cannot be degraded.* Furthermore, its presence is thought to accelerate lysosomal decline, an effect that could promote cell death.

The activity of all three lysosomal pathways of macro-, micro-, and chaperone-mediated autophagy, wanes with age. However, only the mechanism of age-dependent decline in chaperone-mediated autophagy is understood albeit to a limited extent. It is known that the essential "transport" receptor, *LAMP-2A, that allows entrance of proteins into the lysosomes disappears with age.* Consequently, some proteins destined for degradation *cannot enter* the lysosome, cannot be recycled, and so accumulate in the cell. Interestingly, genetically manipulated old mice with an artificial "inducible" gene for liver LAMP-2A (i.e., turned on to make lots of LAMP-2A) display "rejuvenated" liver function with increased presence of this receptor-transporter. These results clearly emphasize the importance of optimal autophagy to maintain cellular health.

Brunk and Terman (2002) proposed that enlarged mt, lysosomal dysfunction, and lipofuscin are interdependent. Reduced fission (reduced mitophagy) and increased fusion of mt create large inefficient mt. Lysosomes attempt to remove these substructures but the presence of ROS and iron-containing complexes within them accelerates lipofuscin formation. In an attempt to degrade lipofuscin, enzymes are wastefully added to the lysosomes in a fruitless effort to destroy lipofuscin. As this overwhelms the lysosomes, crosstalk between lysosomes and mt is postulated to initiate apoptosis. This hypothesis forms the basis of the proposed mitochondrial–lysosomal axis theory of aging by Terman et al. (2010).

> **Box 5.2. More About Autophagy**
> The process of autophagy, especially chaperone-mediated autophagy, is retarded by the hormone, insulin. Insulin influences autophagy through activation of *the negative autophagy regulator and cell nutrient sensor known as TOR (target of rapamycin)*. As insulin levels fall, TOR activation declines and autophagy is enhanced. Improved recycling of defective macromolecules and cell structures eliminates potentially toxic material from the cell and yields an sufficient supply of molecules available for energy or synthesis of new macromolecules. Elevation of autophagy enhances cell survival. Autophagy is maximized in states of low energy intake, for example, CR or starvation, where insulin levels are low and insulin sensitivity is high and conversely autophagy is sluggish in diabetes, especially when insulin levels are poorly regulated.

Peroxisomes Perform Oxidations and Biosynthesize Compounds

The peroxisome is a single membrane organelle arising from the endoplasmic reticulum and other peroxisomes. It is an *oxygen-utilizing organelle, similar to mt, and is thought to work closely with mt* in handling the oxidation of fatty acids. Peroxisomes oxidize many compounds, in particular long-chain fatty acids and inflammatory mediators, and export the fragments for reuse elsewhere in the cell. The oxidation reactions produce the strong oxidant H_2O_2 that is needed in numerous detoxification reactions, for example, alcohol degradation. H_2O_2 levels are controlled by several enzymes (catalase, peroxiredoxins, and glutathione peroxidases) that convert it to oxygen and water. Oxidation of H_2O_2 is critically important since its unwanted presence either in the peroxisome or leakage into the cell cytoplasm will produce serious cellular damage. Box 5.3 gives more information about peroxisomes.

Peroxisomes Lose the Ability to Import Catalase to Degrade Hydrogen Peroxide

The most significant age change in peroxisomes is a *reduction in the importation of key antioxidant enzymes, especially catalase* (see Figure 5.4). Catalase fails to enter the peroxisome and remains in the cytoplasm. Since transportation of catalase into peroxisomes is considered suboptimal even in the young adult cell, peroxisome function is increasingly challenged as the cell ages. One consequence of a catalase deficiency is accumulation of H_2O_2 within the organelle, the presence of which oxidatively damages peroxisome structure and hinders normal breakdown of lipids and polysaccharides. Peroxisomes traumatized in this way

> **Box 5.3. More About Peroxisomes**
> Peroxisomes also carry out the biosynthesis of several key fats such as ether phospholipids (used in myelin), polyunsaturated fatty acids, for example, docosahexaenoic acid (DHA), and bile acids.
> Similar to mt, peroxisomes sense the needs of the cell. How this is achieved is only beginning to be unraveled. Biogenesis, fission, and growth of the peroxisome as well as enzyme importation are some characteristic responses to maintain cellular homeostasis.

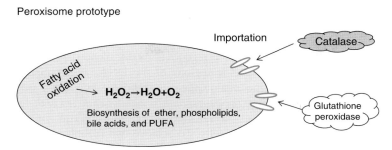

Figure 5.4. Peroxisome prototype. Hydrogen peroxide (H_2O_2) is converted to water (H_2O) and oxygen (O_2); enzymes are imported. PUFA: polyunsaturated fatty acid.

become dysfunctional and paradoxically increase in number in the cell. It is not uncommon to observe increased numbers of *abnormal* peroxisomes in senescent cells from animal models and man.

A second outcome of abnormal peroxisomes is the diffusion of H_2O_2 into the cytoplasm where conversion to other oxidants, such as the hydroxyl radical, is likely and where the unwanted presence of H_2O_2 will change the cellular redox potential to a more oxidized state.

Results of studies that manipulate catalase activity in cells and animals reinforce the importance of adequate amounts of peroxisomal catalase to maintain homeostasis. Scientists have artificially lowered catalase activity in cultured cells with a catalase inhibitor, 3-amino-1,2,4-triazole. Cells inhibited by this chemical display signs of accelerated aging that include mt dysfunction, oxidation of proteins and DNA, reduced peroxisome function, and general disruption of redox potential of the cell. Conversely, modest overexpression of the gene for peroxisomal catalase in mice extends their lifespan about 10% compared to normal mice and expression of a catalase engineered for improved importation delays the onset of senescence in cultured cells.

Nucleus Is the Locus of the Genetic Blueprint for the Cell

The nucleus contains the genetic instructions for the cell phenotype. The genetic instructions are found in the DNA of the nuclear chromosomes, the number of which is species specific. Chromosomes are linear arrangements of chromatin, individual genes tightly coiled and bundled with proteins (histones). Chromatin is either *euchromatic*, which means it is transcriptionally active (genes are actively making RNA to translate into protein), or *heterochromatic*, which means it is transcriptionally inactive (silent genes). Additionally, there exists DNA that is neither euchromatic nor heterochromatic but acts as part of a protective cap (telomere) at the ends of the chromosomes (see mitosis below).

The regulation of gene expression (whether a gene will actually be transcribed and translated into a protein) is complex but of considerable interest since *chromatin remodeling occurs with age*. This *remodeling is termed epigenetic* and refers to changes in gene expression that come about by (i) processes of methylation of DNA (adding a 1-carbon methyl group, $-CH_3$ to the nucleotide base, cytosine),

(ii) acetylation/deacetylation (adding or removing a 2-carbon acetyl group, −CH−CH) from histones, thereby altering chromatin structure to expose silent genes or to silence others, or (iii) by generation of noncoding RNAs (ncRNAs). The noncoding RNAs include regulatory RNAs, which influence genes normally silenced by *transcriptional factors*, proteins that bind to specific regions of the DNA, and routinely control transcription. With epigenetic modulation, the *cell phenotype can change without any alteration in the base sequence of the genes, that is, without changes in the genotype.*

The nucleus is also the focal site for mitosis (cell division), in which one cell divides into two identical, daughter cells. To achieve two identical cells, *the DNA must replicate* (double, make a copy of itself), a feat that necessitates the use of resources throughout the cell in a process termed the cell cycle. Many of the age-associated changes in the cell cycle have been defined and are discussed below.

DNA Experiences Telomere Shortening and Epigenetic Modifications; Proposed Deficiency of Nuclear Lamins

Age-associated changes in the nucleus are abundant and varied. Data support the following changes: (i) telomere shortening, (ii) numerous epigenetic changes that remodel the chromatin, and (iii) structural changes in lamin, a nuclear membrane protein.

Telomeres are long repeats of nucleotides extending from the ends of each chromosome. Telomeres are not genes, (they do not contain information for protein formation), but rather bind multiple layers of proteins to form the telomere complex or "protective cap" at the ends of the chromosomes. Telomeres are likened to the plastic protective tips on shoe laces. Without telomeres and the associated protein complex, chromosomal ends would be vulnerable to degradation or to inadvertent attachment to the adjacent ends of other chromosomes, an event that would prevent cell division (mitosis). Importantly, *telomere size diminishes with each round of cell division* and since mitotic cells divide throughout the lifespan of an organism, telomeres continue to shorten during that time (see Figure 5.5). When the cell reaches a point where a

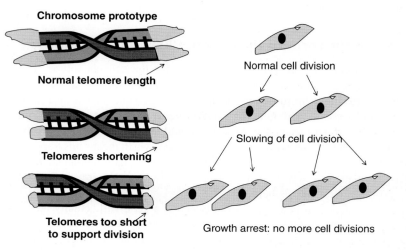

Figure 5.5. Chromosome prototype with gray ends representing telomeres. Telomeres shorten with each round of cell division until too short to safely support DNA replication.

significant number of chromosomes are without protective telomere caps, cell pathways are activated that permanently prohibit future division. This phenomenon provides a possible explanation for the observation that mitotic cells eventually stop dividing and become replicative senescent (see below).

Another aging mechanism in the nucleus is epigenetic modification of the nuclear DNA (nDNA). Epigenetic modification or chromatin remodeling may take the approaches noted above: DNA methylation, histone acetylation, and ncRNA. Although other pathways have been identified that include DNA oxidation, mt ATP output, and signaling between mtDNA and nDNA, histone acetylation in aging is the best studied thus far.

Therefore, *chromatin structure changes with age due in large part to histone acetylation*. Epigenetic influences are potentially enormous. The process of histone acetylation generally "turns on" or activates a gene, and histone deacetylation generally achieves the opposite effect to "turn off" a gene or deactivate it. One group of histone deacetylases (NAD^+-dependent deacylases), called sirtuins, is of particular interest because their enzymatic activity is specifically altered by CR and exercise (demonstrated in animal models). Conversely, histone acetylations are abundant in life-shortening progeroid syndromes. See Box 5.4 for more about epigenetic changes.

A third site of age-associated changes in the nucleus is in the network of proteins, called lamins, situated beneath the double membrane of the nucleus. In the roundworm, aggregation or bunching together of the lamins followed by elongation and fragmentation of the nuclear membrane is a reliable early marker of aging expressed in all cells except neurons. Individuals with Hutchinson–Gilford progeria syndrome (HG) express a mutated protein lamin A called progerin. Cultured HG cells exhibit

---

Box 5.4. More About Epigenetic Changes Related to Aging

With regard to epigenetic changes brought about with methylation, a *reduction in methylation (hypomethylation)* is prominent in nDNA of older cells. This has been attributed to an internal change that reduces DNA methyltransferases activity and additionally to external environmental factors that demethylate DNA, and to DNA oxidation that might structurally prohibit methylation or might alter the cell redox state. A change in redox state would nonspecifically change methylation reactions. These mechanisms potentially adversely affect many genes involved in hormone function, tumor suppression, DNA repair, and cell signaling.

Noncoding RNA (ncRNA) is a significant regulator of gene expression. Among the ncRNA, the microRNAs (miRNAs) are the best understood. MiRNAs influence messenger RNA by promoting its degradation or by repressing its expression (hence no protein is made). miRNAs are abundant and are thought to "fine-tune" gene expression. Some miRNAs have been implicated in aging by their interaction with insulin signaling, a pathway known to regulate the lifespan in animal models. In the roundworm, miRNAs modulate the downstream mediators of the insulin receptor such that KO of certain miRNAs shortens the lifespan and overexpression of certain miRNAs lengthens the lifespan. In mammals (mouse), miRNAs are expressed in a tissue-specific manner throughout the lifespan and miRNAs have been identified that influence all aspects of cell senescence from cell cycle regulators and tumor suppressors to antioxidative enzymes.

distorted nuclear architecture, increased DNA damage, and chromatin remodeling that is thought to cause abnormal differentiation of the stem cell population. This mechanism may explain the HG phenotype that drives accelerated aging and early death (see Chapter 2). There is some evidence that progerin is produced at low levels with age in normal individuals and is associated with chromatin remodeling and DNA damage. Although more needs to be discovered about lamins, it appears that lamins are potentially important regulators of genome stability.

CELLULAR AGING: OBSERVATIONS AND HYPOTHESES

Cells are classified as follows: (i) mitotic cells or cells capable of dividing into two identical daughter cells, such as epithelial cells, (ii) postmitotic cells or differentiated cells, such as neurons generally not capable of dividing, and (iii) stem cells, partially differentiated or pluripotent cells, capable of giving rise to *any* cell type depending on local environmental stimuli (see Table 5.1). *Mitotic, postmitotic, and stem cells are referred to as somatic cells* since they carry the full complement of DNA (46 chromosomes in man for example, species-specific number varies for other organisms) and thus are different from germ-line cells, the sperm and ovum that contain half or a haploid number of chromosomes (23 chromosomes in man). *Somatic cells have a finite lifespan; germline cells are immortal* because hereditary material (DNA) is passed from one generation to the next.

As cells age, changes transpire that produce a *senescence phenotype* in some cells. The senescent cell persists with a phenotype that is dramatically different from that of a presenescent young adult cell. The senescent cell is dysfunctional. It releases numerous factors that negatively impact neighboring cells, and initiates tissue inflammation. Unlike the senescent cell, some cells simply disappear with age and their disappearance decreases tissue size (atrophy). These two distinct cellular adjustments and their proposed initiators are presented below.

The Cell Cycle Is a Tightly Regulated Process of Checklists and Checkpoints

Mitotic cells perform cell replication according to a highly regulated sequence of events, known as the cell cycle. Knowledge of the cell cycle comes from the study of cells in culture under controlled conditions (*in vitro*). The assumption is made that cells in culture behave as cells in their native state, that is, as in their tissue of origin but this must indeed be confirmed by additional experiments in tissues (*ex vivo*) and animals (*in vivo*).

Mitosis is a double-edged sword. On the one hand, it offers cell renewal but on the other hand, the complexity of the process, despite numerous checkpoints and controls, increases the chance for genetic errors. If uncorrected, DNA mistakes set the stage for proliferative arrest, cell death, or uncontrolled proliferation, for example, cancer.

The cell cycle is an intricate process that requires considerable energy, organization, and coordination among the organelles. It is described in Box 5.5.

> **Box 5.5. More About the Cell Cycle**
> The cell cycle consists of separate phases: G_o: rest; G_1: biosynthetic preparation for DNA replication, S: DNA replication; G_2: biosynthetic preparation for physical division into two cells, and M: physical division into two cells. Control factors have been identified that oversee the progress of the cell from G_o through final separation into daughter cells. The transition through the cell cycle from G_o to G_1, G_1 to S, S to G_2, and G_2 to M are tightly regulated. Progression through these phases is largely dependent on a *family of enzymes, the cyclin-dependent kinases (Cdks) that interact in a time-dependent manner with the changing levels of proteins called cyclins*. The mitogenic stimulation of resting cells causes an increase in the level of cyclins sufficient to push the cell from G_o to G_1. Cdk/cyclin complexes interact with other proteins, for example, a protein called *pRB* to activate or inhibit various gene programs necessary for each cycle. Overseeing the process of mitosis are the cell cycle overlords called *cell cycle checkpoints that assure cell progress and fidelity of the copied DNA*. The checkpoint proteins vary with the transition site. A stoppage in G_1 involves activation of the *p53, a tumor suppressor protein* that activates a Cdk inhibitor (p16), one of a family of Cdk inhibitors. Other checkpoints are managed with other kinases and interacting proteins. If cell cycle progression is interrupted, progression ceases until the damage (usually related to DNA) is repaired. If DNA damage is excessive, p53 and pRB activate the pathway to apoptosis. P53 and pRB are also essential players in the regulation of other cell processes that include preservation of cell senescence and autophagy.

Permanent Cell Cycle Arrest (Replicative Senescence) Occurs with Age; Mitotic Cells Express the Senescence-Associated Secretory Phenotype (SASP)

As observed more than 50 years ago (Hayflick and Moorhead 1961), cells in culture divide a finite number of times. The phenomenon is referred to as Hayflick's number and is cell-type specific but importantly it is proportional to the age of the cell donor such that *cells cultured from young donors divide more frequently than cells cultured from older donors*. This observation led to the "mitotic clock" hypothesis of cell aging that hypothesized that at least in replicating cells of the body, an internal clock dictates the number of cell divisions and thus the rate of cell aging. The mitotic clock hypothesis initially suggested a type of "programmed" aging that was subsequently explained by a "nonprogrammed" mechanism of telomere shortening. Figure 5.5 shows a representation of telomere shortening.

Telomeres shorten a finite amount (the degree depends on factors such as oxidative stress) with each round of cell division until a critical number of chromosomes are too short and poorly capped to allow cell division to reliably occur. At this point, a cellular DNA damage response (DDR) is initiated that *permanently arrests the cell cycle* and establishes the state of *replicative senescence in the cell*.

There is now evidence that indicates "cellular senescence is essentially a state of irreversible proliferative arrest caused by stresses that are potentially oncogenic" (Freund et al., 2010). Oncogenic refers to cancer induction. *Telomere shortening is a significant cell stressor*. Other stressors are *tumor-inducing viruses or proteins, severe overt DNA damage, for example, DNA strand breaks, oxidative stress, for example, UV radiation, and aberrant levels of growth factors and hormones*. These stresses initiate many pathways, one of which is DDR. The *DDR encompasses a sophisticated cascade of enzymes and proteins that identifies the DNA damage and attempts to repair it*. If DNA restoration is unattainable, the DDR engages cell cycle mediators to arrest proliferation *or* commit apoptosis. In the case of permanent growth arrest, the metabolic machinery of the cell may change (although the mechanism is not clear) and as a result, the cell secretes factors that (i) seem to preserve the senescent cell (autocrine factors) and (ii) that influence activities in nearby cells (paracrine factors). This *new phenotype is termed the senescence-associated secretory phenotype or SASP*.

The SASP is characterized by an *increase in cell size (volume), reduced metabolic rate, production/secretion of novel substances, and chromatin that condenses in a unique pattern* called senescence-associated heterochromatin foci (SAHF). *SASP secrete 40 or more different proteins* ranging from growth factors and inflammatory/immune mediators to proteases and hormone binding proteins. The lysosomes of the SASP accumulate a new enzyme, *beta-galactosidase* that is of no benefit to the cell but has gained recognition because it is a *stainable biomarker that identifies the SASP* in cultured cells or tissue biopsies.

By virtue of these phenotypic changes, *senescent cells negatively impact adjacent nonsenescent, normal cells*. Proinflammatory factors and degradative enzymes produced by the SASP stress nearby healthy cells, thereby accelerating aging in distal cells but also creating a milieu favorable for disease initiation and progression. It is hypothesized that it is the inflammatory mediators secreted by the accumulating SASP that account in part (another contributor is the aged-immune system) for the persistent low level of chronic inflammation prominent in the elderly, commonly called inflammaging.

All mitotic cells are potential candidates for eventual irreversible replicative or cellular senescence. In other words, *all mitotic cells are likely at some point to stop dividing*. Whether replicative senescence is desirable or detrimental is debated. On the one hand, *replicative senescence prevents tumor formation* (cells are prevented from dividing and hence cannot become tumors), but on the other hand, *replicative senescent cells resist apoptosis by deactivating the cell suicide pathway and thus persist with the novel and potentially disturbing SASP* (see Figure 5.6).

Stem Cells Are Pluripotent and Replenish Missing Cells

Stem cells share with mitotic cells the ability for self-renewal. Unlike mitotic cells, stem cells are pluripotent. This means that under optimal conditions, stem cells are able to differentiate into a specific cell type, for example, muscle or cardiac cell. Stem cells reside in many organs (for example, brain, bone marrow, heart, skeletal muscle, and others) and may also circulate and move to distant organs via blood or lymph. Studies show that stem cells removed from bone marrow (hematopoietic stem cells) and injected into a rat model of myocardial infarction (heart attack) seek out the site of heart damage, where they differentiate into myocardial cells, reduce the area of damage, and improve cardiac function. Small clinical trials in man basically have yielded results similar to those reported in rats. Clearly, improved function through stem cell replacement is a highly prized mechanism that could forestall aging and minimize disease.

It is apparent that despite the tremendous potential of stem cells to "home" to the damaged tissue and replace missing cells, *stem cell activities fail to keep pace with the*

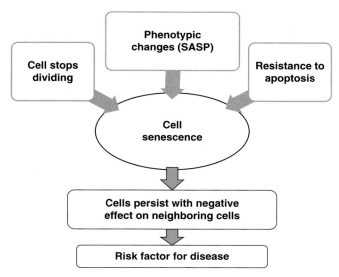
Figure 5.6. Characteristics of cell senescence.

needs of older tissues. It is not known why stem cell replacement capabilities decline with age. It has been proposed that *continuous cell division in the stem cell population leads to replicative senescence* and expression of the SASP similar to other mitotic cells in the body. Since stem cells contain telomerase, the enzyme that restores telomere length and minimizes telomere shortening, replicative senescence via this mechanism is probably unlikely. Whether a deficiency of telomerase arises with time and stress is unknown. Clearly, telomerase is important as exemplified by studies with a mouse model in which telomerase was "knocked out (KO)" specifically in stem cells. The telomerase KO mice exhibited signs of premature aging to include osteoporosis, hair loss, and thymic involution (shrinkage of a major organ of the immune system). These results emphasize the importance of stem cell replacement for survival but whether a telomerase deficiency occurs with age was not demonstrated. Other possibilities, such as dysfunctional mitochondria, accumulation of aggregated protein due to defective autophagy, and activation of the DDR, may contribute to diminished stem cell function.

Additionally, it is hypothesized that the *stem cell environment* is an extremely important factor that could, as it changes in the older organism, adversely affect stem cell function. For example, the hormone estrogen is one of several hormones known to stimulate stem cell differentiation. The effect is mediated by production of endothelial nitric oxide (NO). As the estrogen level declines in menopause, NO levels simultaneously wane. This would adversely limit stem cell function. Additionally, some environmental factors act as negative regulators to depress stem cell function. Examples include elevated blood levels of oxidized LDL (bad cholesterol) and toxic compounds from smoke. Methodologies that place aged stem cells in a milieu comparable to that of a young organism generally report a rejuvenation of the older cells. Future studies seek to define the facilitators of this rejuvenation.

TABLE 5.1. Mitotic and Postmitotic Somatic Cells

Mitotic Cells	Postmitotic Cells
Endothelial cell (inner lining of blood vessels)	Neuron: cell of the central and peripheral nervous system
Epithelial cell (skin, gastrointestinal, respiratory linings)	Skeletal muscle cell: voluntary muscle cell
	Red blood cell (gas exchange); neutrophil; eosinophil; basophil (innate and immune functions)
Osteocyte (bone cell)	Myocardial cell: heart cell
Lymphocyte (T cell; B cell of the immune system)	Endocrine gland cells, for example, ovary, testes, thyroid, parathyroid, pancreas, pituitary, pineal
Fibroblast: connective tissue cell	Macrophage: scavenger cell; immune cell assistant
Stem cells: located in tissues	Renal cell: kidney cell
	Hepatocyte^a: liver cell
	Glia cell^a: auxiliary cell in brain
	Smooth muscle cell^a: contractile cells of blood vessel, gastrointestinal tract, uterus, ureter
	Chondrocyte^a: joint/cartilage cell

^{a}Mitosis induced with injury otherwise postmitotic.

Aging Postmitotic Cells May Activate Apoptosis: Cause of Tissue Atrophy

Postmitotic cells generally do not divide. Myocardial (heart) cells and neurons (brain cells) fall into this category (see Table 5.1). Unlike mitotic cells that stop dividing and exhibit a characteristic SASP, postmitotic cells tend to *disappear* with age. This disappearance is a result of apoptosis. In the young organism, the loss of postmitotic cells is replenished with cells derived from stem cell reserves. As indicated above, current findings suggest that in aged organisms, *cell replacement by stem cells occurs less frequently or not at all*. Therefore, a major outcome of postmitotic cell disappearance is tissue atrophy or shrinkage. *Loss of tissue mass usually results in reduced or impaired tissue function*. An example is sarcopenia (Chapter 7).

Organelle Dysfunction Is the Main Reason Postmitotic Cells Undergo Apoptosis
Several mechanisms have been proposed to explain the loss of postmitotic cells with age. First, some data suggest that *dysfunctional mt* especially in skeletal muscle and myocardial cells contribute to cell extinction. It is proposed that a severe decline in mt function, for example, disrupted electron transport and decline in ATP formation, causes the release of the mt protein, cytochrome c. This protein is an important signal to activate the apoptotic process.

Second, other data suggest that large genomic (DNA) rearrangements observed in aged cardiac tissue and select regions of the brain (studied in mice) may contribute to the disappearance of some postmitotic cells. These rearrangements can be monitored in culture and are inducible by H_2O_2, suggesting a possible role for excess oxidants and free radicals in the death of postmitotic cells. Accumulation of large genomic mutations adversely modifies the expression of a significant number of essential genes. This type of damage exceeds the repair capacity of the DDR. Hence, the outcome is generally apoptosis.

Third, a decline in lysosomal function, for example, autophagy, also contributes to postmitotic cell death. The inability to recycle oxidized or aggregated proteins and damaged organelles permits accumulation of defective organelles and cellular debris that promotes internal damage and leads to apoptosis. This is described above in relation to the mitochondrial–lysosomal axis theory of aging. This theory further proposes that the loss of essential postmitotic cells, for example, neurons of the hypothalamus, leads to system (affecting all organs) dysregulation and a shorter lifespan.

Recent evidence from a mouse model suggests that *some aged postmitotic cells may not disappear, but continue to live with a new phenotype similar to the senescent mitotic cell*. In this altered state, they show some signs of the SASP, for example, express beta-galactosidase and exhibit a nucleus with SAHF expression, and could be expected to contribute to aging and age-related diseases in a manner similar to that described for senescent cells of mitotic origin. This cellular change has not been reported in human postmitotic cells but if confirmed, the presence of SASP in postmitotic cells, such as neurons, might identify a new vulnerability to age-related diseases.

CELL DEATH OCCURS BY AUTOPHAGY, APOPTOSIS, OR NECROSIS

Cell death pathways of autophagy, apoptosis, and necrosis act through the master regulator gene called the tumor suppressor gene (p53). Pathway selection depends on the cell and the conditions that prevail. Autophagy operates to recycle resources and maintain quality control, but under *conditions that excessively activate this process*, for example, low insulin and irradiation, *autophagy leads to cell death* with enlargement of vacuoles and cell content shrinkage.

Apoptosis is programmed cell death, a highly regulated and controlled process introduced above. Specific steps are needed for its activation, commitment, and execution. Activation may be mediated externally through plasma membrane "death" receptors, or internally via major stressors (DNA damage, mutagens). Both routes are assisted by mt membrane changes that eventually stimulate a cascade of degradative enzymes that perform the following activities: caspases digest proteins, endonucleases break down DNA, and multiple enzymes rearrange the cell cytoskeleton and change the cell surface receptors. Together this coordinated degradation of the cell components results in cell shrinkage to a size that can be digested by a phagocytic macrophage (scavenger cell). With apoptosis there is minimal debris and hence few initiators of inflammation. Apoptosis is the likely pathway taken by postmitotic cells. Recall that the SASP resists apoptosis.

Unlike apoptosis, necrosis is an unregulated, chaotic process that results from severe ATP shortage. External factors such as hyperthermia (marked increase in tissue temperature), hypoxia (decrease in oxygen availability), metabolic poisons, cell and tissue trauma, or any particularly noxious stimulus will initiate cell necrosis. In contrast to organized cell shrinkage brought about by apoptosis, the necrotic cell swells allowing leakage of destructive enzymes within the cell and subsequently into the cell's environment. Necrosis attracts other cells such as neutrophils and macrophages that attempt to remove the debris but this "cleanup" produces additional substances that facilitate more tissue destruction.

RELATION OF CELLULAR AGING TO DISEASE

Cell senescence plays a role in several age-associated pathologies. A mechanism for malignancy induction has been proposed by Campisi et al. (2005), who identified the SASP cell as a "bad neighbor." Although the SASP is prevented from becoming a tumor (replicative senescent), it chronically produces and secretes factors that are potent inducers of cell transformation in neighboring cells, the first step in tumor formation. As adjacent cells are continually exposed to substances that modify them, uncontrolled proliferation and migration (metastases) ensue, initiating and supporting neoplastic disease (see Figure 5.7).

Other diseases such as atherosclerosis may be initiated by the SASP cell. Atherosclerosis is a disease that changes the structure of the artery wall. The arterial lesions (plaques) enlarge to the point of rupture and clot formation (heart attack, stroke, and gangrene) or allow parts of the wall to weaken and split (aneurysm). Atherosclerotic lesions tend to develop in vasculature areas of irregular blood flow. It is hypothesized that flow irregularity stresses endothelial cells to the point of replicative senescence and the SASP. Senescent endothelial cells perform poorly, recruit inflammatory cells, and create an environment that favors artery wall remodeling.

Osteoarthritis, a degenerative joint disorder affecting as many as 75% of individuals over 65 years of age may also develop with help of SASP cells. Although the mechanism is not clear, traumatized joints exhibit a preponderance of SASP chondrocytes that resist apoptosis, and produce proinflammatory factors/lytic enzymes. Joint structure is slowly destroyed through SASP-induced persistent inflammation. The prevalence of osteoarthritis

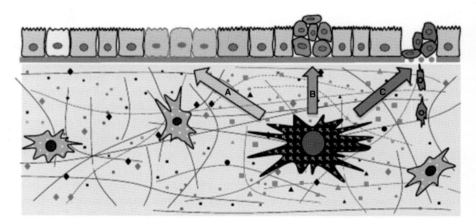

Figure 5.7. Proposed role of senescent cell in development of cancer. Senescent cell releases factors that modify nearby cells (A), promote uncontrolled cell division (B), and enhance invasive movement to other sites (C). (Reprinted with permission Labarge et al. (2012).) (See plate section for color version.)

suggests that some joint stress may be unavoidable, but the onset and severity can be minimized with weight loss and reduction/avoidance of traumatic physical activities.

Dysfunctional mt are postulated to contribute to neuronal loss in Parkinson's disease (PD). Histological findings in PD patients at autopsy show a marked presence of dysfunctional mt in brain sites specifically affected by PD. Dysfunctional mt are potent inducers of apoptosis and could account for the loss of dopaminergic neurons, a hallmark of PD. The cause of mt dysfunction in PD has not been identified. In hereditary PD, mutated genes are responsible for adverse effects on mt biogenesis and function. In cases with no hereditary links, one possible cause may be exposure to toxins, a hypothesis supported by results in animal models.

In sum, several age-related diseases are intertwined with the SASP and dysfunctional mt. Preliminary data suggests that reduced autophagy may contribute to PD and possibly Alzheimer's disease but additional studies are needed.

Summary of Aging of Organelles and Cells

Organelle	Normal Function	Age Change
Mitochondria	Breakdown of nutrients/use of oxygen to produce ATP, carbon dioxide and water	↓ Production of ATP ↑ Formation of ROS → damage/↑ redox potential Fusion > fission
Lysosome	Autophagy (recycling of macromolecules) maintenance of mt health	Slowing of autophagy ↓ Mitophagy ↓ Importation of cell components Accumulation of unwanted toxins → lipofuscin
Nucleus	Master planner of cell activities Directs protein synthesis/controls cell division	Epigenetic change in gene expression (protein synthesis changes) Cell division slows or stops Change in nuclear structure Telomere shortening
Peroxisome	Oxidation of fats and alcohols	↓ Oxidation reactions ↓ Importation of catalase ↑ Hydrogen peroxide
Cells		
Mitotic	Renewal serving numerous functions (barrier, immune, connective, bone)	Replicative senescence (SASP) Sensitive to necrosis
Postmitotic	Nerve, muscle, cardiovascular, metabolic	Apoptosis SASP(?)
Stem	Pluripotent; replacement	Replicative senescence(?) Nonfunctional (lack of supporting milieu)

SUMMARY

Organelles contribute to cell survival in diverse ways: The nucleus contains the DNA "cell blueprint" and directs the synthesis of proteins (transcription/translation); mt utilize oxygen and nutrients to generate power in a transportable energy molecule, ATP (oxidative phosphorylation); lysosomes recycle cell components (autophagy); peroxisomes oxidize and detoxify lipids, acids, and alcohols.

Age changes have been described for several cell organelles. Aging in mt appears as morphological changes, increased mtDNA mutations, and decreased function. Oxidative stress, mtDNA mutations, or altered fission/fusion are possible inducers. Age-associated changes in the nucleus appear as shortened telomeres, remodeled chromatin, and possible defects in the structural lamins. Peroxisomes experience a reduced ability to import the enzyme, catalase. This results in an increase in hydrogen peroxide, whose leakage would damage the cell. Lysosomes are unable to import compounds/organelles destined for degradation and lipofuscin, unrecycled macromolecules and defective mt accumulate and further limit both lysosomal and cell function.

The aging of mitotic cells is termed replicative or cell senescence. Senescent cells are phenotypically altered and as such express the senescence-associated secretory phenotype (SASP). Various factors such as telomere shortening, genotoxicity, and strong mitogenic stimuli produce a SASP characterized by growth arrest, resistance to apoptosis, and production/release of harmful mediators. Senescent cells acutely suppress tumorigenesis; in the long term, however, their persistence causes tissue dysfunction and paradoxically tumorigenesis.

Postmitotic cells mainly disappear with age. As a result, tissues decrease in mass and function. Organelle dysfunction or genomic rearrangements contribute to this disappearance by the mechanism of apoptosis (cell suicide).

The aging of stem cells, needed to replenish absent cells, is poorly understood. Decline in stem cell function has been attributed to dysfunctional organelles, loss of telomerase, and changes in stem cell milieu.

Apoptosis is the complex process of organized cell suicide that allows the cell to disappear without inducing an inflammatory response. In contrast, trauma kills cells by necrosis, a disorganized process guaranteed to generate inflammation.

SASP cells contribute to disease development in several ways: permissive for artery wall remodeling (atherosclerosis), promote tumorigenesis (cancers), accelerate joint destruction (osteoarthritis). Organelle aging (mt, lysosomes) likely contributes to Parkinson's Disease.

CRITICAL THINKING

What role does replicative senescence play in increased vulnerability of the elderly to disease?

Which of the organelles contributes the most to cellular health? Why?

Which of the organelles contributes the most to cellular aging? Why?

Which is the best way for a cell to die? Explain

KEY TERMS

Adenosine triphosphate (ATP) major energy molecule produced in the mitochondria during conversion of oxygen and nutrients to water and carbon dioxide. Energy is released to carry out movement against a gradient, enzyme activity, secretory activity, muscle contraction, nerve conduction, and so on.

Apoptosis cell suicide. It is a highly controlled process that systematically degrades the cell into tiny components. The cell disappears without inducing inflammation.

Autocrine process whereby a cell secretes mediators that act on the secreting cell.

Autophagy the process of recycling of macromolecules, organelles, and membranes in a controlled fashion to optimize reuse. Process carried out by lysosomes.

Biogenesis generation *de novo* of cell organelles; occurs with mitochondria and peroxisomes.

Cell smallest unit of living matter.

Cell cycle description of a complex series of events that must transpire for a resting mitotic cell to divide into two identical daughter cells.

Chaperons proteins that act as protectors to other proteins as in chaperon-mediated autophagy.

Chondrocytes cells lining the joints; subjected to repeated trauma and candidates for replicative senescence.

Chromatin specific subdivision of the chromosomes that contain actively transcribed genes (euchromatin) or silent genes (heterochromatin).

Electron transport chain (ETC) series of enzymes found in the mitochondria that are capable of using electrons from nutrients to convert oxygen to water and carbon dioxide and form ATP.

Euchromatic see chromatin.

Gene sequence of DNA that contains the information to biosynthesize a particular protein.

Genotype all the genes in an organism.

Golgi apparatus organelle of the cell that is involved in production of peroxisomes; also performs a secretory function.

Hayflick's number maximal number of cell divisions that any one cell type may experience. Cells from short-lived organisms have lower maximal number than cells from long-lived organisms.

Heterochromatic see chromatin.

Hydrolytic enzymes enzymes capable of breaking down macromolecules with the addition of water.

Inflammaging term used to describe the presence of low-level chronic inflammatory state in the elderly. It is proposed that this state contributes significantly to aging. Originates from changes within the immune system and from the SASP.

Lysosomes cell organelle that functions to recycle macromolecules, organelles, and membranes by the process of autophagy.

Mitochondria the cell organelle considered the powerhouse of the cell because it produces the high energy molecule, ATP.

Mitophagy autophagy of mitochondria by lysosomes and other pathways.

Mitosis cell division or replication.

Nucleus cell organelle that houses the chromosomes (genetic blueprint) for the cell.

Organelle a substructure of the cell defined by membranes and characteristic function. Examples includes mitochondria, peroxisomes, lysosomes, nucleus, endoplasmic reticulum, and Golgi apparatus.

Osteoarthritis degenerative joint disease involving replicative senescence facilitated by repeated stress on the joints.

Oxidative phosphorylation the process whereby oxygen is converted to ATP with the aid of the electron transport chain in the mitochondria.

p53 protein important cell protein that is a cell cycle gate keeper. If appropriate conditions are not met for normal replication, p53 may prevent replication or initiate apoptosis.

Paracrine process whereby cell secretes mediators that act on neighboring cells.

Peroxisomes cell organelles that perform lipid/alcohol oxidations.

PGC-1a peroxisome proliferator-activated receptor gamma coactivator-1 alpha is a protein that acts as a coactivator of many important gene programs especially those involved in metabolism.

Phenotype all the characteristics of a cell or organism excluding the genotype.

Postmitotic cells that are fully differentiated and can no longer divide.

pRB protein cell protein that prevents abnormal cell division; a gatekeeper for the cell cycle.

Replicative senescence the mechanism whereby cells lose the ability to divide and phenotypically modulate to large, proinflammatory secreting cells that resist apoptosis.

SASP senescence-associated secretory phenotype. Descriptive term to identify aged mitotic cells that no longer divide, resist apoptosis, and secrete unwanted proinflammatory or harmful substances. Also called replicative senescent.

Stem cell pluripotent cells with capacity to develop into any cell type given the appropriate environment.

Telomerase the enzyme with the capability to repair the telomere so that shortening is minimized or prevented.

Telomere the ends of the chromosomes that function as protective regions; with each cell division, the telomere is clipped and becomes shorter until it is too short to support another cell division and replication ceases at that point.

BIBLIOGRAPHY

Reviews

Bratic A, Larsson N-G. 2014. The role of mitochondria in aging. *J. Clin. Invest.* **123**(3):951–957.

Brunk UT, Terman A. 2002. The mitochondrial–lysosomal axis theory of aging: accumulation of damaged mitochondria as a result of imperfect autophagocytosis. *Eur. J. Biochem.* **269**(8):1996–2002.

Calado RT, Dumitriu B. 2013. Telomere dynamics in mice and humans. *Semin. Hematol.* **50**(2):165–174.

Campisi J. 2005. Senescent cells, tumor suppression and organismal aging: good citizens, bad neighbors. *Cell* **120**(4):512–522.

Campisi, J. 2013. Aging, cellular senescence and cancer. *Ann. Rev. Physiol.* **75**:685–705.

Cline SD. 2012. Mitochondrial DNA damage and its consequences for mitochondrial gene expression. *Biochim. Biophys. Acta* **1819** (9–10):979–991.

Freund A, Orjalo AV, Desprez P-Y, Campisi J. 2010. Inflammatory networks during cellular senescence: causes and consequences. *Trends Mol. Med.* **16**(5):238–246.

Galluzzi L, Kepp O, Trojel-Hansen C, Kroemer G. 2012. Mitochondrial control of cellular life, stress, and death. *Circ. Res.* **111**(9):1198–1207.

Giordano CR, Terlecky SR. 2012. Peroxisomes, cell senescence, and rates of aging. *Biochim. Biophys. Acta* **1822**(9):1358–1362.

Gonzalez-Suarez I, Gonzalo S re-alphabetize 2010. Nurturing the genome A-type lamins preserve genomic stability. *Nucleus* **1**(2):129–135. Landes Bioscience to distinquish between Nucleus by Elsevier.

Gorbunova V, Seluanov A, Mao Z, Hine C. 2007. Changes in DNA repair during aging. *Nucleic Acids Res.* **35**(22):7466–7474.

Hayflick L, Moorhead, PS. 1961. The serial cultivation of human diploid cell strains. *Exp. Cell Res.* **25**(3): 585–621.

Kaushik S, Cuervo AM. 2012. Chaperone-mediated autophagy: a unique way to enter the lysosomal world. *Trends Cell Biol.* **22**(8):407–417.

Labarge R-M, Awad P, Campisi J, Desprez P-Y. 2012. Epithelial–mesenchymal transition induced by senescent fibroblasts. *Cancer Microenviron.* **5**(1):39–44.

Lynch-Day MA, Klionsky DJ. 2010. The Cvt pathway as a model for selective autophagy. *FEBS Lett.* **584**(7):1359–1366.

Sena LA, Chandel NS. 2012. Physiological roles of mitochondrial reactive oxygen species. *Mol. Cell* **48**(2):158–167.

Stuart JA, Maddalena LA, Merilovich M, Robb EL. 2014. A midlife crisis for the mitochondrial free radical theory of aging. *Longev. Healthspan* **3**(1):4–18.

Terman A, Kurz T, Navratil M, Arriaga EA, Brunk UT. 2010. Mitochondrial turnover and aging of long-lived postmitotic cells: the mitochondrial–lysosomal axis theory of aging. *Antioxid. Redox Signal.* **12**(4):503–535.

Titorenko, VI, Terlecky SR. 2011. Peroxisome metabolism and cellular aging. *Traffic* **12**(3):252–259.

Experimental

Cortopassi GA, Arnheim N. 1990. Detection of a specific mitochondrial DNA deletion in tissues of older humans. *Nucleic Acids Res.* **18**(23):6927–6933.

Csiszar A, Podlutsky A, Podlutskaya N, Sonntag WE, Merlin SZ, Philipp EER, Doyle K, Davila A, Recchia FA, Ballabh P, Pinto JT, Ungvari Z. 2012. Testing the oxidative stress hypothesis of aging in primate fibroblasts: is there a correlation between species longevity and cellular ROS production? *J. Gerontol. A Biol. Sci. Med. Sci.* **67**(8):841–852.

Safdar A, Hamadeh MJ, Kaczor JJ, Raha S, deBeer J, Tarnopolsky MA. 2010. Aberrant mitochondrial homeostasis in the skeletal muscle of sedentary older adults. *PLoS One* **5**(5):e10778.

Short KR, Bigelow ML, Kahl J, Singh R, Coenen-Schimke J, Raghavakaimal S, Nair KS. 2005. Decline in skeletal muscle mitochondrial function with aging in humans. *Proc. Natl. Acad. Sci. USA* **102**(15):5618–5623.

Tormos KV, Anso E, Hamanaka RB, Eisenbart J, Joseph J, Kalyanaraman B, Chandel NS. 2011. Mitochondrial complex III ROS regulate adipocyte differentiation. *Cell Metab.* **14**(4):537–544.

Yen TC, Chen YS, King KL, Yeh SH, Wei YH. 1989. Liver mitochondrial respiratory functions decline with age. *Biochem. Biophys. Res. Commun.* **165**(3):944–1003.

Yen TC, Su JH, King KL, Wei YH. 1991. Ageing-associated 5 kb deletion in human liver mitochondrial DNA. *Biochem. Biophys. Res. Commun.* **178**(1):124–131.

Zhang C, Cuervo AM. 2008. Restoration of chaperone-mediated autophagy in aging liver improves cellular maintenance and hepatic function. *Nat. Med.* **14**(9):959–965.

SECTION III

ORGAN SYSTEMS: OUTER COVERING AND MOVEMENT: INTEGUMENTARY, SKELETAL MUSCLES, AND SKELETAL SYSTEMS

This section discusses the effects of aging on the outer covering, the integument or skin, and the underlying tissues, skeletal muscles, and bone. It is partly a positional organization since skeletal muscles are attached to the bones and the skin is the enclosing organ. The spatial relation creates serious age "co-dependent" changes. For example, aging muscles (loss of mass and strength) accelerate skeletal changes (loss of bone mass) that predispose to falls and fractures. Additionally but to a lesser extent, aging muscle adds to alterations of skin structure, noticeably wrinkles and sags.

Each tissue excels in a major function: skin (barrier); skeletal muscle (movement); bone (structure). As the maintenance of these tissues declines with age, so do these functions. The *major consequences* are a decrease in wound healing, slowed and unsteady balance and gait, and fragile bones. However, just like cells that perform multiple functions, tissues also participate in many other activities. The skin is home to many cells of the innate and adaptive immune system and is the site of vitamin D production. Muscles are critical for maintenance of overall basal metabolic rate and the regulation of insulin sensitivity and bone is the reservoir for calcium, one of the most important minerals in human physiology. Similarly, a decline in tissue structure and function bodes ill for this array of auxiliary but important functions.

These three tissues share some common elements. First, a variety of dysfunctional cells arise with age: replicative senescent cells (SASP) of skin; apoptotic-prone cells of bone and muscle; dysfunctional myofibers in skeletal muscle; unstable stem cells in muscles and hair follicles. Oxidative stress and inflammatory mediators also accumulate with age in these tissues. Second, although there are several interventions to reduce aging of skin, muscle. and bone, for example, sunscreens, resistance exercise, quality protein consumption, and high impact exercise, the most effective

Human Biological Aging: From Macromolecules to Organ Systems, First Edition. Glenda Bilder.
© 2016 John Wiley & Sons, Inc. Published 2016 by John Wiley & Sons, Inc.

interventions are those that begin in childhood. For example, daily use of sunscreen could prevent ultraviolet radiation-induced aging and skin cancer and an uninterrupted program of vigorous aerobic and resistance exercise begun at a young age could maximize muscle strength and fortify bone density. By maintaining skin, muscle and bone function from an early age onward, the inevitable age-related decrements would have a smaller impact on independence and quality of life.

There are many unresolved issues regarding aging in skin, muscle and bone. Of importance is the lack of consensus on a definition of sarcopenia and dynapenia. Also debated are: procedures to accelerate wound healing in the elderly, the optimal amounts of vitamin D and calcium for bone health in the elderly, nonpharmacological interventions to prevent osteopenia, and effective means to stimulate stem cell function in older adults.

6

AGING OF THE INTEGUMENTARY SYSTEM

OVERVIEW

Unique Aspects of the Integument (Skin)

The integument, commonly called the skin is unique in several ways. It is the largest organ in the human body and because it is the outermost covering, it is constantly assaulted by ultraviolet radiation (UVR) from the sun and by pollutants (smoke, exhaust fumes, and ozone) in the air. Additionally, it is this outer covering, especially the face, that reflects not only our health, beauty, and ethnicity but also our age. As such, considerable effort is expended in maintaining a youthful skin. One final unique aspect is that the skin provides a wealth of valuable services: barrier protection along with temperature regulation, water loss prevention, immunological surveillance, vitamin D production, and sensory input.

Studies on aged skin have encountered several issues. These relate to variability of the skin, studies of small sample size, types of aging, and study focus. First, the assumption is made that all skin is the same, but the variation in thickness, density of glands and hair follicles, fat cell type, and microvascular complexity suggests otherwise. Thus, site location may impact interpretation of age changes. Second, gross observations and measurements must be balanced with cellular and molecular data. Defining underlying mechanisms provides insights into tissue dysfunction. Skin biopsies achieve this goal, but invasiveness limits study size and results from studies with few participants may be misleading. Finally, there is a need to distinguish between environmentally-induced aging and natural or intrinsic aging. Focus has been predominately on environmental damage. Our culturally promoted devotion to

---

Human Biological Aging: From Macromolecules to Organ Systems, First Edition. Glenda Bilder.
© 2016 John Wiley & Sons, Inc. Published 2016 by John Wiley & Sons, Inc.

youthfulness has generated not only a considerable scientific literature on aging of the face but also a multibillion dollar industry of cosmetics and cosmeceuticals (drugs used as cosmetics) that seek prevention, reversal, and amelioration of wrinkled aged skin. In contrast, there are a lesser number of studies addressing skin cancers and a dearth of information on wound healing in elderly skin. The treatment of skin cancers and infections from poor wound healing significantly impact not only health care expenditures but also morbidity and mortality of the elderly.

Skin Aging Results from Extrinsic and Intrinsic Effects

Extrinsic aging refers to age changes in the skin caused by environmental insults. Among the insults, *UVR is considered the most harmful.* Other insults from cigarette smoke and pollution (diesel exhaust, ozone) contribute secondarily to extrinsic aging.

Extrinsically aged skin is tissue that is chronically exposed to the environment. The "unprotected" areas are the face, neck, forearms, and hands. With regard to UVR, skin type affects the extent of damage with fair skin being the most sensitive and dark skin the least. However, dark skin is, nevertheless, vulnerable to UVR damage. *Extrinsic aging is severe and not only produces major wrinkles and sags, called photoaging, but also causes benign/malignant skin cancers* and may possibly accelerate intrinsic aging.

Extrinsic aging *damages skin by production of free radicals and oxidants.* UVR, cigarette smoke, and ozone generate and/or contain an abundance of free radicals and oxidants. As discussed in Chapter 4, oxidative stress damages DNA, peroxidizes membrane lipids, and cross-links proteins. Additionally, oxidants deplete the skin of antioxidant and redox defense mechanisms preventing macromolecular repair. In the presence of continued oxidative stress, the skin becomes dysfunctional.

Intrinsic aging or chronological aging defines age changes in skin areas that are not continuously exposed to the environment. To study intrinsic aging, biogerontologists examine "protected" skin, for example, on the inner upper arms, the buttock, or in comparison to aging of the face, the postauricular (behind the ear) skin. Intrinsic age changes are considered minor compared to extrinsic aging. Clearly, the extent of wrinkling with intrinsic aging is less, but since many other intrinsically-induced age changes such as skin thinning, loss of elasticity, fine lines, decreased pigmentation, and changes in permeability promote tissue dysfunction, intrinsic aging cannot be ignored. Intrinsic alterations are accelerated by decline of hormonal levels especially evident with the dramatic loss of estrogen in menopause.

Skin Layers

Epidermis, Dermis, and Hypodermis Define the Skin
The skin is comprised of three layers: (i) the epidermis or the outermost 8–15 layers of mainly keratinocytes and a lesser number of melanocytes; (ii) the dermis or large matrix-filled layer underneath the epidermis populated with fibroblasts and containing the microvascular system; and (iii) the hypodermis or subcutis, the bottom layer overlaying muscle and filled with adipocytes (fat-containing cells) and connective tissue. Extrinsic and intrinsic aging modify each layer differently (Figure 6.1).

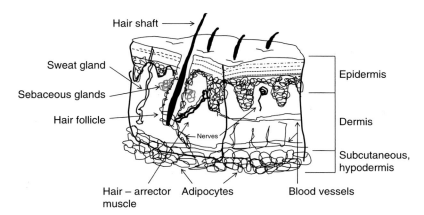

Figure 6.1. Cross section of skin illustrating the three layers and associated structures.

Keratinocytes of the Epidermis Are Continually Renewed from the Basal Layer The topmost layers (stratum corneum) of the epidermis routinely die and flake off, only to be replaced by the cells beneath them that originate from the constantly replicating bottom or basal layer. The basal epithelial cells give rise to the prominent cell type of the epidermis, the keratinocyte, that amass many select proteins, prominent among which is keratin that supplies "waterproofing" to the skin and provides an innate barrier against microbes and toxins. Additional protection is created by the "tight junctions" established between the keratinocytes.

The *basal layer of epithelial cells is uniquely undulating* (hills and valleys) to form a structure that facilitates distribution of nutrients and mediators to the epidermis from the blood supply, located in the dermis.

Melanocytes and Langerhans Cells Provide Protection *Melanocytes containing the sun-protective pigment melanin* (responsible for tanning) are found in the epidermis and supply the keratinocytes with this compound in response to stress, for example, UVR. In addition *antigen-processing cells, called Langerhans cells*, act as sentinels that detect allergens, microbes, and toxins. The epithelial layer also contains *modified epithelial cells that give rise to hair follicles and nails*.

AGING OF THE EPIDERMIS

Extrinsic Aging of the Epidermis

UVR Is the Main Cause of Extrinsic Aging; Pollution Also Contributes
UVR is comprised of UVA (340–400 nm wavelength), UVB (290–315 nm wavelength), and UVC (200–280 nm wavelength) components. Because of its longer wavelength, *UVA penetrates into the dermis* and damages the dermis by multiple mechanisms, noted above. This causes cosmetic facial photoaging. *UVB in contrast penetrates the epidermis*, but no further and is responsible for tanning and sunburn (erythema). Sunburn also harms the skin. Furthermore, UVB is considered an inducer

of skin cancer (photocarcinogenesis). More recently, evidence suggests that UVA may also produce DNA damage sufficient for photocarcinogenesis. UVC does not enter the atmosphere.

Extrinsic Aging Markedly Changes Epidermal Structure The extrinsically aged epidermis appears leathery and blotchy, changes typically seen in heavy smokers or outdoor professionals, for example, farmers. It is characterized by epithelial thickening and a marked flattening of the dermal–epidermal junction. With regard to UVR, low or infrequent exposures stimulate the proliferation of keratinocytes and melanocytes, an effect that *increases production and distribution of the sun-protecting protein, melanin from the melanocytes to the new keratinocytes*. Melanin neutralizes oxidative damage induced by UVR and provides protection from additional damage. *The extent of protection depends on the amount of UVB exposure (degree, duration) and the amount of produced melanin (darkening)*. With chronic UVR exposure, this "tan" protection wanes and epidermal cells are stimulated to divide repeatedly, an effect that thickens the epidermis and produces hyperpigmentation and cancer.

Spotty pigmentations produced by extrinsic aging have been variously called senile lentigo, or sun, liver, or age spots. These are defined flat areas of discoloration (light brown to black) of assorted sizes. Mechanistic studies suggest that melanin production in these lesions is elevated by the UVB-stimulating production of select compounds, for example, the vasoconstrictor, endothelin-1 and the growth factor, stem cell factor. The enhanced interaction between the melanocyte and keratinocyte produced by these mediators produces hyperpigmentation.

Another change is a *reduction in the nourishment of the epidermis* that exacerbates extrinsic damage. This results from the marked flattening of the epidermis–dermis junction. It has been reported that the reduced integrity of the epidermis makes it vulnerable to blistering.

Intrinsic Aging of the Epidermis

Intrinsic Aging Slows Epidermal Activity Unlike the thickened epidermis of extrinsic aging, the *epidermis of intrinsic aging becomes thinner due to the slowdown of mitosis within the basal epithelial layer. This eventually culminates in an accumulation of replicative senescent cells*. Ressler et al., (2006) demonstrated an age-associated increase in senescent epidermal cells (measuring the presence of cyclin-dependent kinase inhibitor, p16, a biomarker of senescent cells) in human biopsy samples from multiple skin locations from birth to 95 years of age.

The slowing of the cell cycle in aged epidermal cells reduces epidermal renewal with age, estimated to be twofold slower compared to that of younger individuals (60 versus 28 days). The presence of a thinner epidermis curtails the effectiveness of the skin as a protective barrier and the slowed renewal of epithelial cells contributes to skin dryness, as the outer cells, infrequently replenished, lose their ability to bind water. Fortunately, the undulating dermal–epidermal boundary is fairly well maintained.

Intrinsic aging of melanocytes tends to produce skin lightening as melanin pigment production declines, *a change that reduces UV protection*. Langerhans cells decline in number, allowing for a decrease in cutaneous immune surveillance. Langerhans cells have the capacity to recognize antigens, for example, pathogens and toxins, and present it to immunocompetent lymphocytes, thus providing an adaptive immunity in the outermost layers of the body. Associated with Langerhans cells are mast cells filled with the amino acid histamine. Activation of mast cells results in the classic "hives" or redness, wheal, and flare response. Both Langerhans and mast cells may play a role in certain skin conditions, for example, atopic dermatitis (a type of eczema).

Aging of the Hair Follicles Leads to Graying
Hair growth occurs in stages of active growth (anagen), transition (catagen), resting phase (telogen), and expulsion of hair shaft with replacement regrowth and start of another cycle. During anagen, the melanocytes in the hair bulb transfer pigment to the hair shaft. *This normal coupling of hair growth and pigmentation occurs for approximately 10 cycles or until 40 years of age.* Thereafter *less pigment is transferred into the hair shaft producing gray and white hair.*

The reason for this change is poorly understood. Limited data show that with age hydrogen peroxide accumulates and reaches millimolar concentrations in hair follicle cells (melanocytes and hair shaft cells). Hydrogen peroxide is a potent oxidant of sulfhydryl groups on methionine-containing proteins. Because of this, key antioxidative enzymes, catalase, methionine sulfoxide reductase systems A and B, and a major melanogenesis-stimulating enzyme tyrosinase are inhibited by the presence of high concentrations of hydrogen peroxide. Other enzymes also are likely oxidized with the outcome of reduced melanocyte function or apoptosis.

The *hair also thins with age, termed senescent alopecia.* This is characterized by a highly variable reduction in hair density, and quality of the hair shaft (generally thinner). The incidence is hard to define. One large subjective study of postmenopausal women (questionnaire analysis) found that diffuse generalized scalp hair loss occurred in 26% of the women; frontal hair loss occurred in 9% of the respondents. However, senescent alopecia is complicated by the high prevalence of another type of hair loss, androgenetic alopecia (male pattern baldness), that affects 70% of elderly men. Female pattern baldness also occurs and affects about 30–40% of older females. Unlike androgenetic alopecia, senescent alopecia is androgen and inflammation independent. As with sun-exposed skin, the scalp is subject to the oxidative effects of UVR. Whereas an exacerbating role for UVR in androgenetic alopecia has been observed, the contribution of extrinsic aging to senescent alopecia remains unknown.

AGING OF THE DERMIS

The Dermis Is Thicker Than the Epidermis and Loosely Configured

The dermis contains *mostly connective tissue proteins, including collagens (80%) and elastins, and polysaccharides called glycosaminoglycans (GAGs).* Dermal proteins are made and secreted by *resident cells identified as fibroblasts*. Also *found in the dermis are a microvascular system, sweat glands, sebaceous or oil-secreting glands, modified muscle fibers attached to hair follicles, and nerve endings sensing pain, temperature, and touch.*

Extrinsic Aging of the Dermis

Deep Wrinkles Result from Extrinsically Induced Changes
In addition to the epidermal changes caused by extrinsic aging (discussed above), the dermis is also dramatically impacted by extrinsic aging. The results are deep wrinkles, sags and telangiectasia (appearance of small blood vessels, capillaries below skin surface). *The photoaged dermis exhibits an abundance of cross-linked collagens and decreased amounts of mature normal collagen.* Excessive amounts of elastin are produced often called solar "elastosis." The elastin is weakened and fragmented due to enzymatic degradation. The *chaotic matrix is attributed to dysfunctional fibroblasts* that fail to produce replacement collagen and additionally secrete degradative enzymes, for example, matrix metalloproteinases (MMPs) that destroy the existing collagen and elastin.

Loss of structural support by collagen and elastin is detrimental in several ways. First, fibroblasts need an orderly matrix to function. Results of cell culture studies show that fibroblasts placed on abnormal substratum become dysfunctional, secrete MMPs, and produce less collagen, an experimental scenario analogous to extrinsic aging. The success of professional dermal fillers is ascribed to the re-establishment of a normal matrix specifically benefitting fibroblasts. This "rejuvenated" environment promotes the much needed "mechanical" stimulus that induces synthesis of new collagen and represses synthesis of MMPs. Second, lack of dermal structure nearly obliterates optimal skin elasticity and strength. As the force of gravity overcomes this weakness, sags and wrinkles appear.

Intrinsic Aging in the Dermis

Modest Dermal Changes Occur with Intrinsic Aging
There are a number of studies that have measured intrinsic changes in the dermis of individuals of different ages (cross-sectional studies). The *replication of fibroblasts slows*, and collagen and elastin levels decline and exhibits evidence of oxidative damage. Structural changes also occur in other matrix compounds, especially those that bind water such as hyaluronic acid and chondroitin sulfate; *water distribution in the dermis becomes uneven with age*. However, overall these changes are *small in magnitude compared to the massive destruction of dermal proteins by extrinsic aging*.

Cosmetically, intrinsic aging of the dermis produces fine lines and wrinkles. Superseding these changes is the intrinsic age-related *decrease in production of sweat and sebum*.

Dermal Glands Are Affected with Intrinsic Aging: Controversial Findings

Although considered highly variable, sebaceous glands of the dermis secrete less sebum with age. In postmenopausal women, sebum production decreases, and sebum

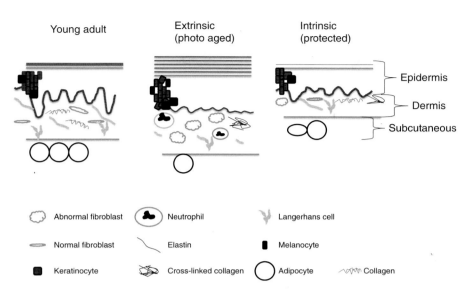

Figure 6.2. Representative changes in the skin due to extrinsic and intrinsic aging. (See plate section for color version.)

replacement time and sebum pore size increase; hormone replacement therapy (HRT) fails to diminish these changes enough to control seborrhea (skin dryness and flaking). In men, sebum production is not changed until the eighties. Gland number stays the same. UVR may cause sebaceous gland hyperplasia, a condition of gland enlargement that is generally benign.

The sweat glands of the skin form sweat made of water, salts, for example, sodium, chloride, potassium, and urea. Sweat acts to cool the surface of the skin by facilitating heat escape. Sweat glands also perform a minor excretory function by secreting the waste product, urea.

There are some data in humans to show that *heat intolerance increases with age and sweat gland number and sweat production decrease with age*. Sweat is critical for cooling the body when core body temperature increases as in exercise or in the presence of a marked elevation in ambient temperature. As the ability to produce sweat declines with age, heat intolerance tends to increase. Some studies show that as a function of age, there is an initial loss of sweating (function and anatomy) in the lower body limbs that with time progresses to the back and upper limbs and finally reaches the head region. There is suggestive evidence that high levels of fitness (as in chronic exercise) slow this progression.

Sensory receptors for touch, pressure, and pain are located in the dermis and gather environmental information on these modalities. As with temperature, information is sent to specific regions of the brain for processing and interpretation. Changes in these sensory receptors are discussed in Chapter 13 (sensory aging).

Changes occur in the cutaneous vascular system. These will be discussed in relation to altered temperature regulation and in more detail in Chapter 9 (cardiovascular aging).

AGING OF THE HYPODERMIS

The Hypodermis Is One of Several Fat Depots

Adipose tissue can be divided into two types: white adipose tissue (WAT) and brown adipose tissue (BAT). Both WAT and BAT are metabolically active. WAT stores lipids as triglycerides for future metabolic use, releases free fatty acids for energy use in other organs, and produces hormones such as leptin and adiponectin that are involved in satiety and insulin sensitivity. BAT thermogenically maintains core temperature during periods of ambient temperature reduction. WAT mass is sizable compared to BAT mass. *WAT is localized to the hypodermis, abdominal deposits (mesenteric and omentum areas), and in the aged in superficial areas (ectopically) on tissues such as liver, skeletal muscle, and bone.* Positron emission tomography plus fluorodeoxyglucose infusion show that BAT previously thought to be present only in the fetus and in the newborn with decreasing presence thereafter is actually present in the adult where it continues to contribute to homeostatic regulation of core temperature. Whether BAT function changes in the elderly is unknown.

The Hypodermis Fat Decreases with Age

The fate of adipose tissue with age has been described for WAT only. There are significant changes in fat depot size, generally beginning anywhere from 40 to 70 years of age, considered times of maximal fat mass accumulation. The *mass of WAT declines with age*. The actual percentage of WAT in the body remains constant due to a simultaneous decrease in lean muscle mass (sarcopenia) and increase in ectopic fat mass (fat localized outside the WAT depots). The *decrease in subcutaneous fat of the hypodermis is a result of a decrease in fat mass and not a decrease in adipose cell number.* A *noticeable age change is a loss of subcutaneous fat in the retro-orbital region* (under the eyes), redistribution to other areas of the face, and then an eventual loss from the face and neck, as depicted in Figure 6.3. These are *effects that reduce support for the dermis and*

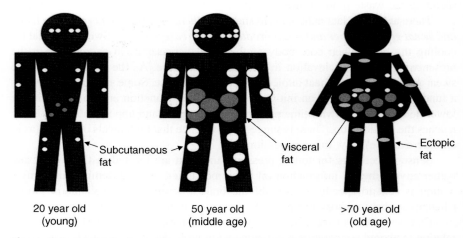

Figure 6.3. Fat distribution with age. (Reprinted with permission from Cartwright et al. (2007).) (See plate section for color version.)

that contribute significantly to wrinkles and sags. Another change is the increase in fat mass of the abdomen (visceral fat increase) and in old age (>70 years of age) there is *an increased presence of fat in unusual locations such as ectopically on bone, skeletal muscle, and liver.* These changes, increase in fat mass in the abdomen and in extra-WAT locations, are *associated with an increased risk of metabolic diseases, for example, type 2 diabetes, hyperlipidemia, and possibly osteoporosis.*

WAT is comprised of fully differentiated cells, called adipocytes, and an abundance (15–50%) of preadipocytes that mature to adipocytes. Results of studies in rats and man show that with age, preadipocytes exert a marked effect on WAT function. Specifically, the ability of preadipocytes to replicate and differentiate into mature adipocytes declines over time. Consequently, *overall function of the WAT is eventually compromised.* Thus, it is proposed that dysfunction of the WAT (especially loss of subcutaneous fat) *reduces the uptake of circulating toxic lipids. This allows for their persistence in the circulation where it is assumed they would induce tissue damage.* Although incompletely understood, the decline of preadipocyte function is strongly influenced by the inflammatory mediatory, tumor necrosis factor alpha (TNFα). The source of TNFα has not been defined.

MENOPAUSE AND SKIN

Menopause Adds to Intrinsic Aging

Estrogen supports skin health in many ways. All cells of the skin display estrogen responsive receptors. In culture, estrogen stimulates proliferation of human keratinocytes, and prevents apoptosis induced by hydrogen peroxide, and formation of inflammatory mediators. In rodents, estrogen increases skin thickness and water content by activating fibroblasts to produce collagen and hyaluronic acid. Other beneficial effects of estrogen on acceleration of wound healing and protection from UVR damage have been shown in various animal models. With menopause, the many actions of estrogen are severely diminished because the estrogen level falls to very low levels and estrogen receptors disappear.

Atrophy and xerosis (dry skin) are the most prominent skin changes associated with menopause. Skin atrophy takes the form of decreased skin thickness brought about by reduced collagen and glycosaminoglycans synthesis by fibroblasts and reduced keratinocyte proliferation due to low levels of estrogen. Skin elasticity also decreases and contributes to wrinkling. Skin alterations are prominent in the first 5 years after menopause and may be reversed or minimized with HRT or topical estrogen (see below).

In addition to estrogen, many other hormones, such as testosterone, growth hormone, and thyroid hormones, influence cells of the skin. Blood levels of these hormones change with age: dramatic fall of growth hormone; modest decline of testosterone, and variable change of thyroid hormones. Changes in these hormones would be expected to impact the health of the skin, but the role that each one plays is vague at present.

CONSEQUENCES OF AGING SKIN

Many Functions of Skin Are in Danger of Decline

The skin provides many benefits. It acts as a protective barrier, and if unbroken, not only *prevents loss of water from the inside to the outside but also prevents invasion from environmental insults* (e.g., pathogens, toxins, and pollutants). An impassable barrier is created by the multilayered epidermis of keratinocytes aided by secreted antimicrobial fatty acids from the sebaceous glands. The vascular capillary network of the skin, sweat production, and hypodermis serve to regulate core temperature in the presence of internal and external temperature stresses. Additionally, the skin is the site of vitamin D production, as well as the location of numerous sensory receptors.

Barrier Function of Skin Is Reduced with Age

A *thin* "functional" epidermis as in intrinsic aging or a *thicker* "dysfunctional" epidermis as in extrinsic aging produces a suboptimal protective barrier.

The epidermis is expected to prevent loss of water and salt and invasion of microbes, allergens, and toxins. These effects depend on the quality of the stratus corneum, sweat and sebum, and immune responsiveness. As evidenced by the age-related increase in susceptibility to xerosis (dry skin), increase in incidence of contact dermatitis, and altered drug permeability, barrier function is reduced with age even in sun-protected skin. Stressful challenges (used experimentally as acetone rubs or tape peals) reveal this vulnerability and an additional prolongation of recovery time from injury in aged skin. Therefore, decreased humidity, solvents, and detergents pose a problem for the aged skin.

Dry Skin Is One Cause of Pruritus Many factors noted above contribute to the presence of dry skin in the elderly. They include a decrease in sebum and sweat production, decreased lipid content of stratum corneum, and the reduced gene expression of aquaporin-3, an essential protein for skin hydration. Whether trans-epidermal water loss changes with age is controversial. Also, measurement of skin pH shows a trend to a more neutral pH that could further dry the skin by activation of proteolytic enzymes.

Dry skin leads to the most common skin complaint of the elderly, pruritus. Chronic pruritus is an "itch lasting for longer than six weeks" (Garibyan et al., 2013). It decreases quality of life due to its negative impact on sleep. Although xerosis is the most common cause of pruritus, many other conditions/factors cause and/or exacerbate it. Among these conditions are immunosenescence (altered immune function), medications, kidney diseases, hematological diseases, neuropathies of damage and inflammation, mites/lice infestation, dermatitis, and psychogenic conditions. Treatment requires an accurate diagnosis and disruption of the itch–scratch cycle. If pruritus is a result of age-related dry skin, an effective protocol includes elimination of harsh soaps/detergents, short warm showers, and application of moisturizers, preferably thick creams (rather than gels or lotions), to retain moisture immediately post shower and throughout the day.

Temperature Regulation Diminishes with Age

The skin assists with temperature regulation. An elevation in ambient temperature elicits a dilation of cutaneous blood vessels (increase blood flow) and increased sweating, processes that promote heat loss and maintenance of core body temperature. With a drop in ambient temperature, cutaneous blood vessels constrict (decrease blood flow), skeletal muscles contract (shivering), and hypodermal fat insulates, processes that conserve and generate heat. These processes are compromised with age. First, with the reduction in subcutaneous fat, there is an increased sensitivity to low temperatures; conversely, as noted above, a reduction in sweat production leads to intolerance of elevated ambient or body temperatures. In studies that measured reflex control of cutaneous blood flow in response to temperature changes, reflex-mediated blood flow declines with age. This deficit is attributed to insufficient sympathetic nerve activity and neurotransmitter release and reduced sensory input from temperature receptors in the skin. Some of these changes are reversed with aerobic training. Others point to the reduction of epidermal/dermal undulations to explain intolerance to temperature changes, clearly compromised with extrinsic aging. This structural change reduces the efficacy of cutaneous blood vessels to regulate heat loss or gain.

Vitamin D Production Declines with Age

Vitamin D is produced in the skin by the action of UVR. The result is D_3 (cholecalciferol), which is subsequently metabolized by the liver to 25(OH)cholecalciferol (25OHD) and thereafter metabolized by the kidneys to 1,25-dihydroxycholecalciferol, the active form of vitamin D. Vitamin D is an essential cofactor that maintains a normal calcium environment (called calcium homeostasis) and its effects are discussed in Chapter 14 (aging of the neuroendocrine system). Figure 6.4 summarizes formation of vitamin D and its general effects.

MacLaughlin and Holick (1985) measured the content of 7-dehydrocholesterol, the precursor to D_3 and the essential compound activated by UVR, in skin from individuals 18–92 years of age and showed that preD_3 declined with age. Both the content of preD_3 and the ability of UVR to convert preD_3 to D_3 fell by more than twofold in this age range. This change is additive with common behaviors of the elderly that include reduced sunlight exposure and decreased consumption of vitamin D-containing foods. Together, a deficiency of 25OHD in the elderly is a likely outcome. Levels of 25OHD below 10–20 ng/ml produce adverse effects associated with vitamin D deficiency such as osteoporosis, cardiovascular disease, and diabetes. According to Bischoff-Ferrari (2014), adequate levels of 30 ng/ml of 25OHD can be achieved with daily consumption of 1600–2000 international units of D_3, although not all vitamin D authorities agree on this value.

Wound Healing Declines with Age

It has been known for nearly a century (see Bentov and Reed (2014) for identification of a 1916 publication) that age and rate of scar formation are inversely related. Wound healing in the elderly requires a longer time compared to young individuals, allowing for an increased risk of infection, worsened by reduced numbers of Langerhans cells.

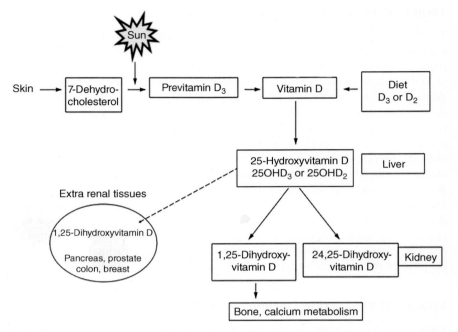

Figure 6.4. Vitamin D metabolism from sun and diet and conversions in select tissues. (Reprinted with permission from Gallagher (2013).)

Surgical site infections (SSIs) are associated with increased mortality, hospital-days, and cost. McGarry et al. (2004) came to this conclusion after a comparison of elderly patients with bacterial (*Staphylococcus aureus*) SSI with elderly without SSI and also with young individuals with bacterial SSI. In the few studies that have directly measured wound healing (injury of thigh skin), the rate of renewal by epithelial cells decreases by 1.9 days and the production of dermal proteins (other than collagen) decreases in older individual (over 65) compared with those 18–55 years of age.

It is proposed that the role of the dermal microcirculation is central to wound healing. It facilitates influx of specialized cells needed to generate an acute inflammatory response and it regulates oxygen tension to prevent microbial growth. In animal models of aging, these changes are evident, but only one study exists that describes cutaneous flow changes with age in man. Tsuchida (1993) reported a 40% drop in blood flow from age 20 to 70 in the deltoid region (upper arm, top of shoulder). Proposals to optimize wound healing and reduce SSI include adequate fluids, temperature regulation, and careful choice of anesthetic. Importantly, additional studies on wound healing in man are needed.

Increased Prevalence of Benign and Malignant Skin Cancers Result from Aged Skin

The most serious effect of aging on the epidermis is induction of skin cancer. "Skin cancer is the most common form of cancer in the United States" (Skin Cancer Foundation). In order of most to least frequently diagnosed skin cancers are basal cell carcinomas, squamous cell carcinomas, and melanomas. According to the American Cancer Association, an estimated 3.5 million (non-melanoma) cancers are diagnosed each year. Data show that repeated exposures to *UVR causes skin cancers. One pathway to tumor induction and progression is via* direct UVR damage of DNA, mutation of p53 tumor suppressor gene, activation of many cancer-facilitating genes and assistance from SASP (senescent) cells. A possible precursor to a squamous cell carcinoma is actinic or solar keratosis. These are precancerous lesions that appear as rough red or pigmented scaly areas of various size and protrusions. Actinic keratosis results from years of sun *exposure. Use of sunscreens (as discussed below) prevents extrinsic aging and associated cancers.*

AGING FACE SYNDROME

The appearance of the face is determined maximally by age and minimally by gender and ethnicity (Figure 6.5). Friedman (2005) defines the aging face (see Figure 6.4). *Structural changes begin in the third decade* with a *lowering of the eyebrows* that appears to reduce the size of the eyes. In the fourth decade, the *eyelids skin weakens* allowing fat movement into this region, *frown lines between the eyebrows appear* (glabellar), the skin folds from nose to mouth (smile lines) become prominent, and the eyebrows continue to descend. In the fifth decade, *forehead wrinkles and crow's-feet appear*; glabellar furrows deepen and vertical lines around the mouth appear. In the

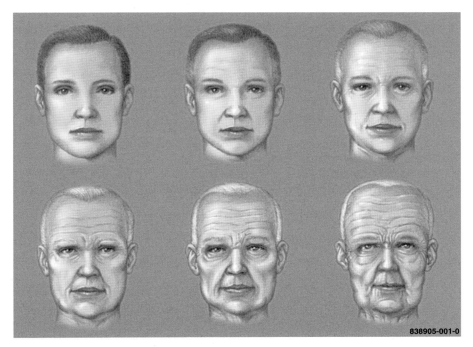

Figure 6.5. The aging face. (Reprinted with permission from Friedman (2005).) (See plate section for color version.)

sixth decade, significant *drooping of the skin around the mouth and neck occurs*, the *nose begins to droop* and the corners of the eyes slant downward; furrows and wrinkles increase, and a downward shift of the mid-face structures produce sagging skin under the eyes. In the seventh decade, the *skin thins and subcutaneous fat disappears*, chin and mouth wrinkles are prominent, eye opening diminishes due to lowered brow and lax eyelids. In the eighth decade, all the *preceding is exaggerated* as the skin continues to thin and the subcutaneous fat disappears further.

Broad-Spectrum Sunscreens Reduce UVR Exposure

The rate and extent of skin remodeling varies from individual to individual, affected by both intrinsic and extrinsic aging. However, extrinsic aging that produces the most dramatic shifts can be significantly minimized by avoidance of environmental insults of UVR and environmental pollutants such as cigarette smoke and ozone. In particular, it is recommended that reduction in *sunlight exposure (sun avoidance, outdoor protective clothes, and sun block usage), should be initiated at an early age*. Until recently, sun block preparations in the United States were effective only against UVB to prevent sunburn. The SPF (sun protection factor) indicates the degree of protection against UVB (the higher the number, the greater the protection up to 50). In June 2011, the FDA ruled that manufacturers of sunscreen could label their products with the words "broad spectrum" if they passed specific FDA-approved *in vitro* tests that showed 90% blockage of UVA and an SPF of at least 15 against UVB. Additionally, an SPF of greater than 50 is to be represented as 50+; SPF values greater than 50 are no more effective than 50. Products that claim water resistance must show evidence of protection during 40 or 80 min of swimming or sweating. The FDA report emphasizes that sunscreens should be applied *generously*, preferably at 2 g/cm^2. Most users apply about 25% of that needed for benefit. In the past, sunscreens that contained compounds such as mexoryl titanium dioxide and oxybenzone were shown to be effective *in vitro* and *in vivo* in absorbing UVA and preventing damaging effects. They remain highly effective by physically blocking UV rays. Many sunscreens now contain different compounds that chemically absorb the rays.

Retinoids as Antiaging Compounds

Retinoids are one of the few topical antiaging compounds that are backed with convincing clinical data (randomized controlled trials (RCT)). The three most prominent retinoids are retinol (vitamin A), retinaldehyde, and retinoic acid (tretinoin). The retinoids act by binding to specific cellular and nuclear receptors to activate a select set of genes, which in turn stimulate epithelial proliferation that thickens the epidermis, tightens the stratum corneum, and stimulates fibroblast to produce GAGs that improve water retention and dermal structure.

Tretinoin was the first to be tested in man for reversal of photoaging. It has been subsequently evaluated in numerous RCTs at different concentrations (0.01–0.1%) for periods of 3 months to 4 years with 20 to over 500 participants. The consistent findings in photoaged skin are a reduction in fine and coarse wrinkling, decreased hyperpigmentation, increased epidermal thickness, and reduced roughness. Although examined in only one study (Kligman et al., 1993), topical tretinoin also reverses

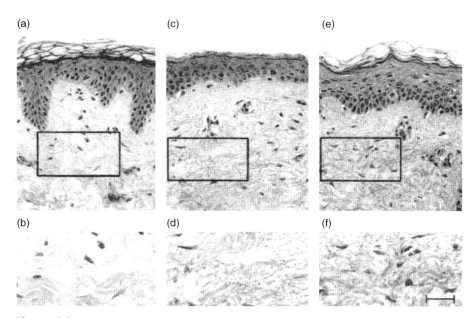

Figure 6.6. Representative histology from study on the effect of topical vitamin A. Daily application of 1% retinol (vitamin A) for 7 days on protected skin of 86 years old is compared with same site histology of young skin. (a and b) 22 years; (c and d) 86 years vehicle; (e and f) 86 years, 1% retinol for 7 days. (Reprinted with permission from Varani et al. (2000).)

intrinsically aged skin. In this study, tretinoin application to the inner thigh of six women for 9 months increased epidermal thickness and enhanced dermal–epidermal undulations, improved melanocyte structure, increased GAG content of the dermis, and stimulated microvascular growth. This is an important study that needs confirmation.

One drawback of tretinoin is production of erythema (redness) that is minimized by concentration reduction. However, the search for compounds without this adverse effect has generated the development of many other retinoid derivatives that compare favorably with tretinoin, considered the gold standard. Vitamin A has been used topically with success in reversing photoaging, although it is unstable, less potent than tretinoin, and must be converted *in vivo* to retinoic acid. Nevertheless, topical application of retinol (7-day application of 1%) to the thigh of 53 individuals 80 years of age or older stimulates fibroblast proliferation, decreases production of proteolytic enzymes (MMPs and collagenases), and increases synthesis of collagen (Figure 6.6).

Other approaches to reducing photoaging are abundant. Traditional techniques include surgery (facelift), skin resurfacing, and botulinum toxin. Newer techniques are nonablative laser rejuvenation and filler materials. These procedures work in different ways. Skin resurfacing with various types of lasers (newer ones penetrating deeper and more selectively) acts by stimulating a wound healing response in the dermis that results in synthesis of more collagen. Botulinum toxin is a neurotoxin that paralyzes the skeletal muscles of the face and when specific muscles are immobilized, wrinkles tend to flatten. Fillers, temporary or "permanent," are injections of collagen

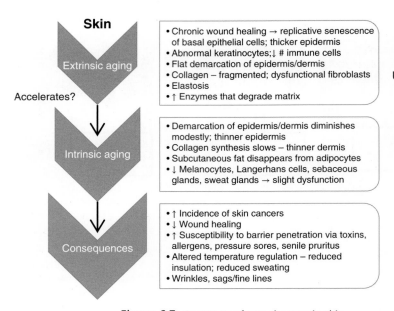

Figure 6.7. Summary of age changes in skin.

or hyaluronic acid-derived products that provide a platform for fibroblast attachment and facilitate fibroblast collagen production and reduction of MMPs.

Oral and Topical Estrogen Reverse Aging of Skin After Menopause

The use of HRT in postmenopausal women is controversial (see Chapter 14). Nevertheless, a preponderance of data from animal studies and observational studies and a few RCTs in women indicate that oral and topical estrogen therapy slow skin aging. These studies generally show that estrogen therapy increases epidermal and dermal thickness and improves skin hydration, collagen content, and elasticity despite differences in estrogen formulations, duration of therapy, site of application (topical only), and size of study. One mechanism for these changes is an estrogen-mediated stimulation of keratinocytes and fibroblast proliferation. Estrogen may possibly improve microcirculation, sebaceous gland secretion, and hair growth, but these are less well documented. Whether wrinkling decreases depends on the duration of therapy and presence of complicating factors such as smoking and photoaging.

SUMMARY

The skin is comprised of three layers: outermost epidermis of keratinocytes and melanocytes; thicker dermis of fibroblasts, matrix proteins, glands, nerves, and microvascular system; and hypodermis of subcutaneous fat. The skin serves as a protective barrier, temperature regulator, immunological sentinel, producer of vitamin D, and sensory relay.

There are two pathways (extrinsic and intrinsic) to skin aging. Extrinsic or environmental (UV radiation, smoking, and pollutants) aging thickens the epidermis and destroys the dermis resulting in major wrinkles/sags and skin cancers.

Intrinsic aging or chronological aging slows keratinocyte replication and decreases fibroblast numbers resulting in epidermal thinning and modestly altered dermal structure. Production of sebum and sweat also decline.

Intrinsic aging predisposes one to xerosis and pruritus, slowed wound healing, reduced production of vitamin D, and suboptimal temperature regulation. Many of these changes are due to loss of hormones, especially estrogen.

Subcutaneous fat diminishes with age and reduces skin support, enhancing wrinkles and sags, and lessens body insulation. Abdominal and ectopic fat mass increase with age and are associated with inflammation and diabetes.

The aging skin can be protected with daily use of sunscreens (broad spectrum, SPF 15), moisturizers, and one of several topical retinoids.

CRITICAL THINKING

Why does extrinsic aging exert such devastation on the skin?

Describe the aging face syndrome if extrinsic aging was completely inhibited.

What is the most serious age change in the skin? why?

How does cellular aging influence tissue aging in the skin?

What are the best procedures to maintain a healthy skin for as long as possible?

KEY TERMS

Actinic keratosis rough and red patches on the skin due to chronic sun exposure that may develop into squamous cell carcinoma.

Adipocytes cells that store fat acids, release fatty acids, and produce metabolic hormones.

BAT brown adipose tissue; cells participate in regulation of body temperature.

Extrinsic aging aging induced by environmental stresses, mainly UVA and UVB radiation, smoke, pollution, and harsh weather.

Fibroblasts connective tissue mitotic cells that produce collagen and elastin to give the dermis form, flexibility, and strength. As they age (replicative senescence), they produce excessive amounts of elastin and degradative enzymes.

Glycoaminoglycans a family of complex polysaccarides and amino acids that are produced in the dermis; assists with water retention and structure.

Homeostasis the constancy of the biological internal environment. The balance of physiological processes to optimize the viability of the organism.

Immunological surveillance physiological function of the immune system to constantly survey the cells of an organism for foreign substances, including cancerous cells.

Intrinsic aging normal chronological aging that excludes extreme, repetitive insults from the environment.

Langerhans cells antigen-processing cells (dendritic cells) found in the epidermis.

Lentigines dark brown skin spots due to aggregations of melanocytes. Result of extensive sun exposure.

Mast cells specialized cells found in the epidermis that contain histamine and are sensitive to various stimuli. When activated, the classic "hive" is evident with redness, edema, and pain.

Metalloproteinase (MMP) degradative enzymes that breakdown collagen and elastin in the connective tissue. Abnormal increase distorts structure and function.

Preadipocyte cell that gives rise to the mature fat cell or adipocytes and influences competency of the adipose tissue.

Photoaging another name for extrinsic aging.

Photocarcinogenesis malignancies induced largely by UVB and UVA radiation with chronic exposure.

Randomized Controlled Trial (RCT) a study design in which the outcomes of a treatment group are compared to those of the nontreatment or control group. Individuals are assigned without bias (randomly) to one of the two groups.

Retinoids family of compounds that includes vitamin A (retinol), retinaldehyde, and retinoic acid.

Sebum the oily, lipid substance produced by the sebaceous glands that adds to the protective quality of the skin.

Solar elastosis component of extrinsic aging that results in excessive production of elastin in the dermis.

TNFalpha signaling molecule that plays a prominent role in inflammatory responses.

UVA ultraviolet radiation of long wavelength that penetrates into the dermis and rapidly ages the skin.

UVB ultraviolet radiation of intermediate wavelength that penetrates the epidermis and produces tanning and sunburn.

WAT white adipose tissue; adipocytes store fats used for energy and influence inflammation; predominant fat type in the body.

Xerosis medical term for dry, rough and peeling skin.

BIBLIOGRAPHY

Review

Bentov I, Reed MJ. 2014. Anesthesia, microcirculation, and wound repair in aging. *Anesthesiology* **120**(3):760–772.

Bischoff-Ferrari HA. 2014. Optimal serum 25-hyrdroxyvitamin D levels for multiple health outcomes. *Adv. Med. Exp. Bio.* **810**:500–525.

Cartwright MJ, Tchkonia T, Kirkland JL. 2007. Aging in adipocytes: potential impact of inherent, depot-specific mechanisms. *Exp. Gerontol.* **42**(6):463–471.

FDA sunscreen monograph, June 16, 2011.

Feng B, Zhang T, Xu H. 2013. Human adipose dynamics and metabolic health. *Ann. N. Y. Acad. Sci.* **1281**:160–177.

Fisher GJ, Varani J, Voorhees JJ. 2008. Looking older: fibroblast collapse and therapeutic implications. *Arch. Dermatol.* **144**(5):666–672.

Friedman O. 2005. Changes associated with the aging face. *Facial Plast. Surg. Clin. North Am.* **13**(3):371–380.

Gallagher JC. 2013. Vitamin D and aging. *Endocrinol. Metab. Clin. North Am.* **42**(2):319–332.

Garibyan L, Chiou AS, Elmariah SB. 2013. Advanced aging skin and itch: addressing an unmet need. *Dermatol. Ther.* **26**(2):92–103.

Gilchrest BA. 2013. Photoaging. *J. Invest. Dermatol.* **133**(E1):E2–E6.

Holick MF. 2014. Sunlight, UV-radiation, vitamin D and cancer: how much sunlight do we need. *Adv. Exp. Med. Biol.* **810**:1–16.

Mukherjee S, Date A, Patravale V, Korting HC, Roeder A, Weindl G. 2006. Retinoids in the treatment of skin aging: an overview of clinical efficacy and safety. *Clin. Interv. Aging* **1**(4):327–348.

Patel T, Yosipovitch G. 2010. The management of chronic pruritus in the elderly. *Skin Therapy Lett.* **15**(8):5–9.

Ramos-e-Silva M, Celem LR, Ramos-e-Silva S, Fucci-da-Costa AP. 2013. Anti-aging cosmetics: facts and controversies. *Clin. Dermatol.* **31**(6):750–758.

Scichilone N, Callari A, Augugliaro G, Marchese M, Togias A, Bellia V. 2011. The impact of age on prevalence of positive skin prick tests and specific IgE tests. *Respir. Med.* **105**(5):651–658.

Sgonc R, Gruber J. 2013. Age-related aspects of cutaneous wound healing: a mini-review. *Gerontology* **59**(2):159–164.

St-Onge M-P, Gallagher D. 2010. Body composition changes with age: the cause or result of changes in metabolic rate and macronutrient oxidation? *Nutrition* **26**:152–155.

Valacchi G, Sticozzi C, Pecorelli A, Cervellati F, Cervellati C, Maioli E. 2012. Cutaneous responses to environmental stressors. *Ann. N. Y. Acad. Sci.* **1271**:75–81.

Waller JM, Maibach HI. 2005. Age and skin structure and function, a quantitative approach (I): blood flow, pH, thickness, and ultrasound echogenicity. *Skin Res. Technol.* **11**(4):221–235.

Waller JM, Maibach HI. 2006. Age and skin structure and function, a quantitative approach (II): protein, glycosaminoglycans, water and lipid content and structure. *Skin Res. Tech.* **12**(3):145–154.

Experimental

Ashcroft GS, Greenwell-Wild T, Horan MA, Wahl SM, Ferguson MW. 1999. Topical estrogen accelerates cutaneous wound healing in aged humans associated with an altered inflammatory response. *Am. J. Pathol.* **155**(4):1137–1146.

Brinkley TE, Hsu F-C, Beavers KM, Church TS, Goodpaster BH, Stafford RS, Pahor M, Kritchevsky SB, Nicklas BJ. 2012. Total and abdominal adiposity are associated with inflammation in older adults using a factor analysis approach. *J. Gerontol. A Biol. Sci. Med. Sci.* **67**(10):1099–1106.

Holowatz LA, Kenney WL. 2010. Peripheral mechanisms of thermoregulatory control of skin blood flow in aged humans. *J. Appl. Physiol.* **109**(5):1538–1544.

Holt DR, Kirk SJ, Regan MC, Hurson M, Lindblad WJ, Barbul A. 1992. Effect of age on wound healing in healthy human beings. *Surgery* **112**(2):293–297.

Inoue Y. 1996. Longitudinal effects of age on heat-activated sweat gland density and output in healthy active older men. *Eur. J. Appl. Physiol. Occup. Physiol.* **74** (1–2):72–77.

Kaye KS, Anderson DJ, Sloane R, Chen LF, Choi Y, Link K, Sexton DJ, Schmader KE. 2009. The effect of surgical site infection on older operative patients. *J. Am. Geriatr. Soc.* **57**(1):46–54.

Kligman AM, Dogadkina D, Lavker RM. 1993. Effects of topical tretinoin on non-sun-exposed protected skin of the elderly. *J. Am. Acad. Dermatol.* **29**(1):25–33.

MacLaughlin J, Holick MF. 1985. Aging decreases the capacity of human skin to produce vitamin D3. *J. Clin. Invest.* **76**(4):1536–1538.

McGarry SA, Engemann JJ, Schmader K, Sexton DJ, Kaye KS. 2004. Surgical-site infection due to *Staphylococcus aureus* among elderly patients: mortality, duration of hospitalization, and cost. *Infect. Control. Hosp. Epidemiol.* **25**(6):461–467.

Maheux R, Naud F, Rioux M, Grenier R, Lemay A, Guy J, Langevin M. 1994. A randomized, double-blind, placebo-controlled study on the effect of conjugated estrogens on skin thickness. *Am. J. Obstet. Gynecol.* **170**:642–649.

Piérard-Franchimont C, Piérard GE. 2002. Postmenopausal aging of the sebaceous follicle: a comparison between women receiving hormone replacement therapy or not. *Dermatology* **204**(1):17–22.

Ressler S, Bartkova J, Niederegger H, Bartek J, Scharffetter-Kochanek K, Jansen-Dürr P, Wlaschek M. 2006. p16INK4A is a robust *in vivo* biomarker of cellular aging in human skin. *Aging Cell* **5**(5):379–389.

Sauerbronn AV, Fonseca AM, Bagnoli VR, Saldiva PH, Pinotti JA. 2000. The effects of systemic hormonal replacement therapy on the skin of postmenopausal women. *Int. J. Gynaecol. Obstet.* **68**:35–41.

Thomas CM, Pierzga JM, Kenney WL. 1999. Aerobic training and cutaneous vasodilation in young and older men. *J. Appl. Physiol.* **86**(5):1676–1686.

Tsuchida Y. 1993. The effect of aging and arteriosclerosis on human skin blood flow. *J. Dermatol. Sci.* **5**(3):175–181.

Varani J, Warner RL, Gharaee-Kermani M, Phan SH, Kang S, Chung JH, Wang ZQ, Datta SC, Fisher GJ, Voorhees JJ. 2000. Vitamin A antagonizes decreased cell growth and elevated collagen-degrading matrix metalloproteinases and stimulates collagen accumulation in naturally aged human skin. *J. Invest. Dermatol.* **114**(3):480–486.

7

AGING OF THE SKELETAL MUSCLE SYSTEM

ORIENTATION TO SKELETAL MUSCLE

Skeletal Muscles Provide Mobility, Strength, Independence, Metabolism, and Thermoregulation

Skeletal, smooth, and cardiac muscles are the three muscle types found in the human body. In adults (30–50 years of age) *skeletal muscle mass comprises about 40% of the total body mass* in men and slightly less in women. Contributing to this mass are over 600 classified skeletal muscles.

Since skeletal muscles are attached to bone (via tendons), their *primary role is to facilitate movement*. Movement of the bones of the skeleton produces voluntary motion and balance. Skeletal muscles of the diaphragm and rib cage enable movement of the chest cavity in respiration and muscles of pharynx (throat region) assist with chewing and swallowing. As skeletal muscle structure and function decline with age, these indispensable physiological activities become difficult. The consequences are serious: *poor balance*, *slowed gait*, *falls*, *exercise intolerance*, *physical disability*, and *loss of independence*.

Skeletal muscles are secondarily important because they *contribute to total body metabolism and uptake of sugar*. A reduction in muscle size or mass reduces metabolism and retards the effects of insulin, changes that lead to an increase in body fat mass, insulin resistance, and risk of type II diabetes (T2D). *Skeletal muscles are also essential for thermoregulation* during exposure to elevated or reduced ambient temperatures. Thermoregulatory adaptation is suboptimal in senescent muscles.

Human Biological Aging: From Macromolecules to Organ Systems, First Edition. Glenda Bilder.
© 2016 John Wiley & Sons, Inc. Published 2016 by John Wiley & Sons, Inc.

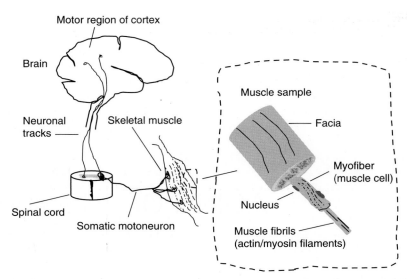

Figure 7.1. Representation of skeletal muscle with somatic motoneuron and connections to cerebral cortex. Inset shows myofiber organization.

Skeletal Muscles Consist of Bundles of Myocytes Innervated with a Motoneuron

Skeletal muscles consist of bundles of *muscle cells called myocytes* (or myofibers) held together with connective tissue. Each myocyte possesses multiple nuclei, up to hundreds of mitochondria (mt) depending on the level of physical activity, and two main contractile proteins, actin and myosin, whose interactive sliding movement produces muscle contraction (shortening). Numerous auxiliary proteins, calcium, and nerve impulse activation assist with muscle contraction. Myocytes are in intimate contact with spinal nerves called somatic motoneurons and respond to nerve impulses from the spinal cord and brain (see Figure 7.1).

SKELETAL MUSCLE SENESCENT PHENOTYPE

Sarcopenia Is the Age-related Loss of Muscle Mass

A gradual age-related loss of lean or fat-free muscle mass (FFM) has been documented to occur in all models of aging (worm, fly, rodent, and monkey) as well as in man. Age-related loss of FFM in man has been assessed in muscle biopsies using histological, biochemical, and molecular probes. Techniques of dual-energy X-ray absorptiometry (DXA), computerized tomography (CT) scans, magnetic resonance imaging (MRI), functional MRI (fMRI), positron emission tomography (PET), and bioelectrical impedance analysis have been employed to study intact muscles. Results from these varied methodologies have shown a *progressive age-dependent decrease in muscle cross-sectional area and/or fiber size or in some studies in man, a decrease in fiber number. This consistent age change in skeletal muscle is termed sarcopenia.*

Baumgartner et al. (1998) proposed an "operational" definition of sarcopenia in which sarcopenia is equal to "2 SD below the mean appendicular muscle mass for healthy young adults." Thus, using (i) a measure of variability around the mean, that is, standard deviation (SD), and (ii) an average of the muscle mass of the arms and legs, that is, appendicular appendages of young subjects (5.45 kg/m^2 for women and 7.26 kg/m^2 for men as measured by DXA), a sex-specific cutoff value below which sarcopenia exists could be determined in the elderly. This criterion not only *identified individuals at risk of balance and gait problems* but also *revealed the prevalence of sarcopenia in individuals older than 80 years of age to exceed 50%*. However, agreement on the actual "below 2 SD" (cutoff) value is not unanimous, nor is their consensus on the best methodology to measure muscle mass. Furthermore, there is little agreement regarding the addition of other parameters (gait and strength) to the definition of sarcopenia. Several studies recommend a more generous cutoff value, a change that reduces the prevalence of sarcopenia in the elderly. Given the debate on the threshold value for sarcopenia, the *prevalence of sarcopenia is generally positioned at 5–13% in individuals 60–70 years of age and 11–50% in those older than 80 years*. A determination of sarcopenia considered "one of the most striking and debilitating age-associated alterations" (Safdar et al., 2010) is important because it has serious consequences for the individual and more importantly it is *amenable to improvement through intervention*.

Rate of Loss of Muscle Mass with Age Is Highly Variable

The *kinetics of age changes* (400 variables) were determined using the best fit linear model applied to a computational analysis of data from 496 studies involving over 53,000 individuals. This analysis indicated that muscle mass declines about 1–2% per year (Sehl and Yates, 2001) in healthy individuals within the age limit of 30–70 years. Longitudinal studies have also determined the rate of mass loss, but the *extent is highly variable*. This is due in part to *confounding factors such as level of physical activity, level of steroidal hormones and growth factors, and diet*.

Myocytes (Type II) Decrease in Number and Size; Fat Content Increases

The decline in muscle mass has been attributed to a loss of myocytes and/or a reduction in myocyte size. An early report of autopsy samples from the thigh muscle (vastus lateralis, most frequently sampled muscle) of healthy individuals, aged 15–83, *attributed sarcopenia to a decrease in myocyte number regardless of type with a preferential size reduction of type II myocytes*. Results of longitudinal and cross-sectional studies basically support these early findings. The selective negative effect on type II myocytes (fast twitch) may be due to a decrease in the function of stem cells (called satellite cells and discussed below). Type II myocytes are characterized as white to red in color, contracting rapidly, utilizing glycolytic enzymes to generate ATP, and fatiguing easily. Type II myocytes are activated with short, high intensity activities. In contrast, type I myocytes are reddish (due to the presence of an oxygen-binding protein, myoglobin), contract slowly, utilize oxidative phosphorylation and hence contain large numbers of mitochondria (mt), and resist fatigue. Type I myocytes facilitate low intensity prolonged activities. One small but interesting 9-year longitudinal study of physically active individuals 71 years of age at study start (Frontera et al., 2008) found a decline in muscle cross-sectional area in the vastus lateralis but no age-related loss of contractile function of isolated myocytes from the vastus lateralis. These findings coupled with those from interventional studies suggest the possibility of a *positive compensatory change induced by physical activity at least in some myofibers.*

As *muscle mass declines, an associated increase in fat mass has been routinely observed.* Lipids accumulate intracellularly (within) and intercellularly (around myocytes). Their presence is associated with altered insulin signaling and inflammation.

Dynapenia Is an Age-Associated Loss of Muscle Strength

Although there has been some reluctance to define loss of muscle strength (dynapenia) as a condition distinct from sarcopenia, Manini and Clark (2012) present several compelling reasons why this is necessary. First, age-related loss of muscle strength *is far greater than can be accounted for by sarcopenia.* It is estimated that sarcopenia can explain only 10% of the loss of muscle strength. Second, *improvement in muscle mass does not result in a parallel increase in muscle strength.* Third, the *association between physical disabilities and dynapenia is statistically stronger* than the association between physical disabilities and sarcopenia. Fourth, *dynapenia but not sarcopenia is statistically associated with increased mortality.* In a recently concluded 13-year study of individuals 53 years of age at study start, Cooper et al. (2014) found a higher mortality rate among individuals who performed poorly on measures of physical capability (hand grip strength, chair rise speed, and standing balance time), all indicators of skeletal muscle strength. Thus, "muscle strength—not simply muscle mass—is a critical factor for determining physical disability and mortality in older adults" (Manini and Clark, 2012).

Rate of Loss of Muscle Strength Is Highly Variable Similar to sarcopenia, a definition of dynapenia is needed to assess its prevalence and develop beneficial interventions. In a recent study by Bohannon and Magasi (2015), grip strength was used to generate a reference value for dynapenia. Based on the t-score (1–2.5 SDs below the average grip strength of 558 individuals, 20–40 years of age), "46.2–87.1% of older men and 50.0–82.4% of older women could be designated as dynapenic." Others, as reviewed by Doherty (2003), measured knee extensor strength in cohorts of different ages and found loss of strength between 20 and 40% in individuals 70–80 years of age and even greater strength loss for those 90 years and older.

Based on the conclusion that muscle weakness is only one of many factors that predict physical disability, Manini and Clark (2012) proposed a "working decision algorithm" to identify dynapenia in the older adult (open for comment at http://dynapenia.blogspot.com.). Briefly, individuals 60 years of age or older are given a knee extensor strength test *if* they have serious risk factors for dynapenia, for example, determined by level of physical activity, lifestyle, malnutrition, obesity, diseases as osteoarthritis or cardiopulmonary, medical history, and self-reported limitations. If risk factors are absent or few in number, grip strength is measured. If grip strength is low, the knee extensor test is administered. Reduced strength on this test indicates dynapenia and the neurological and muscle-based mechanisms for this are explored. Although this is a reasonable and practical approach, defined strength cutoff values are yet to be established and a consensus on this diagnostic scheme is lacking.

Recently, the Foundation of the National Institutes of Health (FNIH) in series of five papers analyzed data from six clinical trials and three observational studies

(>10,000 participants) to generate a definition of *sarcopenia that also included changes in muscle strength and physical performance*. This was compared with criteria proposed by two other groups: International Working Group and the European Working Group for Sarcopenia in Older Persons. Accordingly, FNIH criteria for sarcopenia are (i) gait speed of <0.8 m/s, (ii) grip strength of <26 kg (males) and <16 kg (females), and (iii) appendicular mass (corrected for body mass index) of <0.789 (males) and <0.512 (females). With this definition, the prevalence of sarcopenia is 1.3% (males) and 2.3% (females), 65 years of age and older. FNIH criteria generate the lowest prevalence rates of sarcopenia compared to the above named groups. The FNIH definition of sarcopenia considers dynapenia a subset of sarcopenia. However, given the compelling importance of dynapenia (as noted above), this chapter will treat them separately.

There appears to be no gender preference for dynapenia. However, peak muscle strength is generally greater for men, so *absolute* loss of strength in men is generally higher. Muscles of the upper and lower extremities experience dynapenia. The rate of loss of skeletal muscle strength, according to longitudinal study results, is estimated at 1–3% per year at least at older ages, although this varies with the set cutoff values and factors such as physical activity and diet.

Molecular/Cellular Myocyte Changes with Age

Role of mt, Contractile Proteins, and Motoneurons
The contractile proteins, actin and myosin, interlock in a highly organized pattern to make the complex substructure of the myocyte. In the presence of the high-energy molecule, ATP, the strands (fibrils) of actin and myosin slide past one another to allow fiber shortening (contraction) or generation of force (tension). Since the muscle shortening requires a supply of ATP or other high-energy molecules, myocytes (especially type I) generally contain large numbers of mt to supply the needed ATP. Type II myocytes have fewer mt and generate much of their ATP anaerobically (without oxygen) and thus fatigue readily.

Muscle activity requires a *stimulus from the motoneurons*. Since the ends of the motoneurons fail to touch the muscle fibers directly, the nerve stimulus is relayed to the myocyte via a chemical. In this case, motoneurons release acetylcholine (ACh), a neurotransmitter. ACh diffuses from the motoneuron terminus and acts (binds to a specific receptor) on the muscle fiber to ultimately induce contraction (termed excitation–contraction coupling). A secondary but important activity, the release of calcium from internal stores in and around the muscle fiber, is essential for muscle contraction. Motoneurons receive input from higher centers in the brain (primary motor cortex) (see Figure 7.1).

Aged Myocytes Show Reduced Protein Synthesis, Dysfunctional Mitochondria, and Alterations in Excitation–Contraction Coupling
The *main contractile proteins*, especially myosin heavy chain and mt proteins, *are synthesized at slower rates* in aged myocytes, declining at an estimated rate of 4% per decade. Thus, an age-associated reduction in myocyte protein quantity is anticipated. Associated changes in aged myocytes include reduced mt biogenesis (leading to reduced mt content), decreased enzyme activity, decreased mtDNA copy

number, and decreased ATP production. In addition to the decline in the rate of muscle protein synthesis, the rate of muscle protein degradation also declines due to the low supply of ATP. One inevitable outcome is the accumulation of oxidized proteins.

It is proposed that the *accumulation of mutated mtDNA affects both the quality and quantity of proteins biosynthesized in the mt*, reduces mt function, and may induce apoptosis. Data were recently generated in man that confirmed not only the presence of abnormalities of mt electron transport system (ETS) but also the existence of mt deletion mutations at levels sufficient to induce aberrant myocyte function. The *origin of the mt mutations was not determined and may have been induced by reactive oxygen species* (ROS), possibly from physical inactivity. Corroborating results of longitudinal studies in monkey and rat show the development of sarcopenia and the associated abnormalities of the mt ETS. In sum, suboptimal mt function stresses skeletal myocytes and could contribute to muscle dysfunction.

Several reports in man and a number of animal models suggest that the ability to generate maximal contractile force declines with age. Contraction of the myocyte relies on activation of specific receptors, release of intracellular pools of calcium, sliding of the actin and myosin proteins, and resequestration of calcium. Specifically, the *concentration of the dihydropyridine receptor, essential for myocyte calcium regulation is lower in muscles from older individuals compared to younger individuals*. This loss would produce a reduction in calcium release and hence a decrease in contractile force, possibly accounting for muscle weakness. An uncoupling of excitation–contraction coupling (discussed below) is also important. Other changes such as structural alteration of myosin and abnormal expression of other muscle proteins have been reported in animals and await confirmation in man.

CAUSATIVE FACTORS

Many Factors Influence the Senescent Muscle Phenotype, Especially Physical Inactivity

Physical inactivity appears to be a major lifestyle choice permissive for development of sarcopenia and dynapenia. Other proposed factors such as reduced protein consumption, reduced hormone levels, and diminished myocyte renewal have been studied mostly in relation to sarcopenia.

Physical Inactivity

Since *physical activity significantly impacts myocyte health, sorting out its effects from that of aging represents an arduous challenge.* According to Montero-Fernández and Serra-Rexach (2013), delineation of physical activity, sedentary behavior, and exercise is the necessary first step. Physical activity is "any bodily movement produced by contraction of skeletal muscle that substantially increases energy expenditure." The baseline is sedentary activity of sitting and lying. The next level is low physical activity of standing, slow walking, or lifting light objects. Moderate physical activity is defined as walking at a pace of 3–5 miles/h or moderate level of swimming, and vigorous physical activity means higher intensities of movement, muscle contraction, and oxygen consumption over and above moderate physical activity. Exercise is physical activity that is structured, repetitive, and maintains a level of physical fitness.

How much physical inactivity ensures the senescent muscle phenotype is unknown. A comparison of myocytes (biopsy) from sedentary frail elderly with those from recreationally active elderly revealed "reduced mt function, dysregulation of cellular redox status, and chronic systemic inflammation" (Safdar et al., 2010), changes with potential to create a persistent pro-oxidative and toxic intracellular and intercellular environments. In confirmation, others have shown a reduced mt content in myocytes of sedentary elderly.

Epidemiological studies provide data that a lifetime of physical inactivity (sedentary lifestyle) is associated with a reduced life expectancy with *physically inactive individual experiencing a 30% increase in mortality compared to physically active individuals*. As noted by Booth et al. (2011), in addition to (1) death at a younger age, a lifetime of physical inactivity produces "(2) younger age for onset of physical frailty; (3) fewer years of high quality of life; (4) lowered cardiorespiratory reserve so that stresses, such as major surgery are insufficient to maintain homeostasis within bounds of life; (5) and increased risk of chronic disease." Clearly, physical inactivity results in a plethora of negative changes, including onset of physical frailty. This and many other studies have associated lifelong physical inactivity with sarcopenia and dynapenia. Indeed, some consider that physical inactivity could account for loss of muscle mass and strength independent of any effect of age.

Physical activity and exercise exert *many beneficial cellular effects*. These are enhanced muscle anabolism, reduced oxidative stress, more efficient energy metabolism, reduced muscle inflammation, better mechanical excitation–contraction coupling, minimal muscle denervation, and improved muscle tendon architecture. Additionally, *physical activity can partially slow the progression of sarcopenia related to inactivity and poor nutrition* (Montero-Fernández and Serra-Rexach, 2013).

It is clear that physical activity and exercise minimize and even reverse sarcopenia and dynapenia and additionally offer many other advantages that translate into maintenance of homeostasis, prevention of physical disability, chronic disease risk reduction, and decrease in mortality.

Inadequate Consumption of High-Quality Protein

Balance between Protein Anabolism and Catabolism Preserves Muscle Mass

It is established that *muscle mass is regulated by both mechanical stress, that is, exercise and availability of amino acids (AAs)*. Additionally, exercise and AAs are more effective *together* than either one alone. Both utilize the intracellular signal protein, *mammalian target of rapamycin* (mTORC1), to increase protein anabolism (biosynthesis). Thus, *optimal muscle function depends on adequate physical activity and nutrition*. Additionally, insulin and insulin growth factor-1 (IGF-1) stimulate mTORC1 to achieve normal muscle metabolism.

A number of studies have reported that the anabolic response to an elevation of blood AAs (due to protein consumption as in a meal) is blunted in the elderly and the related deficit in muscle protein biosynthesis contributes to sarcopenia. However, others have found that this "blunting" can be overcome with consumption of higher amounts of "quality" proteins, that is, those containing *essential AAs*. Specifically, *consumption of less than 20 grams (g) of protein per meal is ineffective in stimulating muscle anabolism*. Furthermore, *the consumed protein must contain essential AAs, especially leucine and isoleucine. Leucine stimulates protein synthesis*. Studies (biopsies) show an increase in muscle protein synthesis and activation of mTORC1 and other markers of nutrient signaling after 2 weeks of leucine at 4 g/meal (three meals/day). This protocol appears to reverse the deficit in muscle protein synthesis over the short period. Several months of ingestion of *supplemental essential AAs (15 g/day)* is also beneficial and produces an increase not only in protein synthesis but also lean body mass and muscle IGF-1 levels. Studies of longer duration would be of value.

The effect of other factors on muscle anabolism such as the *level of physical activity and coingestion of other nutrients* are poorly understood, but are also important. It was observed that elderly who performed aerobic exercise prior to protein consumption improve the anabolic effect of AAs due to exercise-induced enhanced blood flow and delivery of AAs to the muscles.

Other Contributing Factors: Neurological, Hormonal, Stem Cell

Postulated Decline in Motoneuron Function
Based on physiology of excitation–contraction coupling, muscle contraction occurs following a nerve impulse. This coupling is thought to deteriorate with age and could contribute to dynapenia. One change of interest is an *age-associated dysfunction of cortical and spinal regulation of motoneurons*. The primary motor cortex modulates the overall recruitment of motoneurons in the spinal cord, facilitating the appropriate response (speed and force) to a voluntary contraction. The contribution of the *cerebral cortex to a maximal voluntary muscle contraction* in the elderly is slightly deficient and as a result there is a *loss of central control* in *certain* muscle groups (knee extensor and elbow flexors, but not ankle dorsiflexor). Studies investigating this issue have generally been small, cross-sectional in design, and report reductions cortical based muscle control of less than 10%. However, it has been emphasized that *even a small central neural reduction from cerebral primary motor neurons and spinal outflow in the elderly would have a significant impact* on an activity as basic as a chair rise that requires a maximal muscular effort.

It is additionally postulated that the *structural interaction of motoneuron terminals with myofibers deteriorates with age*. Therefore, transmission of impulses from nerve to muscle with release of the neurotransmitter, ACh would be compromised. There is some evidence for this from the results of studies with aged rats that describe neuronal disappearance (denervation) and reappearance (reinnervation) of neurons in different fiber bundles and subsequent change in fiber types with loss of type II and replacement with type I fibers. In rats, at least, chronic exercise reverses deteriorative changes by "dendrite restructuring, increased protein synthesis, increased axon transport of proteins, enhanced neuromuscular transmission dynamics, and changes in electrophysiological properties" (Gardiner et al., 2006). Without this stimulus (exercise), motoneurons apparently struggle to innervate newly generated muscle fibers. These findings are important but await confirmation in man.

Reduced Levels of Hormones Impede Maintenance Growth hormone (GH) stimulates protein synthesis in muscle, thereby increasing muscle mass that theoretically could ameliorate sarcopenia. With age, the concentration of GH in the blood declines. GH, although capable of acting alone, also stimulates the production of the auxiliary protein, IGF-1, discussed above as an important mediator of muscle protein synthesis. IGF-1 is required, among other things, for initiating proliferation and differentiation of satellite cells into myocytes, thereby ensuring new muscle formation. Mice that have an abundance of IGF-1 (KI of IGF-1) exhibit muscles of greater mass compared to controls with normal levels of IGF-1.

GH therapy increases muscle mass and decreases fat mass, but has little effect on muscle strength. The combination of GH therapy with resistance exercise, for example, hand weights, fails to increase muscle *strength* above that gained with resistance exercise alone. Several sarcopenia prevention studies examined the effect of growth hormone therapy in combination with testosterone therapy in men or estrogen therapy in women. With validation of elevated hormone levels in the blood, combination therapy increased lean mass, but yielded no improvement in strength in either sex. High doses of GH (600 mg/week generally effective in young subjects to increase muscle mass and strength) are unacceptable in elderly subjects due to activation of quiescent risk factors for diseases, such as prostate cancer, dementia, and diabetes. Lower doses may increase muscle mass but not strength, negating a therapeutic use of GH for the treatment of dynapenia.

Pluripotent Cells Fail to Differentiate As myocytes disappear or become dysfunctional, myocyte replacement is desirable. Replacement mechanisms exist but wane with age. The satellite cell under appropriate conditions, for example, rigorous exercise, is known for its ability to merge with a myocyte and assist with damage repair to achieve a larger, stronger myocyte and hence hypertrophic myofiber. Why it fails in the older muscle is unknown, but it is the focus of ongoing investigations.

In addition to satellite cells, there are several other resident muscle cells that under specific environmental conditions differentiate into myocytes or conversely into adipocytes or fibrotic cells, characteristics of muscle degeneration. Work is under way to fully characterize these various pluripotent cells as to the environmental conditions that epigenetically (turn off or on genes) influence their pluripotent genetic program. It is entirely possible that these mechanisms could be pharmacologically manipulated to provide a therapy in the future for sarcopenia/dynapenia especially for those where exercise is not an option.

Clearly, defining the environmental requirements for stem cell replacement is daunting. Results of parabiosis studies (joining young and old animals such that they share the same circulatory system) have pointed to the need for adequate levels of several hormones, for example, GH and IGF-1, dehydroepiandrosterone (DHEA) and testosterone (males), and estrogen (females). Cytokines such as Il-4 appear to be important according to studies of cultured stem cells. The complexity of the epigenetic changes (see Chapter 5) and the variety of different resident stem cells will no doubt require an array of supportive factors, yet to be discovered.

CONSEQUENCES OF SARCOPENIA/DYNAPENIA

Dynapenia Contributes to Reduced Mobility, Falls, Hospitalization, and Frailty Syndrome

Muscle weakness slows movement, alters gait, and disturbs balance. Abnormal gait and balance lead to falls and fractures that further compromise movement, gait, and balance and create repeated episodes of falls called the "deterioration cycle." At a minimum, falls create a phenomenon called "fear of falling." This is a post-fall mental state of hesitant gait that further slows movement. At its worst, *falls in the elderly require medical attention including hospitalizations that sap financial resources, and may produce additional debilitation and possibly death.*

There are many *reasons for the increased incidence of falls* in the elderly. Primary consideration should be given to a *diagnosis of dynapenia* that requires the initiation of resistance and balance programs. Second, *treatment of specific diseases that compound dynapenia such as osteoporosis, vascular disease with sensory loss, osteoarthritis, and neurological disease* is required. Medications, hearing loss, multitasking, and avoidable environmental hazards should also be considered.

To avoid future falls, a viable plan should be created that includes creation of a safe environment with practical modifications of good lighting, removal of throw rugs and floor clutter, railings on steps, safe bathrooms, and sturdy shoes. The plan should also include identification of medications that reduce balance control (orthostatic hypotension) and increase the risk for falls. Awareness of medication-induced orthostatic hypotension and techniques to minimize it should be taught.

Frailty Syndrome Predicts Increased Risk of Morbidity and Death The frailty syndrome describes a *deteriorative state manifest with serious sarcopenia and dynapenia.* It is characterized with unintentional weight loss, self-reported exhaustion, low grip strength, slow walking speed, and low physical activity. The presence of three of the five conditions predicts a twofold increased risk of morbidity and death, while less than three conditions predicts a twofold risk of frailty progression. The results of the clinical trial, Frailty Intervention Trial (FIT) (Fairhall et al., 2012) showed that multidisciplinary interventions of physical therapy (exercise), nutrition, psychiatric/psychological aids, and socialization reduce frailty and mobility disability. Success is achieved with an individually tailored program.

Other Consequences: Metabolic, Endocrine, and Thermoregulation

Basal Metabolic Rate Declines Due in Part to Sarcopenia Basal metabolic rate (BMR) is one of several components that contribute to an individual's energy expenditure. Other contributors are physical activity and food. However, as much as 60–80% of daily energy expenditure is due to BMR. BMR is the metabolic rate (total oxygen usage) of cells of the body during a state of complete relaxation (preferably sleep). Since a true basal state is difficult to achieve under test conditions, a less stringent state called resting metabolic rate (RMR) is routinely obtained with measurement of oxygen consumption and carbon dioxide production in a peaceful environment (called indirect calorimetry and measured by a metabolic cart in the hospital). RMR is generally 10% higher than BMR.

RMR depends entirely on the lean body mass or FFM to which skeletal muscles contribute nearly half. *RMR declines with age* and in a study by Frisard et al. (2007), the RMR in nonagenarians is down as much as 73% compared to individuals 60–74 years of age. *The age-related fall in RMR has been attributed to sarcopenia but many other factors may be involved.*

In consideration of the physiological axiom of "calories in, calories out" that implies a balance of nutrient ingestion with energy expenditure, a *reduction of RMR predicts a gain in fat deposition (fat weight gain) in the absence of decreasing consumption of food (calories).* This is a serious outcome as elevated fat content induces inflammation and exerts cellular toxicities. However, it remains to be determined *what other effects a low RMR might induce.* This is emphasized in findings by the Baltimore Longitudinal Study (Schrack et al., 2014) that compared two groups of elderly, those with a favorable profiles, for example, absence of physical and cognitive impairments, chronic conditions and comorbidities, and blood profile abnormalities, with those lacking this profile. Surprisingly, the *former had a lower RMR* compared to the latter suggesting *factors other than age and body composition influence RMR.* This adaptation in elderly who are "functional and free of medical conditions" (Schrack et al., 2014) is intriguing and warrants further investigation.

Insulin Resistance Since skeletal muscle is a *main insulin sensitive tissue, essential for glucose utilization*, a reduction in muscle size is a risk factor for the development of insulin resistance and pre-diabetes. A large cross-sectional analysis of more than 14,000 participants (National Health and Nutrition Examination Survey III) found that sarcopenia (determined by bioelectrical impedance) is associated with insulin resistance (determined by the calculated Homeostasis Model Assessment of Insulin Resistance (HOMA-IR) score using fasting insulin and glucose measurements) in both obese and nonobese individuals. The mechanism of this association is unknown, but postulated to be related to infiltrated fat and ensuing growth of an inflammatory milieu. *Visceral fat accumulation arising from a decrease in energy expenditure produces a wealth of inflammatory mediators, for example, TNFα, and interleukin-1beta* (Il-1β) that exacerbates *insulin function* and may accelerate sarcopenia. Also, lipids themselves interfere with insulin signaling. Hyperinsulinemia and eventual development of insulin resistance impair myocyte protein anabolism and mt function that lead to a dysfunctional downward spiral.

Thermoregulation Declines with Sarcopenia In the presence of sarcopenia, thermoregulation is reduced. Not only does sarcopenia alter the thermal properties of the body, but also indirectly affects the blood volume. To adequately handle a heat stress (elevation in ambient temperature), the heart pumps out more blood to the skin to promote heat loss (cooling). Since sarcopenia is associated with reduced blood volume, less blood is pumped by the heart to fill the peripheral blood vessels and less cooling occurs. In a cold environment, the loss of muscle mass limits shivering, a rhythmic contraction of muscles that generates extra heat. Furthermore sarcopenia provides reduced insulation. Additionally, the ability to conserve heat through the mechanism of vasoconstriction of blood vessels in the skin declines with age and adds to poor homeostatic thermoregulation produced by sarcopenia. The summary box summarizes the effects of sarcopenia and dynapenia.

INTERVENTIONS

Exercises

Resistance Exercises Minimize Sarcopenia/Dynapenia Physical inactivity contributes in a major way to sarcopenia/dynapenia. An overwhelming number of longitudinal studies, randomized controlled trials (RCTs), and cross-sectional studies support engagement in resistance exercise to ameliorate sarcopenia and dynapenia. Results of some 14 studies confirm that *resistance exercise in the elderly improves muscle strength*. Whether *lifelong* resistance exercise will completely prevent dynapenia has not been adequately assessed since most studies do not exceed 2 years in duration. Furthermore, there are no studies that have determined whether resistance training acts by *reversing* the maladaptive effects of physical inactivity or *induces* separate optimizing mechanisms.

An increase in muscle strength is apparent within 6–10 weeks of training (three times per week). If resistance exercises are increased on a graded scale and at levels no higher than 80% of one maximal repetition (see four-prong exercise program), these exercises will not increase blood pressure and may be started at any age to achieve a benefit. Improvement in stair climbing and gait velocity is also evident with chronic resistance training. Effects of *resistance exercises that mimic activities of daily living* (ADLs such as personal hygiene, feeding, and dressing) *translate directly to improvements* in ADLs even in sedentary adults.

Effect of Exercise on Myocyte Biology At the molecular level, mechanical stress in the form of resistance exercises produces changes in muscle structure that include *enlargement or hypertrophy of myocytes*, generally type II myofibers, and improvements in muscle content of the contractile protein, myosin heavy chain (MHC). *Muscular strength and cross-sectional area increased significantly in the elderly with a minimum of 6 weeks of resistance training*. Results of muscle biopsies taken from elderly participants after 6 months of continuous resistance exercise show an *altered gene expression* that more closely matches those of younger participants than their own, 6 months earlier. Also, in elderly individuals with type II diabetes (T2D), resistance training produced a reduction in inflammatory mediators. Although both resistance and aerobic exercise can increase muscle mass, long-term resistance exercises induce the largest gains in mass.

As noted by Pasiakos (2012), "Mechanical strain associated with contractile forces generated by exercise is a central physiological driver of protein accretion." It has been shown that a *single bout of resistance exercise increases protein synthesis for up to 48 h*. In the presence of anabolic attenuation to ingested protein observed in some elderly, resistance exercise is a reasonable means to optimizing myocyte protein synthesis.

Chronic Aerobic Exercise Produces Muscle Efficiency, Muscle Endurance, and Many Other Benefits Aerobic exercise actively engages the pulmonary and cardiovascular (CV) systems and necessitates large intakes of oxygen. Aerobic exercise includes running, walking, swimming, cycling, climbing, and dancing. Aerobic exercise improves cardiovascular function with an increase in cardiopulmonary function (VO_2max), an elevation in cardiac output (amount of blood pumped by heart), a increase in arterial compliance (blood vessels are more elastic), and enhancement of endothelial function (directly affects blood flow and health of blood vessels). Other advantages are improved nerve–muscle interaction, weight maintenance, decreased fat deposition, improved cognitive function with decreased anxiety and depression, and reduction in risk factors for CV disease and T2D.

Long-term aerobic exercise additionally improves muscle strength. A 10-year observational study (Crane et al., 2013) compared aerobically active individuals (4 h/week of vigorous exercise) from three age groups (20–39, 40–64, and 65–86) with those less active (no more than 1 h/week of vigorous exercise). Grip and knee extensor strength was greater in the aerobically active individuals and although strength declines with age in both groups, dynapenia was less in aerobically active individuals.

Resistance and aerobic exercises benefit all elderly, even the oldest old and those in chronic care facilities. One case in point are findings that in the lifestyle interventions and independence for elders pilot (LIFE-P), *elderly subjects with potential for disability prevented this outcome and improved their physical performance with 1.2 years of moderate intensity exercise*. This was validated by the short physical performance battery score (determined with tests of walking speed, chair rise, and balance) and improvements in the 400 m walking speed test.

Aerobic Exercise Effects on the Myocyte Muscle contraction *stimulates mitochondrial function, mitochondrial biogenesis*, and also *encourages some aspects of stem cell replacement*. Specifically, aerobic exercise promotes mitochondrial function by "turning on" genes involved in oxidative phosphorylation in type I fibers. Of interest is the study by Parikh et al. (2008) that identified the upregulation (enhancement) of specific genes (muscles biopsies) from young and elderly subjects (70 samples in each age group) performing maximal degrees of aerobic exercise. Upregulated genes in both age groups include genes of the respiratory chain, specifically a component of the NADH dehydrogenase electron shuttle and ATP synthase (enzyme-producing ATP). Interestingly, expression of genes involved in inflammatory processes is simultaneously reduced.

Consumption of Adequate Amounts of Protein Consumption of sufficient quantities of protein is proposed as a means to retard sarcopenia. Blood AAs directly stimulate muscle protein synthesis, an essential activity to achieve muscle homeostasis and improve the quality and quantity of muscle protein. It is estimated that approximately 20% of the elderly population are at risk of consuming less than the recommended daily allowance (RDA) of 0.8 g/kg/day of protein. As shown in Table 7.1, there are many reasons why the elderly fail to consume enough protein. A few of these are reduced energy needs, anorexia, and changes in food preference. Additionally, if the anabolic response to protein consumption is reduced (as deemed possible in elderly), then even *higher amounts of protein are required to stimulate*

TABLE 7.1. Risk Factors for Reduced Protein Intake in Older Adults

Risk Factors	Causes	Consequences
Reduced energy needs	Proportion of energy intake from protein does not change with age	Reduced needs and intake means amount of consumed protein decreases
Physical dependence	Difficulty acquiring and preparing food; 20% homebound elderly get <RDA of protein	Amount of consumed protein decreases
Change in food preference	Carbohydrates and fat-rich foods preferred over protein-rich foods	Protein deficiency coexists with high energy intake
Food insecurity	Limited or uncertain availability of nutritionally adequate and safe foods; high prevalence of elderly receiving congregate, home-delivered meals and other services	Associated with obesity, poor-quality diets low in protein and poorer self-reported health, depression, limitations in ADL and diabetes

Adapted with permission from Volpi et al. (2013).

sufficient protein synthesis. Based on acute protein synthetic rates, an intake of 30–40 g/meal of quality protein (whey, casein, and soybean) yields near-youthful levels of protein synthesis in older individuals. Quality protein sources are important and must contain adequate amounts of leucine, the essential amino acid and prime stimulator of myocyte anabolism.

When combined with resistance exercise, the timing of a protein meal is important but inadequately evaluated in the elderly. One interesting study (Esmarck et al., 2001) measured muscle hypertrophy and strength in 13 subjects (average age of 74 years) who adhered to a resistance program (3 times/week, 12 weeks) and consumed liquid protein (10 g protein, 7 g carbohydrates, and 3 g fats) either immediately or 2 h after completion of exercise. *Hypertrophy and strength increased only in those who consumed protein immediately after exercising.* These fascinating findings need confirmation with a larger group of individuals, a wider age range and different amounts of protein.

FOUR-PRONGED EXERCISE PROGRAM

A four-pronged exercise program is recommended for the elderly and should include *aerobic, progressive resistance, flexibility*, and *balance exercises* (Montero-Fernández and Serra-Rexach, 2013). *Effects depend on intensity, frequency*, and *recovery time*.

Aerobic exercises and their numerous benefits are defined above. Heart rate or the degree of perceived exertion (start at less than optimal and build to "somewhat hard") should be monitored. The American College of Sports Medicine (ACSM) suggests that levels of intensity vary relative to percent of maximal heart rate (220-age) (see Chapter 9 for heart rate calculation controversy). For example, 35–60, 60–80, and 80–90% of maximal heart rate represent low, medium, and high intensity, respectively. Aerobic exercises are performed 3–5 days/week over a continuous period of 20–60 min.

Resistance exercises require muscles to generate a force against an impedance. The impedance is provided either by a machine, a free weight, or one's own body weight as in calisthenics. Progressive resistance exercise may include lifting weights, calisthenics, carrying heavy loads, and heavy gardening. In addition to improvements in mass and strength, there are modest increases in bone mineral density, tendon strength, and lipid profile.

The intensity of the resistance workout is variable: high, medium, or low. According to the ACSM, high-intensity exercises are those that are performed at not less than 70% of one maximal repetition (RM) and with 8–12 repetitions (or two sets of 10 repetitions) at least twice each week. Low intensity starts with 10–40% of 1 RM and one set of 10 repetitions twice a week.

Ideally resistance exercises should include power training (force times speed). This is based on observations that muscle *power declines with age to a greater extent than muscle strength* and *augmentation in muscle power is more strongly correlated with improvements in daily physical necessities* such as arising from a chair. This is a fairly new concept in aging. Power enhancement requires not only a progressive increase in the degree of resistance and number of repetitions, but *also performance speed must increase and rest periods between multiple sets of exercises must decrease.*

Summary of Age Changes in Skeletal Muscle

Phenotype	Contributing Factors	Consequences
Loss of muscle mass (sarcopenia) • Myocyte atrophy • ↓ Type II fibers • ↓ Protein synthesis and degradation • Inadequate myocyte replacement	↓ Physical activity Inadequate consumption—high-quality protein ↓ Levels of hormones	↑ Body fat content ↑ Inflammatory factors Small contribution to dynapenia ↓ RMR Weight gain ↑ Risk of T2D Altered thermogenesis
Loss of muscle strength (dynapenia) • Poor-quality contractile proteins • ↓ mt content • Mutated mtDNA • ↓ Dihydropyridine receptor number • ↓ Central input to motoneurons	Sedentary life style ↓ Physical activity (resistance and aerobic exercise)	Muscle weakness ↓ Gait ↓ Physical performance Exercise intolerance Falls ↑ Morbidity/mortality

As recommended by the ACSM, stretching exercises are to be done two times/week at 10 min/day with a scale of moderate intensity 5–6 on a scale of 0–10. Stretching exercises are applied to muscles in neck, shoulder, elbow, wrist, hip, knee, and ankle. Flexibility exercises are ideally performed concomitant with aerobic and resistance exercise.

To improve balance, the ASCM recommends the use of progressively more difficult balance stances. Possible balance exercises are a series of *one min one foot stances* first without support, second with balance foot placed on an uneven surface, and third with eyes closed. Additionally, dynamic movements that uses toe or heel walking, circular turns, or walking on foam add variety.

Initiation of an exercise program demands certain precautions. Considerations as to comorbidities and mediations, safe attire, nutrition and hydration, knowledge of proper exercise performance, and comfort and safety at an exercise facility are some of the basic ones.

SUMMARY

Skeletal muscles are essential for movement, balance, respiration, and metabolism. Contractions and relaxations of muscles depend on contractile proteins and auxiliary components of the myocyte, interaction with motoneurons, and supportive extracellular environment.

The effect of aging on skeletal muscles is highly variable in large part due to the contribution from physical inactivity, inadequate protein consumption, and decline in hormone (growth factors) levels. Loss of muscle mass (sarcopenia) and loss of muscle strength (dynapenia) are consistently observed in older individuals but a consensus definition for them is lacking. The age-related change in body composition is accompanied by an increase in fat mass. Cellular changes of importance are a decrease in protein synthesis/degradation, mt deficiencies, and reduced excitation–contraction coupling.

Importance has been placed on loss of muscle strength since it results in muscle weakness, reduced gait speed, balance disturbances, exercise intolerance, and falls. These conditions reduce independence, require hospitalization, and promote morbidity and mortality. Dynapenia is slowed by a continuous program of resistance exercise.

The benefits of a four-pronged physical program of aerobic, resistance, balance, and flexibility exercises are incalculable and minimize sarcopenia and dynapenia, produce cardiovascular and cardiopulmonary fitness, generate a favorable lipid profile and reduce risk of cardiovascular disease and T2D, maintain bone density, and reduce mortality.

CRITICAL THINKING

What exercise(s) would you suggest to an older individual with dynapenia? With sarcopenia? With the frailty syndrome? With none of the above?

What is the difference between sarcopenia and dynapenia? Should these conditions be considered separately? Explain. Which one deserves research funding?

How do the following assist with skeletal muscle contraction/relaxation: motoneurons; mitochondria; actin/myosin?

How do the various age-related molecular and cellular changes in the myocyte contribute to loss of muscle mass and strength?

KEY TERMS

Acetylcholine (ACh) chemical mediator called a neurotransmitter released from somatic motoneurons.

Actin a globular protein found in myocytes; component of the thin filament that interacts with myosin in muscle contraction and relaxation.

Anaerobic process that occurs without oxygen and in man leads to the accumulation of lactic acid.

Atrophic reduced in size usually (but not always) due to the loss of cells.

Dihydropyridine receptor skeletal muscle receptor sensitive to the nerve impulse that assists with calcium release in contraction.

Dynapenia progressive age-related loss of skeletal muscle strength.

Electron transport chain (ETC) complexes of proteins in the inner membrane of the mt that are responsible for abstraction of electrons from nutrients to facilitate ATP formation.

Frailty syndrome syndrome with the following symptoms: unintentional weight loss, self reported exhaustion, low grip strength, slow walking speed, and low physical activity.

Growth hormone (GH) a hormone secreted by the anterior pituitary gland that acts in part through the secondary mediator, insulin growth factor-1 (IGF-1). Both GH and IGF-1 assist with formation and maintenance of muscle physiology. In development, they are strong growth (anabolic) promoters. In aging, blood levels decline.

Hypertrophy enlargement in size.

Insulin hormone released from the pancreas in response to elevated blood sugar. It serves to move sugar (glucose) to muscle and liver.

Leucine an essential amino acid (must be consumed in diet) that is instrumental in muscle cell hypertrophy and strength production.

Muscle fiber another name for the muscle cell or myocyte. Single unit of the muscle.

Myocyte individual muscle cell.

Myosin multichain protein that forms the thick filament in the myocyte to interact with actin in muscle contraction.

Parabiosis joining of two organisms such that they share a common blood supply. Used in "rejuvenation" studies by joining a young and an old animal to determine blood factors that slow or retard aging.

Pluripotent cell cell capable of becoming any cell type depending on the presence of growth factors and cell mediators. It is another name for a stem cell.

RDA recommended daily allowance; government suggested quantities of nutrients to prevent gross deficiencies and disease.

Sarcopenia age-related loss of skeletal muscle mass.

Satellite cell one of several stem cells found in muscle. Cells with potential to replace muscle cells that apoptosis.

Testosterone male hormone (steroid) that determines male sex characteristics, including muscle anabolism.

Type I fiber slow twitch muscle fiber, one in which there exist an abundance of mitochondria that enable the muscle fiber to handle endurance activities.

Type II fiber fast twitch fiber that is responsible for speed, but that tires easily; affected by age moreso than Type I fibers.

BIBLIOGRAPHY

Reviews

Booth FW, Laye MJ, Roberts MD. 2011. Lifetime sedentary living accelerates some aspects of secondary aging. *J. Appl. Physiol.* **111**(5):1497–1504.

Clark BC, Manini TM. 2010. Functional consequences of sarcopenia and dynapenia in the elderly. *Curr. Opin. Clin. Nutr. Metab. Care* **13**(3):271–276.

Ciolac EG. 2013. Exercise training as a preventive tool for age-related disorders: a brief review. *Clinics (Sao Paulo)* **68**(5):710–717.

Dam T-T, Peters KW, Fragala M, Cawthon PM, Harris TB, McLean R, Shardell M, Alley DE, Kenny A, Ferrucci L, Guralnik J, Kiel DP, Kritchevsky S, Vassileva MT, Studenski S. 2014. An evidence-based comparison of operational criteria for the presence of sarcopenia. *J. Gerontol. A Biol. Sci. Med. Sci.* **69**(5):584–590.

Doherty, TJ. 2003. Invited review: aging and sarcopenia. *J. Appl. Physiol.* **95**(4):1717–1727.

Gardiner P, Dai Y, Heckman CJ. 2006. Effects of exercise training on alpha-motoneurons. *J. Appl. Physiol.* **101**(4):1228–1236.

Gray H. 2013. *Gray's Anatomy*. London: Arcturus Publishing Limited.

Johnson ML, Robinson MM, Nair KS. 2013. Skeletal muscle aging and the mitochondrion. *Trends Endocrinol. Metab.* **24**(5):247–256.

Konarzewski M, Książek A. 2013. Determinants of intra-specific variation in basal metabolic rate. *J. Comp. Physiol. B* **183**(1):27–41.

Manini TM, Clark BC. 2012. Dynapenia and aging: an update. *J. Gerontol. A Biol. Sci. Med. Sci.* **67A**(1):27–40.

Montero-Fernández N, Serra-Rexach JA. 2013. Role of exercise on sarcopenia in the elderly. *Eur. J. Phys. Rehabil. Med.* **49**(1):131–143.

Nair KS. 2005. Aging muscle. *Am. J. Clin. Nutr.* **81**(5):953–963.

Paddon-Jones D, Rasmussen BB. 2009. Dietary protein recommendations and the prevention of sarcopenia: protein, amino acid metabolism and therapy. *Curr. Opin. Clin. Nutr. Metab. Care* **12**(1):86–90.

Pasiakos SM. 2012. Exercise and amino acid anabolic cell signaling and the regulation of skeletal muscle mass. *Nutrients* **4**(7):740–758.

Volpi E, Campbell WW, Dwyer JT, Johnson MA, Jensen GL, Morley JE, Wolfe RR. 2013. Is the optimal level of protein intake for older adults greater than the recommended dietary allowance? *J. Gerontol. A Biol. Sci. Med. Sci.* **68**(6):677–681.

Experimental

Baumgartner RN, Koehler KM, Gallagher D, Romero L, Heymsfield SB, Ross RR, Garry PJ, Lindeman RD. 1998. Epidemiology of sarcopenia among the elderly in New Mexico. *Am. J. Epidemiol.* **147**(8):755–763.

Bohannon RW, Magasi S. 2015. Identification of dynapenia in older adults through the use of grip strength t-scores. *Muscle Nerve* **51**(1):102–105.

Cawthon PM, Peters KW, Shardell MD, McLean RR, Dam T-TT, Kenny AM, Fragala MS, Harris TB, Kiel DP, Guralnik JM, Ferrucci L, Kritchevsky SB, Vassileva MT, Studenski SA, Alley DE. 2014. Cutpoints for low appendicular lean mass that identify older adults with clinically significant weakness. *J. Gerontol. A Biol. Sci. Med. Sci.* **69**(5): 567–575.

Conboy IM, Conboy MJ, Wagers AJ, Girma ER, Weissman IL, Rando TA. 2005. Rejuvenation of aged progenitor cells by exposure to a young systemic environment. *Nature* **433**(7027):760–764.

Cooper R, Bann D, Wloch EG, Adams JE, Kuh D. 2015. "Skeletal muscle function deficit" in a nationally representative British birth cohort in early old age. *J. Gerontol. A Biol. Sci. Med. Sci.* **70**(5):604–607.

Crane JD, Macneil LG, Tarnopolsky MA. 2013. Long-term aerobic exercise is associated with greater muscle strength throughout the life span. *J. Gerontol. A Biol. Sci. Med. Sci.* **68**(6):631–638.

Dillon EL, Sheffield-Moore M, Paddon-Jones D, Gilkison C, Sanford AP, Casperson SL, Jiang J, Chinkes DL, Urban RJ. 2009. Amino acid supplementation increases lean body mass, basal muscle protein synthesis, and insulin-like growth factor-I expression in older women. *J. Clin. Endocrinol. Metab.* **94**(5):1630–1637.

Esmarck B, Andersen JL, Olsen S, Richter EA, Mizuno M, Kjaer M. 2001. Timing of post exercise protein intake is important for muscle hypertrophy with resistance training in elderly humans. *J. Physiol.* **535** (Part 1):301–311.

Fairhall N, Sherrington C, Kurrle SE, Lord SR, Lockwood K, Cameron ID. 2012. Effect of a multifactorial interdisciplinary intervention on mobility-related disability in frail older people: randomised controlled trial. *BMC Med.* **10**: 120–133.

Frisard MI, Fabre JM, Russell RD, King CM, DeLany JP, Wood RH, Ravussin E, Louisiana Healthy Aging Study. 2007. Physical activity level and physical functionality in nonagenarians compared to individuals aged 60–74 years. *J. Gerontol. A Biol. Sci. Med. Sci.* **62**(7):783–788.

Frontera WR, Reid KF, Phillips EM, Krivickas LS, Hughes VA, Roubenoff R, Fielding RA. 2008. Muscle fiber size and function in elderly humans: a longitudinal study. *J. Appl. Physiol.* **105**(2):637–642.

Goodpaster BH, Chomentowski P, Ward BK, Rossi A, Glynn NW, Delmonico MJ, Kritchevsky SB, Pahor M, Newman AB. 2008. Effects of physical activity on strength and skeletal muscle fat infiltration in older adults: a randomized controlled trial. *J. Appl. Physiol.* **105**(5):1498–1503.

Gordon PL, Vannier E, Hamada K, Layne J, Hurley BF, Roubenoff R, Castaneda-Sceppa C. 2006. Resistance training alters cytokine gene expression in skeletal muscle of adults with type 2 diabetes. *Int. J. Immunopathol. Pharmacol.* **19**(4):739–749.

Katsanos CS, Kobayashi H, Sheffield-Moore M, Aarsland A, Wolfe RR. 2006. A high proportion of leucine is required for optimal stimulation of the rate of muscle protein synthesis by essential amino acids in the elderly. *Am. J. Physiol. Endocrinol. Metab.* **291**(2): E381–E387.

Lambert CP, Sullivan DH, Freeling SA, Lindquist DM, Evans WJ. 2002. Effects of testosterone replacement and/or resistance exercise on the composition of megestrol acetate stimulated weight gain in elderly men: a randomized controlled trial. *J. Clin. Endocrinol. Metab.* **87**(5):2100–2106.

Leenders M, Verdijk LB, van der Hoeven L, van Kranenburg J, Nilwik R, van Loon LJ. 2013. Elderly men and women benefit equally from prolonged resistance-type exercise training. *J. Gerontol. A Biol. Sci. Med. Sci.* **68**(7):769–779.

Lexell J, Taylor CC, Sjöström M. 1988. What is the cause of the ageing atrophy? Total number, size and proportion of different fiber types studied in whole vastus lateralis muscle from 15- to 83-year-old men. *J. Neurol. Sci.* **84** (2–3):275–294.

Mamerow MM, Mettler JA, English KL, Casperson SL, Arentson-Lantz E, Sheffield-Moore M, Layman DK, Paddon-Jones D. 2014. Dietary protein distribution positively influences 24-h muscle protein synthesis in healthy adults. *J. Nutr.* **144**(6):876–880.

Nilwik R, Snijders T, Leenders M, Groen Bart BL, van Kranenburg J, Verdijk LB, van Loon Luc JC. 2013. The decline in skeletal muscle mass with aging is mainly attributed to a reduction in type II muscle fiber size. *Exp. Gerontol.* **48**(5):492–498.

Parikh H, Nilsson E, Ling C, Poulsen P, Almgren P, Nittby H, Eriksson K-F, Vaag A, Groop LC. 2008. Molecular correlates for maximal oxygen uptake and type 1 fibers. *Am. J. Physiol. Endocrinol. Metab.* **294**(6):E1152–E1159.

Safdar A, Hamadeh MJ, Kaczor JJ, Raha S, Debeer J, Tarnopolsky MA. 2010. Aberrant mitochondrial homeostasis in the skeletal muscle of sedentary older adults. *PLoS One* **5**(5): e10778.

Schrack JA, Knuth ND, Simonsick EM, Ferrucci L. 2014. "IDEAL" aging is associated with lower resting metabolic rate: the Baltimore Longitudinal Study of Aging. *J. Am. Geriatr. Soc.* **62**(4):667–672.

Sehl ME, Yates FE. 2001. Kinetics of human aging: I. Rates of senescence between ages 30 and 70 years in healthy people. *J. Gerontol. A Biol. Sci. Med. Sci.* **56**(5):B198–B208.

Verdijk LB, Koopman R, Schaart G, Meijer K, Savelberg HH, van Loon LJ. 2007. Satellite cell content is specifically reduced in type II skeletal muscle fibers in the elderly. *Am. J. Physiol. Endocrinol. Metab.* **292**(1):E151–E157.

Yarasheski KE, Pak-Loduca J, Hasten DL, Obert KA, Brown MB, Sinacore DR. 1999. Resistance exercise training increases mixed muscle protein synthesis rate in frail women and men >/= 76 yr old. *Am. J. Physiol.* **277** (1 Part 1):E118–E125.

8

AGING OF THE SKELETAL SYSTEM

OVERVIEW

Humans have a light yet strong internal support system comprised of skeletal bones. Strength without excess weight is achieved with a mix of flexible proteins and sturdy mineral deposits within the bone matrix. Deceivingly inert in appearance, bones are notably metabolically active and capable of responding to stimuli that vary from mechanical to hormonal. Skeletal bones not only readily adapt to gravity and assorted physical activities but also retains close to 99% of the body's calcium, an indispensable ion that serves many essential functions. Bones also contribute to regulation of blood acid–base balance and blood cell maturation (Table 8.1).

Bone Structure

Bone Is a Mix of Porous and Compact Bone Of the approximately 206 bones in the human skeleton, 80% are macroscopically compact *cortical bones*, for example, skull, jaw, shaft of long bones, and outer edge of joints, and the remaining bones are porous *trabecular (cancellous) bones*, for example, inner aspects of hip, pelvis, and vertebrae. Many skeletal sites have a mixture of these types such as the vertebrae. The periosteum, the fibrous outer covering of bone, is rich in bone cells, blood vessels, and nerves. The periosteum as well as an inner membrane between cortical and trabecular bones, the endosteum, are physical sites of bone adaptation (formation and resorption). *Aging affects cortical bone to a greater extent than trabecular bone* since "70% of all bone loss is cortical" (Seeman, 2013).

Human Biological Aging: From Macromolecules to Organ Systems, First Edition. Glenda Bilder.
© 2016 John Wiley & Sons, Inc. Published 2016 by John Wiley & Sons, Inc.

TABLE 8.1. Bone Functions

- Support
- Movement
- Calcium reservoir
- Acid–base balance
- Blood cell maturation

The microarchitecture of bone consists of internal concentric structures, some evolving as cartilage and others arising as a result of modeling and reshaping. In cortical bone these units (osteons with a central Haversian canal) are tightly packed together and consist of bone materials that include minerals (calcium and phosphate), matrix proteins (type I collagen, proteoglycans, and other proteins), sensory bone cells called osteocytes, nerves, and a small blood supply. In trabecular bone, these units are similar, but appear as plates and are more difficult to clearly define. The most interesting feature of bone is the *impressive complex of channels (canaliculi)* that contain the dendritic projections of the osteocytes that communicate with projections from neighboring osteocytes as well as with bone surface cells. Extracellular fluid baths the canaliculi. It is *this complex that responds to mechanical loads or deformations termed customary strain stimulus* (CSS). This is discussed below. The major components of bone are illustrated in Figure 8.1 and Table 8.2.

Modeling and Remodeling Processes

Bone Senses and Responds to Mechanical Strain
The concept of bone as a mechanostat was proposed more than 25 years ago by Frost (1987). Frost's theory

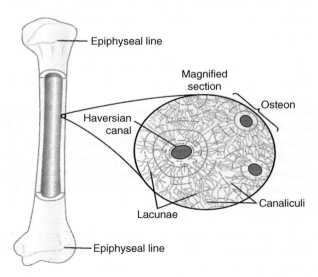

Figure 8.1. Structure of bone. (Reprinted with permission from Guyton and Hall, 2006.) Guyton AC, Hall JE. 2006. *Textbook of Medical Physiology,* 11th ed. Philadelphia: Elsevier Saunders, pp. 983. (See plate section for color version.)

TABLE 8.2. Bone Microarchitecture

Types	Cells	Components
Cortical • Compact Trabecular • Porous	Osteoblast • Forms bone • Produces matrix of collagens; calcium-phosphate hydroxide salts Osteocyte • Mechanosensor • Coordinates activities Osteoclast • Resorption of bone • Produces enzymes to breakdown matrix Progenitor cells • Future bone cells	Osteon • bone unit with inner canal Periosteum • outer bone forming region of cortical bone Endosteum • inner bone forming region of cortical bone Canaliculi • interconnecting network of channels

states that *bones adjust to different magnitudes of mechanical strain throughout life*. Mechanical strain above a threshold amount (magnitude) initiates bone formation and mechanical strain below the threshold magnitude induces bone loss (resorption). Skerry (2006) extended this theory to include multiple mechanical thresholds (no one mechanostat) because "the skeleton responds to a complex strain stimulus, made up of numerous different parameters of which peak magnitude is only one." The term "customary strain stimulus" replaces mechanical threshold. *Customary strain stimulus varies throughout the skeleton and is modified by genetics, gender, disease, drugs, and, of course, age.*

Modeling and Remodeling Establish Bone Strength

The composition and structure of bone is developed through the *"final common pathway" of modeling and remodeling processes.*

Modeling describes bone formation and bone resorption as separate entities in time and place and is the process responsible for bone growth from birth to young adulthood. *Remodeling requires the coupling of bone resorption with bone formation such that bone resorption occurs first followed by bone formation or "filling-in" thereafter.* Remodeling refashions bone by repairing microdamage generated by high strain stimuli or strain above the CSS and by removing bone when the strain falls below the CSS.

Bone Cells (Osteoblasts, Osteoclasts, and Osteocytes) with Different Functions Create the Density and Geometry of Bone

During growth, both modeling and remodeling participate to create bone structure and composition. Bone areas that receive high levels of strain (due to developing muscle strength) create bone through stimulation of *osteoblasts* that secrete a variety of collagens and the minerals (calcium phosphate salts) to establish the extracellular

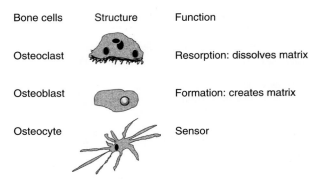

Figure 8.2. Bone cells: appearance and function.

bone matrix; bone areas that receive less strain undergo bone resorption (dissolution) carried out by *osteoclasts* that dissolve the matrix through release of numerous lysosomal proteases and hydrogen ions (Figure 8.2). Examples of reduced bone strain are limb disuse (cast), bed rest, or space flight.

Remodeling adds importantly to bone growth. This coupled process begins with bone resorption by the osteoclasts, followed by bone formation with osteoblasts and the formation of the osteon with entrapment of the osteoblast that becomes a permanent resident cell, the *osteocyte* (Figure 8.2). Remodeling not only continues construction of an impressive sensing network of the osteocytes noted above but also allows adaptation to strain over and above that generated by normal skeletal muscle strength. This is evident in the superior density and quality of bone that receives high impact strain most commonly experienced by elite athletes. A study by Schipilow et al. (2013) reported greater mineral density and bone strength in the weight-bearing distal tibia (lower larger leg bone) of female alpine skiers and soccer players compared to bone measurements from nonathletes.

INEVITABLE LOSS OF BONE

Changes in Mechanical Strain

With Age, Modeling Nearly Stops and Remodeling Is Stimulated by Strains Below the CSS Frost (1997) postulated *that bone modeling ceases at age 30*, although some data suggest it continues throughout adulthood but at a very slow rate. Importantly, the "bone size, architecture and mass attained during growth determine the relevance of bone loss during advancing age" (Seeman, 2008). This is a crucial observation because *peak bone mass attained at the end of growth determines as much as 60% of the risk of developing osteoporosis* in later years (Gunter et al., 2012). Thus, it is imperative that children and adolescents engage in continuous vigorous physical activities during development to ensure optimal bone density and strength in young adulthood thereby reducing the risk of osteoporosis in later years.

In the adult, it is primarily bone remodeling that determines bone quality since the contribution from bone modeling is minimal. As muscle mass and strength decline (as with sarcopenia, dynapenia) and physical activity decreases generally by choice,

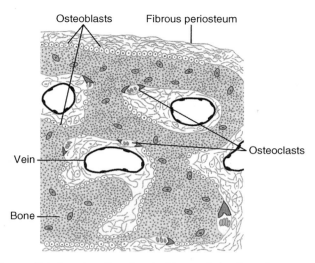

Figure 8.3. Bone cells in location. (Reprinted with permission from Guyton and Hall (2006).) Guyton AC, Hall JE. 2006. *Textbook of Medical Physiology,* 11th ed. Philadelphia: Elsevier Saunders, 982 pp. (See plate section for color version.)

mechanical strains lessen. As mechanical strain falls below CSS, remodeling, bone loss and fragility occur. This scenario is *considered by some to be the major reason for bone loss with age.* However, *strain-dependent bone loss is exacerbated by the lower levels of regulatory hormones and lifestyle choice* such as smoking and diet.

Bone Cell Functions Deteriorate with Age Remodeling is initiated by microdamage or perturbations in the canaliculi that stimulates apoptosis in the sensing cell, the osteocyte. Thereafter, newly generated osteoclasts remove bone and create bone cavities, which are subsequently filled in by an assembly of osteoblasts (see Figure 8.3). Together these cells embody the bone basic multicellular unit (BMU) that not only renews bone but can also potentially strengthen it if bone resorption and bone formation are equal.

According to Seeman (2008), with age, *bone formation (osteoblastic function) declines and bone resorption (osteoclastic function) stays the same or decreases.* This is *accelerated with loss of estrogen.* After menopause (see Chapter 14) there is an increased rate of remodeling, thus creating more BMUs whose resorption in the presence of decreased bone formation accelerates bone loss. Additionally, an enhanced remodeling rate removes dense mineralization and replaces it with less dense material. Three to five years after menopause, bone formation "catches up" with the initial accelerated resorption and the rate of bone loss slows. However, the presence of increased number of BMUs causes a continual loss of bone for years. The characteristics of modeling and remodeling are summarized in Table 8.3.

Trabecular Structure Disappears and Cortical Structure Becomes More Like Trabecular Bone; Gender Differences Initially accelerated remodeling occurs on trabecular bone due to its larger surface area, but as this surface

TABLE 8.3. Characteristics of Bone Processes

Modeling	Remodeling
• Bone formation and bone resorption are independent activities; uncoupled	• Bone resorption and bone formation are dependent activities; coupled
• Process used in bone growth; slows or absent in adulthood	• Process to maximize bone strength; altered in adulthood
• Responds to mechanical strains above and below CSS; above = bone formation; below = bone resorption	• Responds to mechanical strains above and below CSS; above = remodeling rate increases to strengthen bone; below = remodeling rate also increases with bone loss
• Effect in elderly—absent	• Effect in elderly—bone formation < bone resorption results in bone loss

CSS: customary strain stimulus.

area shrinks, *remodeling transpires on cortical surfaces that become more trabecular-like in appearance.* Areas deep in bone, unaffected by remodeling, suffer oxidative damage, for example, cross-linked collagens, and lots of AGEs due to absence of bone renewal.

Bone loss in men is less than in women for several reasons. The male skeleton is larger and men do not experience the dramatic reduction in estrogen innate to women. Additionally, since bone loss is due to decreased bone formation within the BMU, trabecular thinning occurs first that basically preserves cortical bone, the bone type needed for bone strength. Thus, risk of fractures is less in men than in women. However, fracture risk is comparable in men and women where the two sexes are of the same age and with the same bone mineral density (BMD).

Large Epidemiological Studies Show Bone Loss in Men and Women, but Greater Loss in Women

Imaging techniques to assess bone status have been used in cross-sectional and longitudinal studies in man. The dual energy X-ray absorptiometry (DXA) measures BMD in a two-dimensional format (aBMD). It is used as a screening tool for osteoporosis, but does not provide information on bone quality (architecture and strength). Studies reviewed by Clarke and Khosla (2010) performed over the past 25 years on *aBMD data found the rate of bone loss in the vertebrae, pelvis, and ultra distal wrist is especially rapid immediately following menopause.* Thereafter, bone loss slows. No sudden similar bone loss was experienced by men, although aBMD declined slowly after age 40. Figure 8.4 is a composite assessment of cortical and trabecular bone loss based on these data.

Quantitative computed tomography (QCT) is another imaging technique used in the evaluation of bone changes with age. Unlike DXA, *QCT provides a volumetric measurement of BMD* (vBMD) and can be used on the spine and hips (cQCT) or the appendages (pQCT). These data are also amenable to calculation of bone strength. It is a useful research tool because it reveals more about bone quality than the DXA and is sometimes used clinically for a diagnosis in puzzling bone loss conditions.

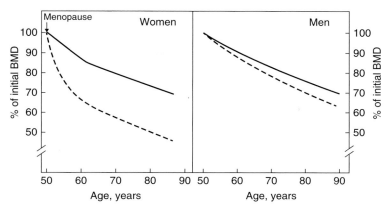

Figure 8.4. Patterns of age-related bone loss in women and men. Dashed lines represent trabecular bone and solid lines, cortical bone. The figure is based on multiple cross-sectional and longitudinal studies using DXA. (Reproduced with permission from Khosla and Riggs (2005).) Khosla S, Riggs BL. 2005. Pathophysiology of age-related bone loss and osteoporosis. *Endocrinol. Metab. Clin. North Am.* 34(4):1015–1030.

In a large cross-sectional study of over 600 participants (20–92 years of age), *QCT analysis of various bones (lumbar spine, femoral neck, distal radius, and distal tibia) showed a complexity of bone changes*, some of which Riggs et al. (2004) considered beneficial (increased cross-sectional area of the vertebrae and proximal femur due to outward periosteal and endocortical remodeling) and others were detrimental (55% loss of central trabecular bone, 24% loss of peripheral trabecular bone, 25% loss of cortical bone in women, and lesser percent changes in men), generally beginning in middle age. Riggs et al. (2004) concluded that the *loss of both trabecular and cortical bones and the small skeletal size of women predispose them to a high incidence of fragility fractures* (those arising from normal activities).

Nonmechanical Factors: Hormones, Vitamin D, and Calcium

Loss of Estrogen Contributes Significantly to Bone Loss in Both Elderly Women and Men

Many hormones contribute to bone health. They include sex steroids, adrenal steroids (AS), insulin and parathyroid hormones; they operate in different ways. Estrogen (estradiol, E2) is considered to be *the most significant systemic regulator of bone for both women and men*. The effects of E2 on bone are complex and incompletely understood. In the main E2 slows bone remodeling rate by preservation (prevents apoptosis) of osteocytes and osteoblasts and eliminates (increases apoptosis) osteoclasts. These *pleiotropic osteogenic actions of E2 diminish dramatically in menopause as E2 levels plunge to near zero*. The result as discussed above is augmented bone remodeling and bone mineral loss, changes that are reversed with hormone replacement therapy (E2 and progesterone preparations) during and after menopause.

A deficiency of E2 in men accounts in large part (estimated 70%) for bone loss with age. Several cross-sectional studies, a 4-year longitudinal study, and an experimental manipulative study with sex steroids concluded that *a decline in E2, not testosterone (T), is associated with bone loss*. E2 arises from T (via an aromatase enzyme) and as T levels decline with age (see Chapter 14), albeit not as dramatically as E2 in women, E2 levels in men will also decline and hence influence bone health. Secondarily, T directly affects remodeling by prevention of resorption and enhancement of bone formation. However, these effects are modest. So as T declines with age, T-supporting activities on bone deteriorate somewhat, but *by far the larger effect is from the reduction in E2*.

Vitamin D and Calcium Deficiency Activate Parathyroid Hormone-Directed Bone Resorption

Blood calcium plays a role in bone remodeling. Bone is a singular reservoir of bound calcium (mineral salts of calcium phosphate). The remaining 1% of calcium circulates in the blood, half of which is bound to protein complexes and the remainder is ionized and complex-free. *Ionized calcium is tightly regulated by the parathyroid hormone (PTH), vitamin D, and itself and their respective receptors*. PTH is produced by the parathyroid glands (four small glands situated on top of the thyroid gland); vitamin D is generated by the action of UV radiation on the skin (see Chapter 6) and is obtained in the diet. Calcium is a dietary essential.

Ionized calcium is vital for nerve conduction, enzyme activity, muscle contraction, and secretory functions, to name but a few, and hence a homeostatic mechanism maintains its concentration within a very narrow range. For example, a reduction in ionized calcium below this range stimulates the parathyroid glands to produce and release PTH. PTH activates osteoclasts to dissolve bone, an effect that releases calcium into the blood. PTH also acts on the kidney to enhance reabsorption of calcium (prevent its urinary loss) and additionally to produce more vitamin D. Vitamin D acts on the GI tract to absorb additional dietary calcium. Thus, PTH, vitamin D, and calcium act in concert to restore ionized calcium to a physiologically acceptable concentration.

To maintain calcium homeostasis without theft of bone calcium, intake of calcium and its intestinal absorption must be optimal. This requires an adequate amount of vitamin D to facilitate calcium transfer across the gut. However, many elderly (males and females) not only do not consume enough calcium (intake <700–800 mg/day) but are also vitamin D deficient (less that 20–40 ng/ml of 25-OH vitamin D_3) (see Chapter 6). Normal levels of vitamin D and calcium promote normal mineralization; *a vitamin D/calcium deficiency in adults produces inadequately mineralized matrix and weak bone structure that leads to increase risk of fractures*.

Vitamin D deficiency is associated with hyperparathyroidism (elevated secretion of PTH). In both sexes, levels of PTH have been shown to increase with age, possibly the result of vitamin D and calcium deficiency. In light of this, the daily calcium recommendations (Institutes of Medicine, 2010) for women >51 years of age and men >70 years of age is 1200 mg/day. Vitamin D consumption of at least 800 international units (IU) is also recommended. As reviewed by Bauer (2013), these recommendations are based on data from a number of clinical trials that show a modest benefit in fracture reduction with calcium and vitamin D supplements or calcium alone. In consideration of the variability of supplement absorption and the potential of kidney

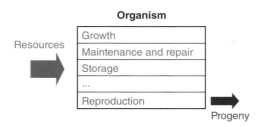

Figure 3.3. Choices that organisms faced according to Darwin's natural selection tenets. (Reproduced with permission from Kirkwood (2005).)

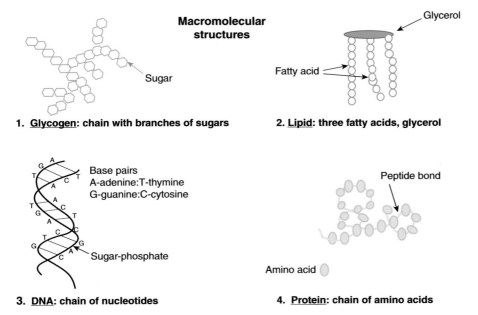

1. **Glycogen**: chain with branches of sugars
2. **Lipid**: three fatty acids, glycerol
3. **DNA**: chain of nucleotides
4. **Protein**: chain of amino acids

Figure II.3. Representative structures of macromolecules.

Human Biological Aging: From Macromolecules to Organ Systems, First Edition. Glenda Bilder.
© 2016 John Wiley & Sons, Inc. Published 2016 by John Wiley & Sons, Inc.

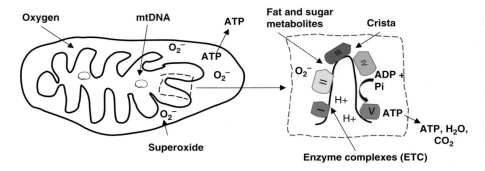

Enzyme complexes facilitate use of oxygen and electrons/protons from nutrients to generate ATP, carbon dioxide, and water

Figure 5.1. Mitochondrion prototype. Components are mitochondrial DNA (mtDNA), electron transport chain (ETC) with enzyme complexes, free radicals (superoxide, O_2^-), ATP production (carbon dioxide, CO_2 and water, H_2O), and hydrogen ion gradient (H^+).

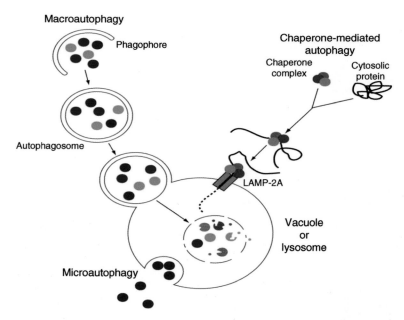

Figure 5.3. Illustration of the three pathways of autophagy in the cell. (Reprinted with permission from Lynch-Day and Klionsky (2010).)

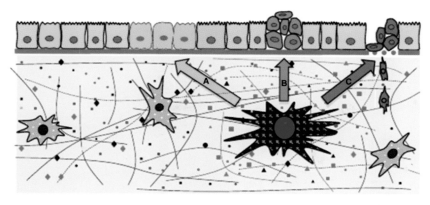

Figure 5.7. Proposed role of senescent cell in development of cancer. Senescent cell releases factors that modify nearby cells (A), promote uncontrolled cell division (B), and enhance invasive movement to other sites (C). (Reprinted with permission Labarge et al. (2012).)

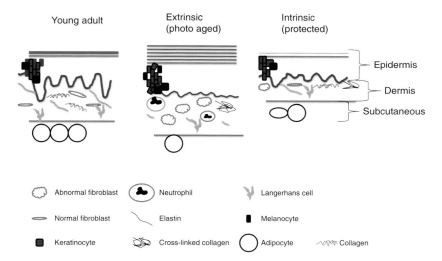

Figure 6.2. Representative changes in the skin due to extrinsic and intrinsic aging.

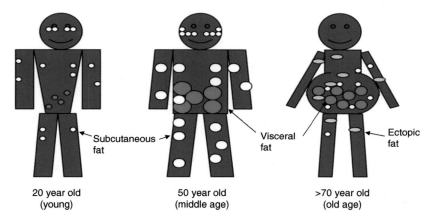

Figure 6.3. Fat distribution with age. (Reprinted with permission from Cartwright et al. (2007).)

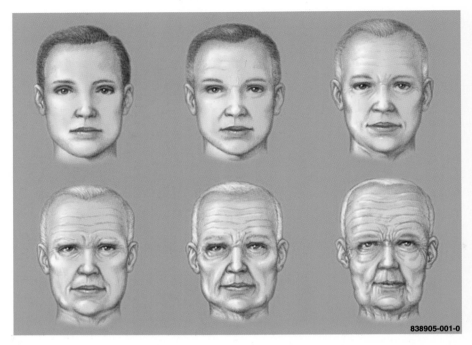

<u>Figure 6.5.</u> The aging face. (Reprinted with permission from Friedman (2005).)

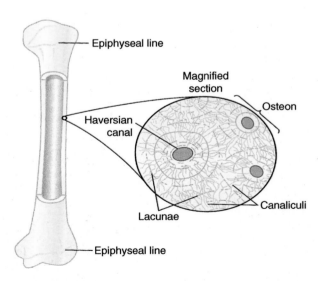

<u>Figure 8.1.</u> Structure of bone. (Reprinted with permission from Guyton and Hall, 2006.) Guyton AC, Hall JE. 2006. *Textbook of Medical Physiology,* 11th ed. Philadelphia: Elsevier Saunders, pp. 983.

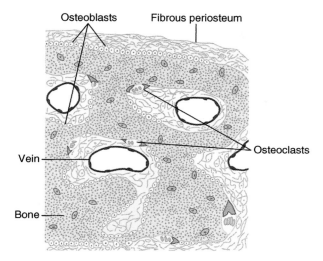

Figure 8.3. Bone cells in location. (Reprinted with permission from Guyton and Hall (2006).) Guyton AC, Hall JE. 2006. *Textbook of Medical Physiology,* 11th ed. Philadelphia: Elsevier Saunders, 982 pp.

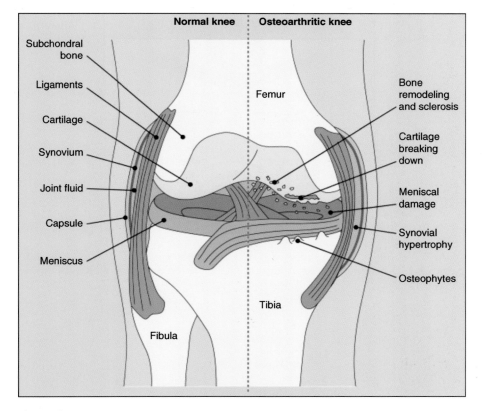

Figure 8.5. Osteoarthritis of the knee joint. (Reprinted with permission from Hunter and Felson (2006).)

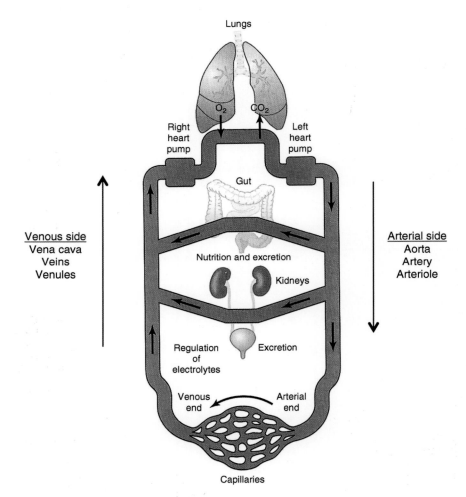

Figure 9.1. Overview of the circulatory system. (Reprinted with permission from Guyton and Hall (2006).)

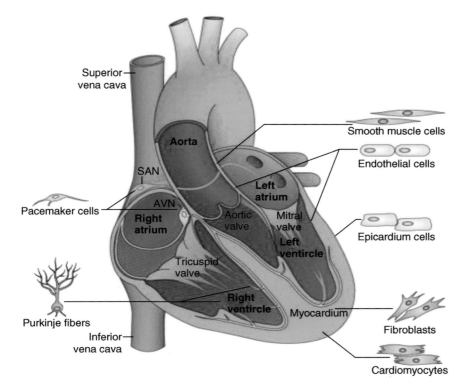

Figure 9.2. Cell types found in the heart. (Reproduction with permission from Xin et al. (2013).)

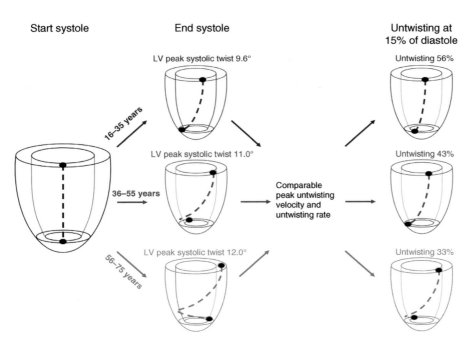

Figure 9.4. Age-associated changes in left ventricular (LV) twisting and untwisting during contraction and relaxation of the heart muscle. (Reprinted with permission from van Dalen et al. (2010).)

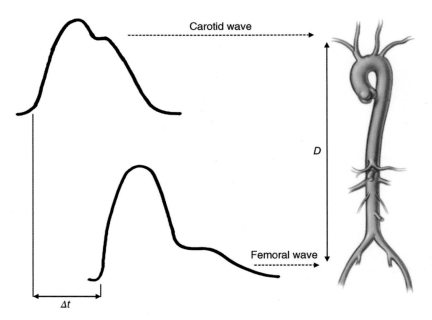

Figure 9.6. Pulse wave velocity determination. Transit time is estimated by the foot-to-foot method. The foot of the wave is defined at the end of diastole, when the steep rise of the waveform begins. The transit time is the time of travel of the foot of the wave (Δt) over a known distance (D). (Reprinted with permission from Calabia et al. (2011).)

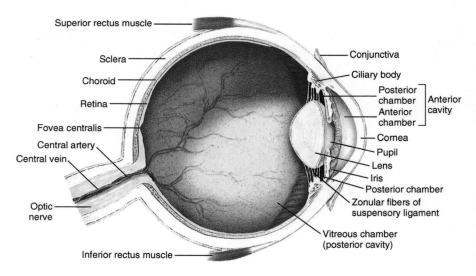

Figure 13.1. Anatomy of the eye. Fovea centralis is the central portion of eye and functions to provide clear central vision. (Reprinted with permission from Barett et al. (2012).)

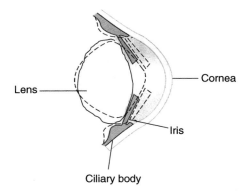

Figure 13.2. Accommodation. Dotted line illustrates changes in lens shape with accommodation; iris also constricts. (Reprinted with permission from Barett et al. (2012).)

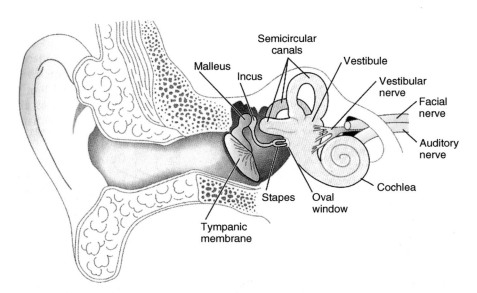

Figure 13.3. Anatomy of the ear. (Reprinted with permission from Costanzo (2006).)

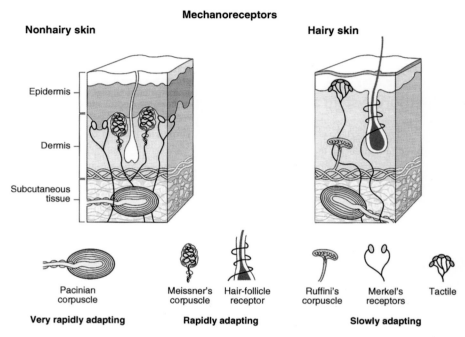

Figure 13.4. Skin mechanoreceptors. (Reprinted with permission from Costanzo (2006).)

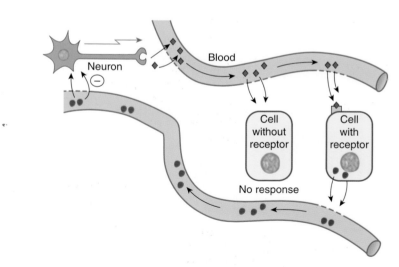

Figure 14.1. Negative feedback inhibition. Hypothalamic releasing hormone stimulates select pituitary cells (those with responsive receptors) to release the tropic hormone that inhibits release of the hypothalamic hormone (negative feedback loop). (Reprinted with permission from Patrick Bilder, PhD.)

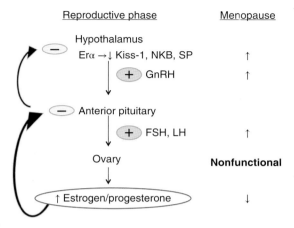

Figure 14.3. Regulatory changes in the hypothalamus–pituitary–ovarian axis during menopause. KiSS-1: kisspeptin; NKB: neurokinin B; Erα: estrogen receptor alpha; GnRH: gonadotropin-releasing hormone; FSH: follicle-stimulating hormone; LH: luteinizing hormone; SP: substance P. Hypothalamic mediators are discussed in Rance (2009).

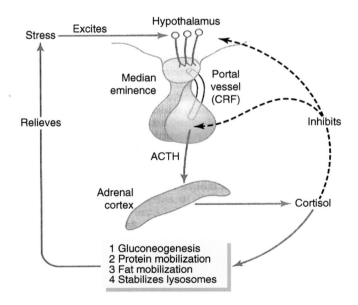

Figure 14.5. Regulation of the hypothalamic–pituitary–adrenal axis. (Reproduced with permission from Guyton and Hall (2006, p. 956).)

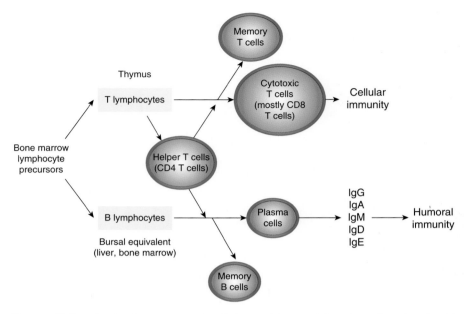

Figure 15.3. Adaptive immunity with cellular and humoral functions. (Reprinted with permission from Barrett et al. (2012).)

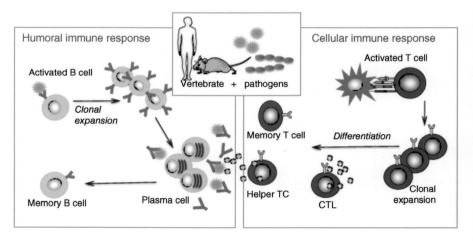

Figure 15.4. Clonal expansion is essential for immune response. (Reprinted with permission from Müller et al. (2013).)

stone formation, dietary sources of calcium and vitamin D (sun source) are preferable to over-the-counter supplements.

Growth Hormone Levels Decline with Age: A Change that May Influence Bone Function Although growth hormone (GH) (in association with insulin growth factor-1, the mediator of many of the effects of GH) is a major player in longitudinal bone growth in children and adolescents, its role in bone maintenance in the adult and certainly in the elderly is less well understood. Adults with confirmed GH deficiency exhibit lower BMD and as a group have a higher risk of fractures than those without the deficiency. Additionally, research suggests that aged human bone cells in culture are less responsive to GH/IGF-1. This coupled with a 5–20-fold drop in serum GH levels with age has led biogerontologists to speculate that both the reduced plasma levels of GH (also IGF-1) and abnormal cell signaling contribute to bone loss.

Insulin Resistance May Exacerbate Bone Loss Insulin is an osteogenic hormone and assists with maintenance of bone calcium. It is gender indifferent, but its beneficial effects are hampered in the presence of insulin resistance arising from type II diabetes (T2D). Bone loss is an expected outcome of poorly controlled T2D. The incidence of hip fracture risk is elevated in patients with T2D. DXA scans fail to detect diabetes-dependent bone changes for unknown reasons. One explanation is that *insulin resistance facilitates a decrease in bone quality rather than bone density*, possibly by acceleration of AGE (advanced glycation end product) formation or interaction with IGF-1. As emphasized by Yamaguchi and Sugimoto (2011), surrogate markers of increased fracture risk need to be identified for patients with T2D.

Elevated Levels of Adrenal Steroids Enhance Bone Loss Adrenal steroids such as cortisol are released during psychological and physical stress. High levels of AS promote bone calcium loss by several mechanisms: lowering calcium and increasing PTH levels, directly promoting bone resorption, and directly decreasing bone formation. Endocrine hypersecretion in Cushing's disease or the use of pharmacological doses of AS produces osteoporosis, a disease of severe bone loss (see below). Additionally, age affects the regulation of AS such that AS levels remain elevated for longer periods of time. This is intertwined with numerous factors, for example, sex-steroid milieu, inflammation, obesity, type of stress, dementia, and depression (for others, see Veldhuis et al. (2013)).

Smoking Exerts Numerous Effects to Augment Bone Loss Smoking is an independent risk factor for osteoporosis and is associated with lower BMD and increased risk of fracture compared to no smoking. The mechanism for this is not clear, although it is known that smoking generates a plethora of free radicals and oxidants (noted in other chapters), decreases absorption of calcium, increases adrenal steroid production, and also accelerates the onset of menopause. Negative effects on bone appear to be dose dependent and partially reversible with smoking cessation.

A summary of mechanical and nonmechanical factors that influence bone loss is given in Table 8.4.

TABLE 8.4. Factors Influencing Bone Loss in the Elderly

---

1. Mechanical strain less than CSS
    - Less modeling
    - Increased remodeling rate with resorption > formation
2. Menopause
    - Loss of estrogen removes osteogenic effects; remodeling rate increases; resorption > formation

   Estrogen decrease in men
    - Reduction in testosterone (precursor) yields fall in estrogen; effects as in women but to lesser degree
3. Vitamin D/calcium deficiency
    - Increase levels of parathyroid hormone (bone resorption)
4. Decrease growth hormone (IGF-1)
    - Loss of osteogenic effects
5. Increase in adrenal steroids (stress, glucocorticoid use)
    - Bone loss through multiple mechanisms
6. Smoking
    - Radicals and oxidants implicated in matrix deterioration; effects additive with 2,3,5 above

---

CSS: customary strain stimulus.

OSTEOPENIA; OSTEOPOROSIS

Extent of Bone Loss Defined by DXA and FRAX

The DXA scan detects BMD through the use of low-dose radiation and has been identified by the World Health Organization (WHO) as the method of choice to detect osteoporosis in postmenopausal women and men older than 50 years of age. "DXA allows accurate diagnosis of osteoporosis, estimation of fracture risk and monitoring of patients undergoing treatment" (El Maghraoui and Roux, 2008). In the elderly population, definitions of normal, osteopenia, and osteoporosis are established with *calculation of the T-score*. A measured BMD (g/cm^2) generally of the spine and hip is compared with a gender-specific normal value (averaged from BMD of 30 year olds). The delta is divided by the standard deviation (extent of variability around the mean) of the normal 30-year-old average. According to the WHO, *T*-scores signify the following: normal (−1 to +4), osteopenia (−1 to −2.5), osteoporosis (−2.5 to −4 or more), and severe osteoporosis (−2.5 plus fragility fractures). See Box 8.1 for other methods.

Osteopenia becomes the DXA-based medical term for bone loss somewhere between normal and osteoporosis. This degree of bone loss is thought to precede osteoporosis, the disease well known for its increased fracture risk. Nevertheless, fractures occur with osteopenia such that the WHO developed the Fracture Risk Assessment Tool (FRAX) algorithm. Obtained from population-based studies, *the FRAX appraises other factors such as* "age, gender, body weight and body mass index, a history of fracture(s), hip fracture(s) in parents, current smoking, excessive alcohol intake, rheumatoid arthritis, glucocorticoid use, and other forms of secondary osteoporosis" (Eriksen, 2012). Although vitamin D deficiency and bone metabolism markers are not included in FRAX, FRAX combined with BMD allows 10-year risk calculation for determination of need for pharmacotherapy. According to guidelines by the National

Osteoporosis Foundation (NOF), individuals should receive drug therapy if their 10-year hip fracture probability is 3% and their major fracture probability is 20%.

Pharmacotherapy for Osteoporosis Reduces Fracture Risk There are 9 million elderly with osteoporosis and 48 million with osteopenia according to NOF estimates (2013). The health care burden is significant and expected to increase. The consequences of osteoporosis are serious. *Twenty-five percent of women over 60 years of age with a hip fracture die within 1 year of the fracture emphasizing a high risk of mortality postfracture.* Furthermore, the prospect of morbidity from osteoporosis of the spine is high. Compression of the spine fosters a reduction in height, a curved spine, altered posture and gait, and an increased incidence of falls. Back pain often accompanies these structural changes. Other outcomes include loss of independence and associated depression.

The therapy for osteoporosis is directed toward increasing bone mass and density to achieve an increase in bone strength and a reduction in fractures. Vitamin D and calcium levels are determined and if deficient, brought within the normal range. Although estrogen replacement therapy is an effective treatment of osteoporosis, it remains controversial due to potential side effects (see Chapter 14). Selective estrogen drugs (selective estrogen receptor modulators (SERMs)) that target the bone and not breast or uterine tissue are useful, but carry a risk of blood clots. Drugs of choice are those without any estrogenic effects classified as antiresorptive or bisphosphonates (prevent bone resorption, possibly by prevention of osteocyte apoptosis). Recently, evidence of atypical fractures with long-term use and the rare event of osteonecrosis of the jaw may limit their use to shorter durations or in select populations. Both SERMs and bisphosphonates are taken orally and are highly efficacious. They have been shown to increase BMD and reduce fracture risk 30–50%. Other drugs such as a human recombinant PTH (PTH fragment) and a humanized monoclonal antibody (blocker of an osteoclast receptor) are also efficacious, but must be administered by injection.

NONPHARMACOLOGICAL INTERVENTIONS

Most Fractures Result from Falls: Ways to Prevent

The statistics indicate that 90% of hip fracture and all of wrist fractures result from falls (Eriksen, 2012). Since the majority of falls occur in the home, suggested revisions that reduce fall incidence include the following: secure rugs, remove clutter, improve lighting, add hand rails on steps, reduce use of multiple medications including sleep aids and tranquilizers, receive training on fall-induced orthostatic hypotension (lightheadedness on standing from sitting or reclining position), correct sensory loss (sight, hearing), avoid multitasking, and improve balance and reflex rate with targeted exercises. Numerous trials have shown the benefit of well-designed multicomponent exercise programs that emphasize balance training as well as strength and endurance in the reduction of falls and the reduction of injuries from falls in the community-dwelling elderly and in the physically frail in nursing homes.

Value of Exercise for Bone Loss Prevention Is Debated

There is a need for nonpharmacological procedures to prevent osteopenia, osteoporosis, and the associated risk of fractures. It is clear that large mechanical strain stimulates bone remodeling and improves bone strength in children and young adults. What remains to be determined is the mechanical strain (magnitude, frequency) that would strengthen the skeleton of the elderly sufficiently to prevent fractures. As noted above, muscle weakness and physical inactivity remove most of the needed mechanical strain and accelerate bone loss remodeling.

There exist many studies of high-impact exercise programs for the elderly, but the results thus far are less than stellar. A meta-analysis of randomized controlled trials (RCTs) that assessed the effect of exercise on bone loss and that were published prior to October 2009 concluded that "exercise can significantly enhance bone strength at loaded sites in children but not in adults" (Nikander et al., 2010). Another meta-analysis (Hamilton et al., 2010) of bone mass (DXA) and geometry (pQCT) data of elderly subjects enrolled in different study designs (RCT, cross-sectional, and prospective) found that "exercise effects appear to be modest, site-specific, and preferentially influence cortical rather than trabecular components of bone." The best results were associated with high-impact loading exercises and continued compliance, objectives difficult to achieve in the elderly.

RCTs completed in 2010 and thereafter (see Table 8.5) show that high-impact exercises, for example, squat jumps, hopping, or other weight-bearing *impact exercises, produce small gains in BMD (~2–5%) in areas scanned (lumbar spine, femur neck) or prevent BMD loss in these regions compared to nonexercising controls.* In these more recent studies, exercises were performed for 6–18 months

TABLE 8.5. Interventional Studies for Prevention of Bone Loss, 2010 to Present

Specifics	Exercise Variation	Results
Number of participants	High impact	Lumbar spine (LS)
• 35–180; men and women	• Multicomponent weight bearing exercise	• ↑ BMD 0.4–1.8% in men
Age range (years)	• Hopping, jumping, squat	Femoral neck (FN)
• 50+	Resistance exercise	• ↑ BMD 0.5–2.8%
Duration	• High velocity, progressive	Control BMD
• 6–18 months	• Low and high intensity	• ↓ 0.9–1.2%
• 1 study:12 years	Duration	12-year study
Number of studies	• Every day (jumping); 1–2–3×/week	• Less bone loss in LS and FN
• 12 RCT		Resistance exercise alone
		• ↑ Lean body mass, fitness, limb strength and balance

BMD: bone mineral density; RCT: randomized controlled trials; studies referenced in bibliography.

and generally with a low number of participants (largest study included 150 participants). Although a combination of preserved BMD and lower bone turnover (as measured by Winter-Stone et al. (2011)) favors a reduction in fracture risk, it remains elusive as to how much BMD and geometric remodeling are required for a significant decrease in fracture number. Whether modest gains in BMD would continue to accrue if high-impact exercises were continued beyond 18 months is unknown. Resistance exercises (three times weekly) alone do not change BMD, although resistance exercises offer other benefits of improved muscle strength and balance and favorable body composition (see Chapter 6).

As shown from animal studies, an increase in periosteal bone thickness (widening out of cortical bone) translates into bone strength. Application of intermittently applied mechanical stress with an increasing rate of deformation is an effective stimulator of periosteal bone formation in animal models. Whether this format can be translated into useful exercises for the elderly and also produce stronger bones needs to be determined.

The rather modest effects of high-impact exercise on bone density and architecture in man, whether due to poor compliance or less than optimal mechanical impact, have spurred research into the evaluation of another approach: high-frequency, low-magnitude mechanical stimulation (LMM) to prevent bone loss. As reviewed by Ozcivici et al. (2010), LMM successfully improves bone quantity and quality in several animal models by the induction of extremely small deformations in bone matrix, which through multiple mechanisms favor bone formation. In a small study, 32 postmenopausal women treated daily with LLM (two, 10 min periods) for 1 year exhibited a 0.04% increase in BMD. The BMD loss of 2.14% in the control group led Rubin et al. (2004) to conclude that LMM prevents bone loss. The follow-up, a larger (200 participants, 60 years and older, with osteopenia) and longer (2 years) randomized, double-blind, sham-controlled trial, called Vibration to Improve Bone Density in Elderly Subjects (VIBES), is ongoing. The objective includes evaluation of LLM efficacy on BMD (DXA), bone geometry (QCT), bone turnover (markers), balance stability (balance tests), muscle strength (hand grip), and the number of self-reported falls.

Whole-body vibration (WBV), vibration of magnitudes many fold higher than LMM used for muscle building, has also received attention as a way to retard bone loss. Thus far WBV and variations thereof have yielded mixed results and additional studies comparable to VIBES are needed.

AGING IN JOINTS: OSTEOARTHRITIS

Osteoarthritis, an Inflammatory, Destructive, Painful Joint Disorder, Has Many Causes

Osteoarthritis (OA), considered the most common joint disorder in the United States (Zhang and Jordan, 2010), is characterized by joint pain and a specific radiological identity. Joint pain may arise for many reasons, but for a diagnosis of OA, there must be evidence of the following: *extra bone growth called osteophytosis, narrowing of the joint space, extra bone growth along the joint line, and formation of subchondral cysts*. These changes are illustrated in Figure 8.5.

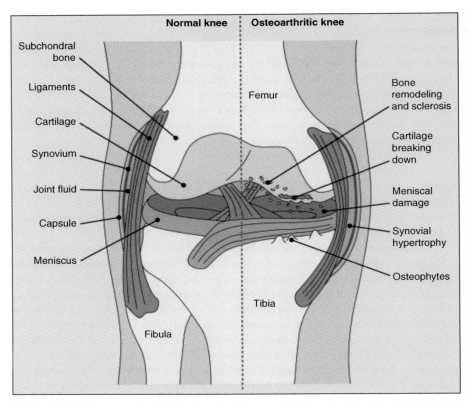

Figure 8.5. Osteoarthritis of the knee joint. (Reprinted with permission from Hunter and Felson (2006).) (See plate section for color version.)

Estimated prevalence of OA has increased over the years and varies according to joint site and age. The highest prevalence is attributed to knee OA (27.8%, 45 and older; 37.4%, 65 and older). The second and third most common sites for OA are the hip and the hand, respectively.

Although *the main risk factor for OA is age, it is not the only risk factor*. Indeed, an abundance of additional risk factors known to contribute to OA has obscured an accurate definition of the role of aging in this common joint disorder. Compounding risk factors are as follows: obesity (especially for knee OA), past trauma (meniscal tear, anterior cruciate ligament damage around knee), occupation (repetitive joint usage), genetics (heritability of ∼50%), race/ethnicity (prevalence and joint selection varies by race), diet (vitamin D deficiency <33 ng/ml of 25-OH vitamin D_3 associated with OA; others of vitamin K and selenium deficiency possible), physical activity and sports (generally that of elite athletes), misalignment (normal hip–knee–ankle alignment necessary for load distribution), musculoskeletal aging of dynapenia (not clear whether joint damage fosters muscle disuse or whether muscle weakness permits joint damage) and osteopenia, and congenital abnormalities (rare).

It has been proposed that the age-dependent changes in articular cartilage (ends of bone) represent an example of failure of homeostatic joint maintenance. Since OA is rare

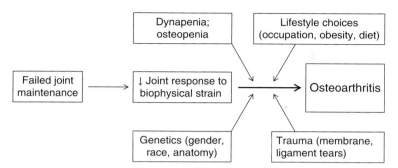

Figure 8.6. Multifactorial etiology of osteoarthritis.

in young individuals whose joints experience the same degree of stress as elderly joints, it is reasoned that older joints fail to adequately respond to joint stress. Thus far there is *no convincing explanation for this failure*. It is known that in OA animal models, the number of chondrocytes (resident cells of cartilage) decreases and some chondrocytes express characteristics of stress-induced senescence identical to the destructive senescence-associated secretory phenotype (SASP) (see Chapter 5). These cellular alterations would produce a weakened and thinned matrix of cross-linked collagens, changes that diminish biophysical stress-resistant properties of the joint and possibly provide the foundation for OA. *Many individuals never suffer from OA* suggesting that age changes *per se* do not cause OA. However, the age changes, although poorly defined in man, are thought to contribute to a *suboptimal maintenance that fails when other risk factors (noted above) are present*. Many of the major risk factors are given in Figure 8.6.

With the use of newer imaging techniques (magnetic resonance imaging (MRI) and ultrasound), it is clear that OA involves changes in all aspects of the joint: synovial fluid, articular cartilage, subchondral cartilage, bone, bone marrow, membranes, and ligaments. It is now evident that *inflammation plays a role in OA*.

Box 8.1. More About Bone Loss Detection

In a recent retrospective study (multidetector computed tomography study) (Samelson et al., 2012) of second- and third-generation cohorts of the Framingham Heart Study (690 participants, 40–87 years of age), a comparison of vBMD in lumbar and thoracic vertebrae of men and women found that vBMD and strength in both sexes declined more in lumbar vertebrae compared to thoracic and that this change was greater in women than men.

To determine the dynamic state of bone function, serum and urine markers of bone formation and resorption are measured. It has been argued that determination of these biomarkers provides more reliable information of bone activity compared to that obtained for DXA scans. Additionally, the measurement of biomarkers is less hazardous (no radiation exposure). Biomarkers are becoming more popular in tracking the effect of chronic exercise on bone physiology. Serum markers indicative of bone formation are osteocalcin, a bone growth factor, and alkaline phosphatase, an enzyme. Urinary markers of bone resorption are breakdown products of bone collagen such as deoxypyridinoline and N-telopeptide.

Synovitis (inflammation of the synovial lining of the joint) is common and associated with pain of the knee and hand.

There is no cure for OA. Current therapy relieves the symptoms. Acetaminophen is recommended as the best choice for long-term analgesic treatment of the mild to moderate pain of OA. For severe cases or those not responding to acetaminophen, nonsteroidal anti-inflammatory drugs (NSAIDs) have been of value. However, side effects of NSAIDs, for example, the risk of adverse GI bleeding, are of concern. Other pharmacological therapies are intra-articular injections of steroids or hyaluronic acids and use of opioids. Nonpharmacological approaches include physiotherapy, walking aids, and acupuncture. Severe cases require joint replacement (arthroplasty), fragment removal, and fused joints. Frontier research is focused on preventing bone loss, blocking pain centrally (in the brain), and most importantly suppressing joint inflammation.

SUMMARY

Skeletal bones comprise a dynamic system that constantly adjusts to changes in external mechanical strain and internal hormones. The processes of modeling (bone growth) and remodeling (bone strengthening) are accomplished through the activities of several cell types: osteoclast, osteoblast, and osteocyte. Age-related bone loss is multifactorial and attributed to an age-related decrease in mechanical strain below the customary strain stimulus intertwined with estrogen deficiency (menopause, lesser loss in men) and secondarily to vitamin D and calcium deficiency, and others (smoking and insulin resistance).

As defined by DXA analysis (T-score) and related tests, osteopenia signifies a modest loss of BMD and osteoporosis represents a condition of serious bone loss associated with an increased risk of fractures. Although several drugs have shown benefit in slowing bone loss in osteoporosis and reducing fracture risk, a search for nonpharmacological interventions continues. Very modest effects on reduction in BMD loss with high-impact exercises and supplemental vitamin D (800 IU/day or higher + 1200 mg calcium/day) have been reported.

Osteoarthritis is a painful joint disorder of the knees, hip, and hands involving radiographic evidence of bone and joint destruction. Failed joint maintenance compounded by genetics, musculoskeletal aging, and lifestyle choices have been identified as causes.

CRITICAL THINKING

What is osteopenia? What advice can you give to prevent it?

What bone processes change with age? What can or should one do about them?

Why is osteoporosis considered an insidious disease?

Why is it not particularly surprising that high-impact exercise has only a modest effect on bone loss prevention?

What is(are) the best way(s) to avoid osteoarthritis?

In addition to recommendations given in the chapter, develop additional strategies to prevent falls.

KEY TERMS

Basic multicellular unit (BMU) group of cells required for bone remodeling.

Bone mineral density (BMD) the amount of bone mineral content determined by DXA scan.

Calcium the mineral essential for muscle, nerve, and secretory functions. Blood levels of ionized calcium are tightly controlled by the action of the parathyroid hormone. Calcium is stored in bone; it must be consumed daily.

Cortical bone compact or dense bone, generally the outermost portions of bones. Most abundant bone type.

Customary Strain Stimulus (CSS) mechanical threshold required for bone homeostasis. Influenced by many factors such as bone type/location, hormones, age, environmental stimuli.

Dual energy X-ray absorptiometry (DXA) radiological scan used to measure mineral content of bone.

Fracture risk assessment tool (FRAX) an analysis that includes risk factors for osteoporosis to include age, gender, body weight and body mass index, a history of fracture, lifestyle factors, and so on.

Low-magnitude mechanical stimulation (LMM) currently under study as a non-pharmacological nonexercise intervention to prevent bone loss.

Osteoblast bone cell that forms bone structure (osteogenesis).

Osteocalcin protein produced by osteoblasts; marker of bone formation.

Osteoclast bone cell that degrades or reabsorbs bone minerals (osteolysis).

Osteocyte resident bone cell found in high numbers that senses changes in bone deformations and signals bone resorption or formation; susceptible to apoptosis.

Osteon the basic unit of compact bone containing the osteocyte, canaliculi, and the Haversian canal.

Osteopenia the degree of bone loss greater than normal, but not as great as in osteoporosis. BMD T-score of -1 to -2.5.

Osteoporosis a type of bone disease in which there is loss of BMD greater than 2.5 standard deviations below the mean (bone mineral density average of individuals $\sim$30 years of age). This loss of BMD increases the risk for fractures. The disease requires drug therapy to increase BMD.

Parathyroid hormone (PTH) hormone produced by the parathyroid gland. It regulates the level of calcium in the blood.

Periosteum outer region of cortical bone; site of bone formation.

Quantitative computerized tomography (QCT) imaging technique used to determine bone quality (geometry and strength).

***T*-score** calculated value obtained by comparison of BMD of elderly with that of a normal value (average value from cohort of 30 year olds).

Trabecular bone bone that is loosely constructed representing a "sponge-like" appearance; usually found in the central cavity of long bones.

Vitamin D vitamin D_3 is the provitamin that is synthesized in the skin by the action of UV irradiation. The active form of vitamin D acts throughout the body. Its major function is maintenance of bone health by effects on calcium absorption and bone cell function.

BIBLIOGRAPHY

Review

Bauer DC. 2013. Clinical practice: calcium supplements and fracture prevention. *N. Engl. J. Med.* **369**(16):1537–1543.

Clarke BL, Khosla S. 2010. Physiology of bone loss. *Radiol. Clin. North Am.* **48**(3):483–495.

El-Khoury F, Cassou B, Charles M-A, Dargent-Molina P. 2013. The effect of fall prevention exercise programmes on fall induced injuries in community dwelling older adults: systematic review and meta-analysis of randomised controlled trials. *BMJ* **347**: f6234.

El Maghraoui A, Roux C. 2008. DXA scanning in clinical practice. *QJM* **101**(8): 605–617.

Eriksen EF. 2012. Treatment of osteopenia. *Rev. Endocr. Metab. Disord.* **13**(3):209–223.

Frost HM. 1987. Bone "mass" and the "mechanostat": a proposal. *Anat. Rec.* **219**(1):1–9.

Frost HM. 1997. On our age-related bone loss: insights from a new paradigm. *J. Bone Miner. Res.* **12**(10):1539–1546.

Gunter KB, Almstedt HC, Janz KF. 2012. Physical activity in childhood may be the key to optimizing lifespan skeletal health. *Exerc. Sport Sci. Rev.* **40**(1):13–21.

Hamilton CJ, Swan VJ, Jamal SA. 2010. The effects of exercise and physical activity participation on bone mass and geometry in postmenopausal women: a systematic review of pQCT studies. *Osteoporos. Int.* **21**(1):11–23.

Honig S, Rajapakse CS, Chang G. 2013. Current treatment approaches to osteoporosis- 2013. *Bull. Hosp. Jt. Dis.* **71**(3):184–188.

Hunter DJ, Felson DT. 2006. Osteoarthritis. *BMJ* **332**(7542):639–642.

Institute of Medicine. 2010. *Dietary Reference Intakes for Calcium and Vitamin D*. (Eds Ross CA, Taylor CL, Yaktine AL, Del Valle HB), Committee report from the Institute of Medicine of the National Academies. Washington DC: The National Academies Press.

Kassem M, Marie PJ. 2011. Senescence-associated intrinsic mechanisms of osteoblast dysfunctions. *Aging Cell* **10**(2):191–197.

Khosla S, Oursler MJ, Monroe DG. 2012. Estrogen and the skeleton. *Trends Endocrinol. Metab.* **23**(11):576–581.

Loeser RF. 2013. Aging processes and the development of osteoarthritis. *Curr. Opin. Rheumatol.* **25**(1):108–113.

Moncada LV. 2011. Management of falls in older persons: a prescription for prevention. *Am. Fam. Physician* **84**(11):1267–1276.

Nikander R, Sievänen H, Heinonen A, Daly RM, Uusi-Rasi K, Kannus P. 2010. Targeted exercise against osteoporosis: a systematic review and meta-analysis for optimising bone strength throughout life. *BMC Med.* **8**: 47–63.

Ozcivici E, Luu YK, Adler B, Qin Y-X, Rubin J, Judex S, Rubin CT. 2010. Mechanical signals as anabolic agents in bone. *Nat. Rev. Rheumatol.* **6**(1):50–59.

Seeman E. 2008. Structural basis of growth-related gain and age-related loss of bone strength. *Rheumatology (Oxford)* **47** (Suppl. 4):iv2–iv8.

Seeman E. 2013. Age- and menopause-related bone loss compromise cortical and trabecular microstructure. *J. Gerontol. A Biol. Sci. Med. Sci.* **68**(10):1218–1225.

Skerry TM. 2006. One mechanostat or many? Modifications of the site-specific response of bone to mechanical loading by nature and nurture. *J. Musculoskelet. Neuronal Interact.* **6**(2):122–127.

Wenham CY, Conaghan PG. 2013. New horizons in osteoarthritis. *Age Ageing* **42**(3):272–278.

Wong PK, Christie JJ, Wark JD. 2007. The effects of smoking on bone health. *Clin. Sci. (Lond.)* **113**(5):233–241.

Yamaguchi T, Sugimoto T. 2011. Bone metabolism and fracture risk in type 2 diabetes mellitus [Review]. *Endocr. J.* **58**(8):613–624.

Zhang Y, Jordan JM. 2010. Epidemiology of osteoarthritis. *Clin. Geriatr. Med.* **26**(3):355–369.

Experimental

Falahati-Nini A, Riggs BL, Atkinson EJ, O'Fallon WM, Eastell R, Khosla S. 2000 Relative contributions of testosterone and estrogen in regulating bone resorption and formation in normal elderly men. *J. Clin. Invest.* **106**(12):1553–1560.

Khosla S, Melton, LJ 3rd, Robb RA, Camp JJ, Atkinson EJ, Oberg AL, Rouleau PA, Riggs BL. 2005. Relation of volumetric BMD and structural parameters at different skeletal sites to sex steroid levels in man. *J. Bone Miner. Res.* **20**(5):730–740.

Khosla S, Riggs BL, Robb RA, Camp JJ, Achenbach SJ, Oberg AL, Rouleau PA, Melton, LJ 3rd. 2005. Relationship of volumetric bone density and structural parameters at different skeletal sites to sex steroid levels in women. *J. Clin. Endocrinol. Metab.* **90**(9):5096–5103.

Riggs BL, Melton, LJ 3rd, Robb RA, Camp JJ, Atkinson EJ, Peterson JM, Rouleau PA, McCollough CH, Bouxsein ML, Khosla S. 2004. Population-based study of age and sex differences in bone volumetric density, size, geometry, and structure at different skeletal sites. *J. Bone Miner. Res.* **19**(12):1945–1954.

Rubin C, Recker R, Cullen D, Ryaby J, McCabe J, McLeod K. 2004. Prevention of postmenopausal bone loss by a low-magnitude, high-frequency mechanical stimuli: a clinical trial assessing compliance, efficacy, and safety. *J. Bone Miner. Res.* **19**(3):343–351.

Samelson EJ, Christiansen BA, Demissie S, Broe KE, Louie-Gao Q, Cupples LA, Roberts BJ, Manoharam R, D'Agostino J, Lang T, Kiel DP, Bouxsein ML. 2012. QCT measures of bone strength at the thoracic and lumbar spine: the Framingham Study. *J. Bone Miner. Res.* **27**(3):654–663.

Schipilow JD, Macdonald HM, Liphardt AM, Kan M, Boyd SK. 2013. Bone micro-architecture, estimated bone strength, and the muscle–bone interaction in elite athletes: an HR-pQCT study. *Bone* **56**(2):281–289.

Veldhuis JD, Sharma A, Roelfsema F. 2013. Age-dependent and gender-dependent regulation of hypothalamic–adrenocorticotropic–adrenal axis. *Endocrinol. Metab. Clin. North Am.* **42**(2):201–225.

Exercise and bone loss prevention studies

Allison SJ, Folland JP, Rennie WJ, Summers GD, Brooke-Wavell K. 2013. High impact exercise increased femoral neck bone mineral density in older men: a randomised unilateral intervention. *Bone* **53**(2):321–328.

Basat H, Esmaeilzadeh S, Eskiyurt N. 2013. The effects of strengthening and high-impact exercises on bone metabolism and quality of life in postmenopausal women: a randomized controlled trial. *J. Back Musculoskelet. Rehabil.* **26**(4):427–435.

Bemben DA, Bemben MG. 2011. Dose–response effect of 40 weeks of resistance training on bone mineral density in older adults. *Osteoporos. Int.* **22**(1):179–186.

Bolton KL, Egerton T, Wark J, Wee E, Matthews B, Kelly A, Craven R, Kantor S, Bennell KL. 2012. Effects of exercise on bone density and falls risk factors in post-menopausal women with osteopenia: a randomised controlled trial. *J. Sci. Med. Sport* **15**(2):102–109.

Gianoudis J, Bailey CA, Ebeling PR, Nowson CA, Sanders KM, Hill K, Daly RM. 2014. Effects of a targeted multimodal exercise program incorporating high-speed power training on falls and fracture risk factors in older adults: a community-based randomized controlled trial. *J. Bone Miner. Res.* **29**(1):182–191.

Kemmler W, von Stengel S, Bebenek M, Engelke K, Hentschke C, Kalender WA. 2012. Exercise and fractures in postmenopausal women: 12-year results of the Erlangen Fitness and Osteoporosis Prevention Study (EFOPS). *Osteoporos. Int.* **23**(4):1267–1276.

Kukuljan S, Nowson CA, Sanders KM, Nicholson GC, Seibel MJ, Salmon J, Daly RM. 2011. Independent and combined effects of calcium–vitamin D_3 and exercise on bone structure and strength in older men: an 18-month factorial design randomized controlled trial. *J. Clin. Endocrinol. Metab.* **96**(4):955–963.

Marques EA, Mota J, Viana JL, Tuna D, Figueiredo P, Guimarães JT, Carvalho J. 2013. Response of bone mineral density, inflammatory cytokines, and biochemical bone markers to a 32-week combined loading exercise programme in older men and women. *Arch. Gerontol. Geriatr.* **57**(2):226–233.

Marques EA, Wanderley F, Machado L, Sousa F, Viana JL, Moreira-Gonçalves D, Moreira P, Mota J, Carvalho J. 2011. Effects of resistance and aerobic exercise on physical function, bone mineral density, OPG and RANKL in older women. *Exp. Gerontol.* **46**(7):524–532.

Multanen J, Nieminen MT, Häkkinen A, Kujala UM, Jämsä T, Kautiainen H, Lammentausta E, Ahola R, Selänne H, Ojala R, Kiviranta I, Heinonen A. 2014. Effects of high-impact training on bone and articular cartilage: 12-month randomized controlled quantitative MRI study. *J. Bone Miner. Res.* **29**(1):192–201.

Whiteford J, Ackland TR, Dhaliwal SS, James AP, Woodhouse JJ, Price R, Prince RL, Kerr DA. 2010. Effects of a 1-year randomized controlled trial of resistance training on lower limb bone and muscle structure and function in older men. *Osteoporos. Int.* **21**(9):1529–1536.

Winters-Stone KM, Dobek J, Nail L, Bennett JA, Leo MC, Naik A, Schwartz A. 2011. Strength training stops bone loss and builds muscle in postmenopausal breast cancer survivors: a randomized, controlled trial. *Breast Cancer Res. Treat.* **127**(2):447–456.

SECTION IV

INTERNAL ORGAN SYSTEMS: CARDIOVASCULAR, PULMONARY, GASTROINTESTINAL, AND URINARY SYSTEMS

INTRODUCTION

The next three chapters discuss aging in several internal organs of the body. Although each organ system performs unique functions, an interdependence among them is both obvious and necessary. For example, oxygen is retrieved by the pulmonary system and distributed throughout the body by the cardiovascular system. Nutrients are generated by the gastrointestinal (GI) system and in conjunction with oxygen are utilized by the cells. Metabolic waste products are either excreted by the urinary and GI systems or expelled as carbon dioxide by the lungs. Deterioration of function in any one system eventually secondarily impacts and impedes the others.

At rest (sedentary), functionality of internal organ systems appears untouched by aging. In contrast, functional loss becomes obvious in the face of energy demands above rest (for example, exercise, surgery). Stress of modest and vigorous physical activity creates serious problems due to age-related deficits in cardiac, pulmonary, and renal reserve.

Organ remodeling is a recurrent theme in aging of internal organs, a pattern evident in the heart, the blood vessels, kidneys and chest cavity structures, fostered in large part by senescent cells. Neuronal dysfunction also factors into the aging of the cardiac, pulmonary, and GI systems. Many age changes, although not diseases in themselves, pose a vulnerability to disease. For example, intimal medial thickening predisposes to coronary artery disease, arterial stiffness may progress to systolic hypertension, deficits in diastolic contraction underlie diastolic heart failure, and nephron damage advances to chronic kidney disease. Although aging may prime the tissue milieu for disease onset,

---

Human Biological Aging: From Macromolecules to Organ Systems, First Edition. Glenda Bilder.
© 2016 John Wiley & Sons, Inc. Published 2016 by John Wiley & Sons, Inc.

pathological changes need not occur. Many modifiable factors play a role in disease initiation and progression.

One pathway to minimize the contribution of aging to disease development is with a program of chronic exercise (aerobic and resistance). Interventional studies are presented in this section that demonstrate improvements in cardiac and pulmonary reserve, improved vascular health, and reduction in other disease-related risk factors such as abnormal lipid profile. Management of GI and urinary system-associated conditions of stress incontinence, chronic constipation, and possibly nocturia also benefit from both general and muscle-specific exercises.

9

AGING OF THE CARDIOVASCULAR SYSTEM

OVERVIEW

The cardiovascular (CV) system delivers oxygen and nutrients to every tissue in the body. This essential function is performed by the heart (acting as a pump) and vasculature (network of blood vessels). This is a *closed system* in which blood (red blood cells, white blood cells, platelets, proteins, ions, and water) is pumped by and through the heart and distributed to the lungs (pulmonary system) and to the rest of the body (distant or peripheral tissues, for example, liver, brain, kidneys, and skeletal muscle) and returned back to the heart for another distribution. The exchange of nutrients and gases within the tissues facilitates cellular respiration and metabolism (see Figure 9.1 and Chapter 5).

The association between *optimal CV function and low morbidity, mortality, and longevity* is well established. Thus, *CV maintenance or preservation is key to successful aging*. Additionally, in yet poorly understood ways, aging of the CV system is a prominent risk factor for CV diseases (see Table 9.1 for subcategories) suggesting an urgency to slow CV aging. Indeed, the major cause of death in individuals 65 years of age and older is cardiovascular disease accounting for 26.5% of all deaths in the United States as of 2010 (National Vital Statistics Report, 2013). In individuals older than 85 years of age, mortality due to CV disease is 30.8%. Within both age groups, mortality due to CV is independent of gender and race (Caucasian, African-American, and Hispanic).

Human Biological Aging: From Macromolecules to Organ Systems, First Edition. Glenda Bilder.
© 2016 John Wiley & Sons, Inc. Published 2016 by John Wiley & Sons, Inc.

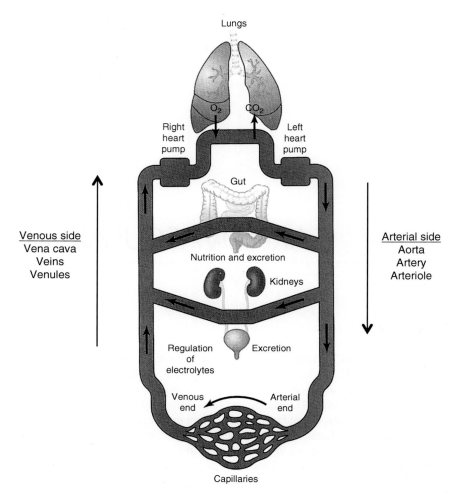

Figure 9.1. Overview of the circulatory system. (Reprinted with permission from Guyton and Hall (2006).) (See plate section for color version.)

To elucidate age changes in the CV system, structure/function in the elderly is compared with that in the young (cross-sectional study design) or measured over several years (longitudinal study design). Complex imaging in combination with standard devices have accelerated the understanding of CV aging. These clinical tools are detailed in Box 9.1.

Blood pressure, blood flow, and cardiac function (contractility, heart rate, and conductivity) must be maintained within a narrow or homeostatic range for survival. Normal values are given in Table 9.2. As the organism ages, the CV system struggles with homeostatic maintenance.

TABLE 9.1. Disease Terminology

Cardiovascular disease (heart disease)

1. Coronary heart disease (atherosclerosis- progressive inflammatory disease of the artery wall that results in structural remodeling)

   Myocardial infarction (MI): heart attack or sudden deprivation of oxygen to heart due to ruptured plaque and blood clot
   Angina: chest pain
   Stroke: prolonged oxygen/nutrient deprivation to the brain

2. Arrhythmias: irregular heart rhythm
3. Congenital (birth) heart defects
4. Valvular heart disease: affects aortic or mitral valve
5. Cardiac infections: bacterial, viral, and parasitic

Source: Mayo Foundation for Medical Education and Research, 2014. http://www.mayoclinic.org/diseases-conditions/heart-disease/basics/definition/CON-20034056.

Box 9.1. Clinical Tools to Study CV Aging
Several valuable non-invasive approaches are: (1) an echocardiogram, an instrument that uses sound waves to generate a 2- or 3-D images of the heart; (2) speckle-tracking echocardiography is echocardiography that analyzes the "speckle" patterns generated with cardiac motion and strain; (3) cardiac magnetic resonance imaging (CMRI), an imaging device that uses a magnetic field and radio wave pulse energy to generate a detailed view of the heart; and (4) computed tomography (CT), imaging that employs serial x-rays to obtain a 3-D image. An invasive method, radionuclide venticulography, takes nuclear scans of the heart following radiolabelling of red blood cells. The analyses from these instruments and techniques generate cardiac images during rest and stress-induced activity to determine contractility (pumping ability), coronary blood flow, and/or valve function. Other cardiac functions such as conductivity (flow of nerve impulse across the heart) is assessed by an electrocardiogram (ECG) that captures cardiac electrical activity. The waves of electrical activity recorded on an ECG signify the time and character of key cardiac events. ECG results are useful in the assessment of coronary artery disease, abnormal rhythms and age changes in heart rate. Conventional and indispensible devices are the sphygmomanometer (pressure cuff) to measure systolic and diastolic pressure generated during the cardiac cycle of contraction (systole) and relaxation (diastole), respectively, and the plethysmography (volume detector) or ultrasound probe to record blood flow in the forearm to determine endothelial function.

TABLE 9.2. Cardiovascular Homeostatic Values

Variable	Range	Measurement
Blood pressure	120/80 mmHg or slightly lower	Sphygmomanometer
Heart rate	65–75 beats/min	Pulse rate; ECG
Blood flow	Variable depending on artery or vein	Brachial artery ultrasound/plethysmography
Cardiac output	5 L/min	Indicator dilution; pulse contour; echo-doppler ultrasound
Ejection fraction	65%	ECG; computerized tomography; MRI
Contraction/relaxation (systole)/diastole)	At 70 beats/min, 0.27 s in systole; 0.53 s in diastole	Echocardiogram

ECG: electrocardiogram; MRI: magnetic resonance imaging; Hg: mercury; mm: millimeter; min: minutes.

HEART

Structure: Matrix, Cells, Valves

The Fibroblast is the Most Abundant Cell Type: Role in Matrix Turnover The heart is a complex tissue comprised of four chambers and many different cell types: cardiomyocytes (CMs), fibroblasts, pacemaker cells, Purkinje cells, and endothelial cells (see Figure 9.2). All of these cells are susceptible to aging. The most abundant cell type (>50%) in the heart is the fibroblast. Traditionally, fibroblasts were deemed passive cells dispersed among the CMs and tasked with maintenance of extracellular matrix (ECM), a fibrillar network formed from collagens (predominately type I) and many other molecules, for example, proteoglycans, proteases, and growth factors. Supported by recent data, fibroblasts not only regulate EMC homeostasis (producing and degrading matrix proteins such as collagen) but also connect with CMs to assist with biochemical, metabolic, and electrophysiological functions. Their contribution to CV function is exceedingly important.

Collagen Fibrosis Due to Aberrant Fibroblasts; Cross-Linkage Stiffens the Ventricles There is solid evidence that the *aged heart displays evidence of fibrosis, an increase in collagen formation in and around the cardiac cells and blood vessels*. This is due to marked changes observed in the resident fibroblast. With stress, fibroblasts phenotypically modulate, divide, move around, and most significantly produce more collagen, hence the fibrosis. Fibroblasts are generally directed to this state by increased stress on the heart (higher pressures) and through changes in growth factors, neurotransmitters, and hormones. *Fibrosis is one step to cardiac remodeling that may eventually lead to heart failure.*

There is also evidence that cardiac collagen becomes cross-linked with advanced glycation end product (AGE) adduct formation (see Chapter 4). The *presence of AGE-induced collagen cross-linkage makes the collagen stiffer and less elastic than unmodified collagen*. The result is an age-associated increase in ventricular stiffness,

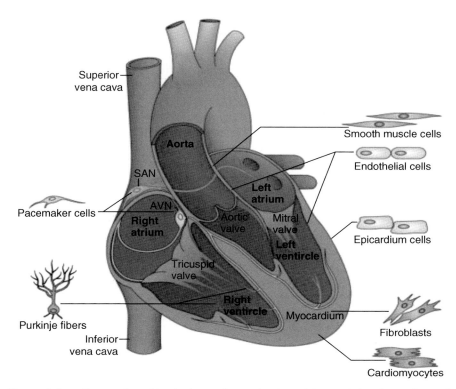

Figure 9.2. Cell types found in the heart. (Reproduction with permission from Xin et al. (2013).) (See plate section for color version.)

a change thought to contribute to diastolic dysfunction (see below). Supportive data from a number of cross-sectional studies, and one 4-year longitudinal study of 788 subjects, average age of 60 years (Borlaug et al., 2013), using comprehensive echocardiography showed that despite the use of antihypertensive medications, the stiffness of the heart during contraction and relaxation increased 14 and 8%, respectively, in the presence and absence of CV disease. In this study, the age-associated ventricular stiffness was more prominent in women. The reason for the gender difference is not known.

Left Ventricular Hypertrophy Is Debated; Heart Appears to Remodel Its Shape An increase in left ventricular (LV) mass implies cardiac remodeling that may signal early signs of heart failure. Investigations as to whether age *per se* induces LV hypertrophy (enlargement or increase in mass) and hence encourages heart failure are worthwhile endeavors. There have been many studies with this objective, mostly using echocardiography, but more recently cardiac magnetic resonance imaging (CMRI). However, the answer remains elusive. While echocardiography studies (79–501 participants across a broad age range) from 1977 to the present have found a positive association between left ventricular hypertrophy and age (i.e., LV mass increases with age), these same studies also found other factors (body mass index, male gender, systolic blood pressure, abdominal circumference, and mitral valve dysfunction) to be associated with LV hypertrophy. Of these, systolic blood pressure was the best predictor of increased LV mass. Interestingly, Gando et al. (2010) comparing physically fit and unfit pre- and postmenopausal women found that the higher the level of cardiopulmonary fitness, the lower the degree of LV hypertrophy, a relation found only in the older but not younger women. Also, Fiechter et al. (2013) using CMRI technique reported a decrease in LV mass in healthy individuals, from 20 to >70 years of age, asymptomatic for cardiovascular disease. *Thus, age per se appears to be a weak (if at all) inducer of ventricular mass enlargement.*

Although ventricular mass remains constant or may decrease with age, a fascinating CMRI study by Hees et al. (2002) observed in over 300 healthy subjects (average age of 56 years) that *both genders experience an increase in wall thickening with age*. In essence, the study results show that with age *the shape of the heart remodels from elliptical to spheroid because the interventricular septum thickens more than the opposing free wall.* Thus, in agreement with others, *the heart remodels in both men and women with age*.

CMs Are Muscle Cells that Contract and Relax to Pump the Blood
CMs contain alternating filaments of contractile proteins, actin and myosin, that slide past each other during contraction, and additionally rely heavily on firm regulation of calcium pools and an adequate supply of ATP from mitochondria. In these aspects, CMs are similar to skeletal myofibers (Chapter 7). However, in contrast, CMs are tightly connected to one another through intercalated disks that ensure a coordinated (in unison) contraction/relaxation from each heart chamber, and are additionally connected to a web of fibroblasts and anchored to the ECM.

Renewal of CMs Declines with Age; Senescent CMs Increase
CMs have long been considered postmitotic cells (fully differentiated following birth) and thus deemed unable to replicate or renew themselves. In support of this, histological data of human hearts (17–90 years of age) from individuals who died of non-cardiovascular conditions *showed a significant loss of CMs (as much as 30%) with age in the* LV and somewhat lower loss in the right ventricle (RV) (Olivetti et al., 1991), a change apparent only in men (Olivetti et al., 1995). Furthermore, the loss of CMs was associated with hypertrophy (enlargement) of the remaining CMs, postulated as an adaptation to a relative elevation in pressure.

More recently, this *view of cardiac stasis, death, and hypertrophy has been radically revised and replaced by the concept of significant CM renewal.* Bergmann et al. (2009) measured the incorporation of atmospheric radioactive carbon (^{14}C) (from nuclear bomb detonations between 1955 and 1963) into the DNA of CMs. This allowed assessment of DNA proliferation in CMs that indicated *CMs continue to divide well into adulthood.* From these measurements and associated theoretical modeling, Bergmann et al. (2009) concluded that ventricular CMs renew themselves at a rate of 1% per year beginning at age 20. *This renewal rate declines to 0.3%/year at age 75* and implies that *half of the CMs are replaced within one's lifespan.* The idea of the heart as a static tissue with no renewal potential is no longer tenable. Furthermore, these findings boost hope for future "therapeutic regeneration" to ameliorate heart disease.

Although the ^{14}C study has been criticized for its assumptions, results from *several other studies show an even greater regenerative capacity of the heart*. In a small study of eight terminal cancer patients who received the radiosensitizer iododeoxyuridine as therapy, the incorporation of this compound into the DNA of actively dividing cells revealed *that an average 22% of the CMs* was replaced with fully differentiated (contractile) CMs. A second larger study (74 hearts from individuals 19–104 years of age) that used sophisticated labeling techniques to identify human cardiac stem cells at various stages of development as well as senescent cells identified the following cell types in the heart: (a) naive cardiac stem cell, (b) cardiac stem cell in various stages of development into CMs, and (c) senescent CMs (Kajstura et al., 2010). According to this work, "from 20 to 100 years of age the myocyte compartment was *replaced completely 15 times in females and 11 times in males*." Interestingly, senescent CMs (determined by the senescent marker p16^{INK4a}) are found in hearts from individuals of all ages, but the *number of senescent CMs increases with age and is greater in hearts from males (80% at 104 years of age) compared to females (45% at 102 years of age)*. The pool of competent cardiac stem cells that develop into new CMs increases with age (renewal greater in women than men). However, the critically important finding is that the *time to onset of senescence is shorter in cells of older hearts compared to those of younger hearts (14 years at age 20 compared to 2 years at age 90)*. This means that cardiac stem cells of older hearts divide less frequently before expressing the senescent phenotype. The reason for this age-associated acceleration of senescence is unknown, but evidence implicates telomere shortening. An increased rate of senescent CM formation and an increased number of senescent CMs create a vulnerability for this tissue. Senescent cells are sensitive to death by apoptosis, necrosis, and autophagy, none of which is favorable to the heart. Thus, *repair capacity decreases and vulnerability to stress increases in the older heart*. An unanswered question is why cardiac renewal favors the female, apparent after 50 years of age. A hormonal explanation is not unreasonable, but the mechanism for this is unknown.

The discrepancy between recent study results (cardiac renewal) and early histological findings (no renewal and CM hypertrophy) may possibly relate to the different methodologies and perhaps to the more stringent exclusion criteria used in recent studies to remove the contribution from heart disease.

Aging of Aortic and Mitral Valves May Progress to Valve Disease

The aortic heart valve (valve located between LV and aorta) and mitral heart valve (valve located between the left atria and LV) (see Figure 9.2) in more than 40% of individuals in their 70s have echocardiographic evidence of fibrosis and calcification. These changes in the aortic valve, termed aortic valve sclerosis, are exacerbated by comorbidities of hypertension, end-stage renal disease, LV hypertrophy, and smoking. Other prominent changes, for example, calcification in the mitral valve, specifically mitral annular calcification (MAC) (mineralization of the valve ring), as studied by Barasch et al. (2006) in 3929 community dwellers, are strongly associated with CV disease (atherosclerosis) rather than aging *per se*.

A life-threatening valvular change is aortic valve stenosis. The Jerusalem Longitudinal Cohort study (Leibowitz et al., 2013) of 497 participants found that a small percentage of elderly (up to 8.5% >85 years of age) have this condition, essentially a stiffening of the aortic valve flaps. It is a serious condition that increases the probability of death by fourfold within 5 years. How aging predisposes the valves to pathological fibrosis and mineralization is unknown, but proposed initiators are altered fluid shear stress (mechanical force of the blood on the cells lining the valves) and inflammation, both highly associated with aging.

Function: Diastolic Relaxation, Maximal Heart Rate in Exercise

Cardiac Cycle Includes Contraction (Systole) and Relaxation (Diastole)
The function of the heart is to pump blood to the lungs and peripheral tissues. For each heart beat, the heart contracts (systole) and ejects a portion of its content (ejection fraction (EF)), relaxes (diastole), and fills with blood for the next beat. The ejected blood pushes against the vasculature creating a systolic pressure that travels forward as a wave and is detected as the pulse. During diastole, the pressure falls and the recoil of the blood vessels aid forward flow of the blood. These changes are depicted in Figure 9.3.

In this way, the ventricles expel blood at a rate of ~5 l/min, with a heart rate of 65–75 beats/min. The amount of blood in the ventricles at the end of diastole and immediately prior to systole is the end diastolic volume (EDV) and the heart ejects on average about 65% of the EDV. The amount of blood remaining at the end of systole is the end systolic volume (ESV). Many factors regulate cardiac output and arterial blood pressure, which are indicated in Table 9.3. Age *per se* may influence arterial pressure, peripheral resistance, heart rate and cardiac relaxation (discussed below) and, as described earlier, age also predisposes the heart to a shape change and decreased compliance.

Myocardial Relaxation Slows with Age: Problem with the Twist
The *prominent age-associated functional changes in the human heart are (i) a decrease in early diastolic filling rate and (ii) an altered cardiac (heart rate) response to exercise (or stress)*. The first of these relates to relaxation of the ventricles. It was observed that *the time it takes to fill the ventricles in early diastole increases with age*. This is explained by a deviation in the twist and torsion of the heart independent of disease. Basically, as shown by a prospective speckle-tracking echocardiography study (van Dalen et al., 2010) of 75 subjects (16–75 years of age) and a retrospective CMRI study of 183 subject in four age groups (from 20 to >70 years of age) (Fiechter et al., 2013),

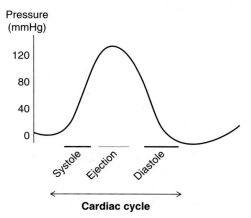

Figure 9.3. Aortic pressure changes during the cardiac cycle.

TABLE 9.3. Arterial Blood Pressure Regulation

Arterial blood pressure (mmHg)

1. Cardiac output (L/min)
   a. Stroke volume (ml/beat)
      - Myocyte contractility
      - Myocardial stretch
   b. Heart rate (beats/min)
2. Peripheral vascular resistance (dyn×s/cm^5)
   - Vessel length
   - Blood viscosity
   - Lumen diameter (constriction/dilation)

mm Hg: millimeters of mercury; L/min: liters per minute; ml/beat: milliliters per beat; dyn·s/cm^5: dynes seconds per centimeter to the fifth power.

the older heart twists too much in systole (contraction) and consequently is deficient in untwisting in diastole (relaxation).

Cardiac myofiber tracts appear to play a significant role in the twisting of the ventricular in systole. As described by Lumens et al. (2006), the external subepicardial fibers are aligned obliquely in a left-handed spiral around the ventricle, the inner subendocardial fibers are oriented in a right-handed spiral, and the midwall myofibers are oriented horizontally. Although the subendocardial fibers are stimulated to contract first, they are secondarily overwhelmed by the subepicardial counter-contraction. In this respect, van Dalen et al. (2010) showed that the heart twists 9.6° in the young (16–35 years of age) and 12° in the older subjects (56–75 years of age) (see Figure 9.4). Additionally, since the rate and velocity of diastolic untwisting do not change with age, the *excess systolic twist in older subjects creates a diastolic delay relative to systole*. It is unlikely that fibrotic changes in the ventricles cause the diastolic delay since diastolic velocity and rate are not changed with age, but rather only diminished relative to the systolic twist peak. It is postulated that the increased systolic twisting is due to weakening of the subendocardial myofibers that permits a greater contractile override by the subepicardial myofibers. CMRI data for this in healthy elderly have been provided by Lumens et al. (2006).

This *diastolic modification is considered a possible precursor to diastolic heart failure* (DHF), which is the most common type of heart failure in the elderly. Zouein et al. (2013) describe DHF as heart failure with a normal EF concomitant with diastolic dysfunction. The pathophysiology of DHF includes other changes such as ventricular remodeling and stiffness, oxidative stress, inflammation, endothelial dysfunction, and a high incidence of comorbidities, for example, obesity, diabetes, and hypertension. Many of these changes are already evident in small degrees in normal aged hearts. It is noted that DHF differs from classic or systolic heart failure that exhibits a reduced EF due to compromised systolic function.

This diastolic change promotes the development of atrial hypertrophy. Near the end of diastolic filling, the atria actively push blood into the ventricles. As age-associated diastolic dysfunction progresses, the contractility of the atria needs to increase. This change may eventually induce atrial enlargement. Atrial hypertrophy is

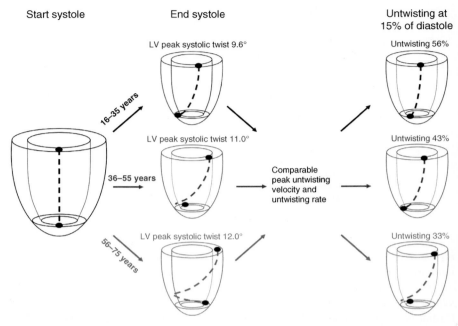

Figure 9.4. Age-associated changes in left ventricular (LV) twisting and untwisting during contraction and relaxation of the heart muscle. (Reprinted with permission from van Dalen et al. (2010).) (See plate section for color version.)

frequently associated with atrial arrhythmias (abnormal heart beats), the incidence of which increases with age.

Cardiac Response to Exercise Declines with Age, Primarily Due to Reduced Augmentation of HR

Exercise increases the uptake of oxygen above resting levels, a change termed VO_2max that requires not only efficient extraction of oxygen by the peripheral tissues but necessitates cooperation from multiple systems, for example, the CV and the pulmonary system (Chapter 10), the autonomic nervous system, skeletal muscle system (Chapter 7). Thus, the impact of age on the contribution of the CV system to exercise is only part of a more complex picture in which age changes potentially affect all of the aforementioned systems.

Many cross-sectional studies have assessed the effect of age on the CV response to aerobic exercise. For example, Fleg et al. (1995) conducted a large exercise study on 200 participants (22–86 years of age) rigorously screened to ensure the absence of heart disease. Cardiac function measured by radionuclide ventriculography during rest and maximal cycling exercise was significantly influenced by age. Specifically, at rest (sitting), HR declines and systolic blood pressure increases with age. Although men outperform women at all ages, maximal cycle work rate declines (approximately 40%) with age in both sexes. The *exercise-induced increase in EF, HR, and cardiac index (CO related to body surface area) was less in older participants than younger ones.* There are also age-related increases in end systolic volume (expressed as index relative to body surface area) and total systemic (peripheral) vascular resistance

(TSVR), a measure of arteriole function. With the exception of an age-associated higher end diastolic volume in men at peak exercise, the aforementioned age-related changes were similar in men and women.

Correia et al. (2002) examined the effect of age on submaximal exercise for a longer period of time (in contrast to most other studies, including Fleg et al. (1995), that studied maximal exercise, generally for no more than 12 min). The idea was to approximate "usual aerobic activities of everyday living" and the "classic endurance training regimens." Forty subjects recruited from the BLSA exercised (graded treadmill) for approximately 81 min at 70% of maximal calculated VO_2max. Comparison of the hemodynamic response to exercise in the two age groups (average 37 and 66 years) at 10 min (as in previous studies) with the exercise response to prolonged exercise revealed that cardiac performance increases at all ages during prolonged exercise, a result of an increase in HR, an increase in cardiac contractility (increase in stoke volume and decrease in ESV), and an increase in EF. However, *the cardiac augmentation stimulated by exercise was significantly less in older individuals than in younger ones*. These changes could not be explained by differences in effort between the two groups since both exercised at the same percent of VO_2max, nor were they due to other possible confounding influences from blood pressure, TSVR, and arterial elasticity since these were similar at the early and late time points. Thus, it was concluded that *cardiac performance reserve declines with age*. Although these studies were done with untrained subjects, as reviewed by Tanaka and Seals (2008), "master" or "elite" athletes, performing at exceptionally high levels, also experience a similar blunted response to exercise generally after the age of 60—the extent of a dampened cardiac response is however less in trained individuals compared to untrained individuals.

There are several *serious consequences of reduced cardiac reserve*: (i) the rate of *work output decreases* under stress (calculated from cycle exercise to be as much as 40% for sedentary elderly); (ii) an *increase in exercise intolerance* (lower lactate threshold) that generates fatigue to limit further activity; (iii) a *reduction in independence* and quality of life (QOL) related to lack of energy; (iv) *predisposition to tissue hypoxia* (low oxygen) and cellular damage that eventually diminishes organ function; and (v) *reduced ability to respond to stress*, for example, surgery. In contrast, aerobic training significantly improves the response of the body to physical activity at all ages (see Interventions section below).

Control of Heart Rate During exercise, the elderly basically fail to increase their HR and stroke volume (SV) (blood pumped per beat) to the same extent as a young person. Whether SV or HR contributes the most to this deficit is debated. However, many interesting studies have focused on an age-dependent effect on HR as discussed below.

HR is influenced by three factors: (i) the intrinsic beating of the cardiac pacemaker cells in the sinoatrial (SA) node, (ii) beta-adrenergic stimulation from the sympathetic nervous system (SNS) nerves, and (iii) muscarinic stimulation from the parasympathetic nervous system (PNS) nerves. The latter two comprise the outflow of the autonomic nervous system (ANS). Any one or all three could be responsible for the age-associated decreased resting (seated) HR and for the reduced rise in maximal HR with exercise.

The contribution from the PNS to exercise is generally quite small and hence its effect on exercise-induced HR changes with age is considered least likely of the three possible regulators. In contrast, the SNS and intrinsic pacemaker function come into play during exercise and are prime candidates.

Based on results, mostly from animal models, an age-related diminution of beta-adrenergic stimulation to the heart is observed and is explained by a reduction in number and/or sensitivity of specific beta-adrenergic receptors (βrs). βrs located on the myocyte plasma membrane respond to norepinephrine (NE), the neurotransmitter released from SNS nerves in response to stimulation as in exercise. As SNS nerves are activated, the released NE binds to βrs (lock and key fashion) and sets into motion cellular changes that increase HR, force of ventricular contraction, and conduction velocity of electrical activity through the heart. In the presence of a reduced number of βrs, an attenuated HR (as well as contractility and conductivity) response to NE occurs. Adding to determinations in animal models, observations in 26 explanted human hearts, 1–72 years of age, demonstrated that *βrs responsiveness is severely impaired with age and is attributable to the disappearance of βrs and their associated dysfunction*. Additionally, in a study comparing exercising subjects (28–72 years of age), half of which were infused with propranolol, a drug that blocks βrs and prevents the effects of NE, Fleg et al. (1994) showed that the presence of βrs blockage obliterates the age-associated blunting of the response to exercise. Both studies concluded that beta-adrenergic input to the heart declines with age.

A second proposal, however, also convincingly shows that an age-associated decrease in intrinsic heart rate (that due to the spontaneous activity of special pacemaker cells in the heart (sinoatrial nodal cells)) is responsible for the decline in resting HR and reduced HR response to exercise. Christou and Seals (2008) measured exercise-induced maximal HR in 30 individuals (22–65 years of age) and in a separate experiment determined intrinsic heart rate (pharmacological blockade) and the heart rate response to isoproterenol (a drug that mimics NE and stimulates the βrs). In this study, *an age-associated decrease in intrinsic HR reduction accounted for 70% of the exercise-induced reduction in maximal HR in the elderly*. Beta-adrenergic responsiveness also declined with age, but its contribution was minor. Earlier studies using pharmacological manipulation also showed slowing of intrinsic SA nodal rate with age and concluded that intrinsic HR declines ∼0.6–0.8 beats/min/year. The cellular mechanism for this slowing is not known, but possibilities include a decrease in the number of SA nodal cells and/or a corruption of SA nodal milieu with unwanted fat and collagen. Similar structural changes are found within the conductive fibers (His-Purkinje cells) and may contribute to age-associated rhythm disorders, for example, premature atrial beats, paroxysmal supraventricular tachycardia (short burst of fast heart beats) that may progress to a more serious and life-threatening condition of atrial fibrillation (irregular and fast HR).

Preconditioning Benefit Lessens with Age The phenomenon called preconditioning, extensively investigated in animal models, refers to a brief period of cardiac ischemia (oxygen deprivation) that protects against a larger future ischemic insult occurring either within hours or within days. Thus, there are two windows of protection. As reviewed by Juhaszova et al. (2005), this phenomenon has also been documented in humans. Specifically, a short bout of ischemia (e.g., angina) prior to a heart attack reduces the size of the heart damage and/or yields a better prognosis. However, as with animal models, the *benefits of preconditioning decrease with age*. The *short-term protection is lost in elderly patients*. Study results show that angina immediately prior to a heart attack is not protective, whereas angina a week before a heart attack relates to better outcomes (decreased incidence of arrhythmia, heart failure, and deaths). Other studies suggest that the *requisite threshold to activate preconditioning increases with age*. Why the response to preconditioning diminishes with age is unclear, but in studies with aged animals, deficiencies in mitochondria function and lower levels of nitric oxide (an essential vasodilator produced by the endothelium) account for this loss.

VASCULATURE

The vasculature is comprised of blood vessels with variable wall thickness, composition, and internal (lumen) diameter. Blood that is ejected from the LV flows first through the thick-walled *aorta* into the large- and medium-size *arteries* and then to the smaller *arterioles* and finally to thousands of thin-walled *capillaries* located within the tissues. From the tissue capillaries, the blood moves back to the heart through the venous system of very small *venules*, larger *veins*, and finally the largest *superior* and *inferior vena cava* that connect with the right atria and finally the RV. Blood ejected from the RV courses into the lungs via the pulmonary artery and returns to the heart via the pulmonary vein (see Figure 9.1).

The main function of the arterial circuit (aorta–arteries–arterioles–capillaries) is to supply the peripheral tissues with nutrients, for example, sugars, fats, proteins, hormones, ions, various small molecules, and oxygen. Waste products, for example, urea, nitrogen-related products, lactic acid, and other metabolic breakdown products and carbon dioxide are carried away from the tissues through the venules–veins–vena cava to the heart for exchange in the lungs or excretion by the kidneys. The exchange of oxygen for carbon dioxide and nutrients for waste products occurs at the capillaries, structures that are one cell thick to allow rapid diffusion of material to and from the tissues. To reiterate, these processes facilitate internal or cellular respiration (oxidative phosphorylation) (see Chapter 5). External respiration occurs in the lungs (see Chapter 10).

Wall Remodeling

Three Layers to the Vasculature Excluding the single-layer capillaries, all other blood vessels exhibit three discrete layers: (i) tunica intima, the innermost layer of (a) endothelial cells, (b) subendothelial or basement membrane layer, and (c) elastic tissue (lamina interna); (ii) tunica media, or middle layer with variable layers of smooth muscle cells (SMCs), matrix of collagen and elastin fibers, and an elastic tissue layer (lamina externa); and (iii) adventitia, outer covering of loose connective tissue with nerves, fibroblasts, and a blood supply (vasa vasorum) in the case of large arteries. Veins and venules are similar to arteries and arterioles in composition, but are generally thinner and with internal valves in some veins (Figure 9.5).

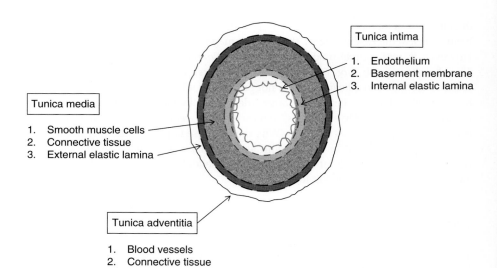

Figure 9.5. Cell layers of the large artery.

Matrix Deterioration Leads to Arterial Stiffness *Aged arteries exhibit a stiffness* not present in arteries of young individuals. This is explained by several matrix changes: an excess of collagen, a surplus of elastin fragments (elastocalcinosis), and AGE formation.

The arterial tunica media of large arteries, for example, aorta, carotid progressively accumulates abnormal collagen and elastin. The consequence is a physical alteration that makes them less compliant (less stretchable). Furthermore, a condition termed *elastocalcinosis* develops with influx of calcium and fragmentation of elastin fibers. Elastocalcinosis, previously missed in autopsy specimens due to histological processing techniques, has been identified with newer procedures. As reviewed by Atkinson (2008), elastin fibers are likely damaged by calcium mainly through inflammatory processes, and secondarily by oxidant stress, elevated plasma cholesterol, and metabolic dysfunction exacerbated by diseases of renal failure and diabetes. Therefore, it is postulated that a reduction in artery wall inflammation would minimize matrix destruction.

Arterial stiffness is a predictor of future hypertension. *Arterial stiffness is detected by an increase in systolic blood pressure, a widening of the pulse pressure, and by an increase in pulse wave velocity (PWV).* PWV is the rate of movement of the pressure wave produced by each cardiac systole. The increased systolic pressure generates the wave on impact with the arterial wall. The wave continues forward to the peripheral tissues and then reflects backward to the heart during diastole to summate with the pressure of ventricular filling (normal outcome). With compliant or elastic arteries, the pulse wave is small in width and the wave moves slowly through the system. In contrast, in stiff arteries, the pulse wave meets the less compliant vessel wall and reacts much like water pressure through a rigid pipe. *This causes the pulse wave to widen, to move more rapidly (increase in PWV), and to return to the heart earlier and earlier.* Measurements of PWV in cross-sectional studies of >500 subjects

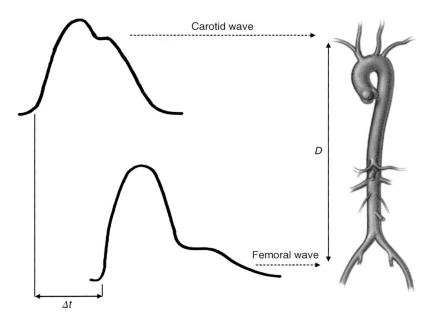

Figure 9.6. Pulse wave velocity determination. Transit time is estimated by the foot-to-foot method. The foot of the wave is defined at the end of diastole, when the steep rise of the waveform begins. The transit time is the time of travel of the foot of the wave (Δt) over a known distance (D). (Reprinted with permission from Calabia et al. (2011).) (See plate section for color version.)

over a wide age range found that beginning in middle age, the aortic PWV arrives in early systole (not diastole as in young) and in older subjects, PWV arrives in late systole, an effect that adds to the afterload (pressure against which the heart must pump) and elevates systolic pressure, decreases diastolic pressure, and widens the pulse pressure.

One of several serious consequences of arterial stiffness and increased PWV is that the *peripheral tissues, for example, kidneys and liver, receive a greater than expected force. This is speculated to contribute to tissue damage and chronic inflammation.* Second, the altered PWV promotes development of intimal–medial thickness (IMT) (see below), a type of arterial remodeling that is positively associated with the occurrence of systolic hypertension (systolic pressure >140 mmHg and diastolic pressure >90 mmHg). Systolic hypertension is a risk factor for atherosclerosis and stroke. Because of the interdependency among large artery rigidity, thickness, and elevated pulse pressure, *systolic hypertension has become the most common form of hypertension in the elderly* (Figure 9.6).

Intimal–Medial Thickness Increases with Age

The tunica intima normally contains a few (or none) smooth muscle cells. However, with age, this layer *tends to accumulate smooth muscle cells yielding a new state of intimal–medial thickness* (see Figure 9.7). In addition to age, IMT is also associated with dyslipidemia (high ratio of total cholesterol to HDL), elevated systolic blood pressure, and coronary

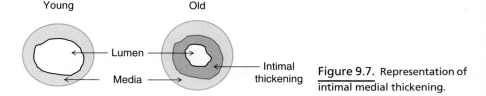

Figure 9.7. Representation of intimal medial thickening.

heart disease (CAD). To prioritize these factors, results from the BLSA showed that IMT is minimal in healthy elderly without CAD; of intermediate size in elderly asymptomatic for CAD, but with proof of relevant ECG changes; and of the highest magnitude in those with overt CAD. Accordingly, IMT is considered a risk factor for future cardiovascular disease with the *extent of artery wall thickening relating directly to the level of risk*, that is, the thicker the artery wall, the greater the risk for cardiovascular disease. However, the absolute disease risk from vascular thickening is significantly less compared to other risk factors, for example, hypercholesterolemia, diabetes, and smoking.

Although some evidence of IMT is expected in healthy aged arteries, wall thickening is associated with other negative aspects of vascular aging that include endothelial dysfunction, artery wall stiffness, and changes in pulse wave velocity. Results of a small study of thoracic aortas from sudden death victims (20 year olds compared to 65 year olds) without CV disease revealed not only relatively greater IMT in the elderly but also confirmation of inflammation, presence of abnormal collagens, abundance of degradative enzymes, and expression of a potent vasoconstrictor and inflammatory inducer, angiotensin, its receptor, and supporting proteins. It is proposed that the *aged artery wall is somewhat thicker and contains an undercurrent of inflammation.*

Vasculature Dysfunction

Blood flow and pressure are managed through the counter-force of arteriolar resistance determined in part by arteriolar lumen diameter. Blood vessels regulate pressure (tone) in the CV system by virtue of (i) the elasticity of the large arteries and veins and (ii) largely by the relaxation (vasorelaxation and dilation) and contraction (vasoconstriction and contraction) of the arterioles.

The elasticity of large arteries decreases with age as described above. Little is known about the aging of the venous system, but the low venous pressure suggests that aging may be minimal. In contrast, arteriolar aging is evident. Arterioles vary their lumen diameter by vasoconstriction and vasorelaxation in response to neurotransmitters from the ANS (e.g., norepinephrine and acetylcholine), circulating epinephrine and adrenal steroids, and a plethora of other factors produced locally by the endothelium and other tissues.

Irregularities in arteriolar function (vasoreactivity) contribute to hypertension, although a debate exists as to whether this is the mechanism of hypertension in the elderly as explained above (recall etiology of systolic hypertension with large artery stiffness).

Endothelial Cell Becomes Dysfunctional; Loss of Vasodilator Endothelial cells play a pivotal role in the health of blood vessels. Endothelial cells form the endothelium, the internal covering of the vasculature. The endothelium acts as a barrier and regulates permeability of substances into the surrounding tissues. It regulates the growth of new blood vessels (angiogenesis), prevents clot formation, and establishes the response to inflammation. Some of these functions are due to production of endothelium-derived relaxing factor (EDRF), also known as nitric oxide (NO). NO is important for three reasons: It relaxes blood vessels to allow increased blood flow and hence is a valuable vasodilator; it combines with oxidants to neutralize their oxidative damage and thus acts as an antioxidant; it inhibits a substance called endothelin that constricts blood vessels, indirectly promoting vasodilation.

Endothelial function is assessed by plethysmography or ultrasound (see Box 9.1). Both approaches have their limitations, but are the best to date to determine the *in vivo* status of endothelial function. As determined by these methods, *endothelial function deteriorates with age*. Dysfunctional endothelial cells (an example of replicative senescence) exhibit a reduced ability to perform cardinal functions: regulating blood flow, regulating angiogenesis, depressing inflammation, and overseeing vessel wall structure and associated functions such as permeability. Vascular homeostasis declines due to a reduction in NO production. Gerhard et al. (1996) studied 119 healthy individuals (19–69 years of age) and observed an age-associated decrease in endothelium-dependent vasodilation beginning as early as 40 years of age. Endothelium-independent vasodilation (smooth muscle cell contraction without aid of endothelial NO) however was not affected by age. Results of numerous studies have since confirmed these findings. As NO production by endothelial cells declines, vasoconstrictors, for example, endothelin, angiotensin, and others, are free to act unopposed. As a result, blood vessels more readily constrict, blood pressure increases, and other effects, for example, artery wall thickness and inflammation, are promoted. *Loss of NO is a loss of vascular homeostasis*. The compromised blood flow to the tissues sets the stage for end organ damage.

Smooth Muscle Cells Regulate TSVR and Fill the Media SMCs are important cells within the vasculature since this cell type (i) contracts and relaxes to decrease or increase arterial blood flow, respectively, changes that determine TSVR, and (ii) provides structural support for the large arteries, but with minimal contractile function. Similar to skeletal and cardiac myocytes, the SMC contains the necessary contractile proteins, actin and myosin albeit configured loosely. Contraction relies heavily on calcium dynamics and ATP to create actin–myosin cross-bridges, but unlike other muscle types, SMC can also maintain constant tension with minimal use of ATP.

Medial Smooth Muscle Cells Migrate and Modulate The influence of age on SMCs has been largely studied in animal models. Results with animals show that *diverse stresses* for example, endothelial dysfunction, excessive tension, oxidative stress, or toxins to medial SMCs, *phenotypically alter them*. The new phenotype exhibits little of the contractile actin–myosin proteins, initiates cell division and migrates to distal sites. IMT is a product of phenotypically modified SMCs in which

SMCs migrate from the media to the intima. Recent work (Ferlosio et al., 2012) of immunostained aortas from young and old human donors show an increased number of phenotypically modified cells in aged aortas. These cells express a stem cell-like phenotype and as shown from animal studies, this is a cell type more likely to divide, migrate, and remodel the artery.

Baroreflex Sensitivity Decreases with Age The baroreflex is a mechanism by which the ANS acutely regulates blood pressure. Receptors in the large arteries (carotid and aortic) sense a change in blood pressure. An increase in blood pressure activates the baroreceptors to send impulses to brain centers that in turn stimulate the PNS (vagus nerve) to lower HR and vasodilate the arterioles, changes that lower blood pressure. Conversely, a drop in blood pressure reduces the activity of the baroreceptors and generates a reflex to increase SNS activation that increases HR, CO, and vasoconstriction, changes to reverse the fall in blood pressure. *With age, the sensitivity of this homeostasis-producing reflex to either a rise or a fall in blood pressure diminishes*. In older women, the baroreceptor sensitivity to high blood pressure is less than in men, possibly predisposing them to hypertension. Furthermore, results of a large population study (Rotterdam study of >2000 subjects) (Mattace-Raso et al., 2007) and a smaller study (25 subjects) (James and Potter, 1999) confirm that orthostatic hypotension (light-headedness on sudden standing) is related to an age-dependent decrease in sensitivity of the baroreflex to low blood pressure. Several studies point to large artery remodeling and in particular, loss of arterial compliance as a mechanism for the sensitivity reduction. Furthermore, exercise (3 months, daily, 40–45 min of walk or walk–jog at 60–85% maximal heart rate) (Monahan et al., 2001) improves not only the arterial compliance but also the sensitivity to this important CV maintenance mechanism.

AGE AS RISK FACTOR FOR CV DISEASE

As already described, the CV system experiences a number of age changes (see Summary Box). CAD is the most prevalent disease and the major cause of death in the elderly. Among the many age changes, *endothelial dysfunction is considered an important factor that contributes to a CV disease vulnerability*. Thorin and Thorin-Trescases (2009) proposed that the senescent endothelial cell is the prominent proatherogenic element fostering atherosclerosis. The senescent endothelial cell permits influx of oxidized lipids and toxins, thereby facilitating chronic wall inflammation, damage, and eventual atherosclerotic plaque formation, the hallmark of atherosclerosis. Wang et al. (2010) argue that with time the vessel wall, itself, takes on an "age-associated secretory phenotype" analogous to that described for the senescent cell (see Chapter 5).

Strait and Lakatta (2012) propose that age-associated changes in structure, function, and repair of the heart provide little resistance to the increased incidence of CV diseases such as HF. Specifically, the stiffening of the heart and larger blood vessels elevates systolic pressure, exacerbates diastolic dysfunction, and creates more work for the heart. The result is inadequate cardiac reserve during any stress. The repair capacity of both the endothelium and the heart itself, both with an abundance of senescent cells cannot rebound sufficiently with ischemic events. The outcome is tissue damage. This is worsened with comorbidities, for example, diabetes and chronic hypertension.

PRESENT AND FUTURE INTERVENTIONS: AEROBIC EXERCISE, PHARMACOTHERAPY, AND CALORIC RESTRICTION

Several interventions have shown benefit in retarding the above-described age changes in the cardiovascular system. CV improvement in the elderly has resulted from (i) chronic aerobic exercise or (ii) pharmacological therapy. The effects of caloric restriction on CV aging are impressive, but have only been studied in animal models.

Chronic Aerobic Exercise Minimizes the Effect of Age on CV Structure and Function

The advantages of aerobic exercise in the elderly have been drawn from results of epidemiological and observational studies and clinical trials. Epidemiological studies point to a lower incidence of CV disease in exercising individuals compared to sedentary individuals and observational results show that endurance-trained elderly have higher VO_2max (can consume more oxygen and do more work), preserved LV compliance (no stiffness) compared to those without training. Generally, physically conditioned elderly have a higher LV EF due to a decrease in ESV (more forceful contraction).

Data from clinical trials indicate that sedentary elderly engaged in physical conditioning improve their VO_2max by 16.3% (meta-analysis of 41 trials of 2012 subjects, 60 years and older) (Huang et al., 2005). Additionally, several studies that measured aortic compliance before and after aerobic exercise consistently report improved compliance or prevention of the age-related loss of arterial elasticity. Other studies demonstrate that endothelial dysfunction of sedentary elderly can be reversed with 3 months of aerobic exercise (walking) and that chronic exercise (elderly athletes) prevents age-associated endothelial dysfunction.

Furthermore, risk factors for CV and other diseases are reduced with physical conditioning. Study results show a decrease in blood pressure in the elderly with systolic hypertension, a more favorable lipid profile (increase in HDL), and improved insulin sensitivity. Other benefits are a small increase in bone mineral density or a slowing of bone loss (see Chapter 8) and reduction of depression, especially that provoked by a diagnosis of CV disease.

Fletcher et al. (2013) in the American Heart Association (AHA) scientific statement on exercise testing and training provide general guidelines for endurance and resistance exercises for the elderly. Endurance training must be performed more than five times a week, for more than 30 min, and at an intensity of 55–90% of maximum predicted HR, 12–16 on perceived exertion scale (Borg scale equal to "somewhat hard to hard") or 40–80% of VO_2max. The maximum predicted HR (220-age) although recommended by the AHA, has been criticized for lack of validation. A meta-analysis of the HR response to exercise in over 18,000 subjects plus a prospective study that measured HR changes to incremental treadmill exercise in

500 subjects with different degrees of physical conditioning (Tanaka et al., 2001) found exercise-induced maximal HR fits the formula of (208-0.7 × age) better than (220-age). The difference is relevant because the target HR is lower with the original formula and hence reduces both the intensity level and the benefit.

Despite numerous exercise-induced benefits, the number of endurance-trained elderly is small. Several explanations have been given (Fleg, 2012) that include (i) elderly mindset associating exercise with the "young," (ii) comorbidities, for example, osteoarthritis that could limit certain exercises, and (iii) failure of physicians to recommend and encourage a physical conditioning program for their elderly patients. Clearly, exercise is not just for the young, physical limitations can be overcome by appropriate exercise selection and encouragement by physicians can be improved with better awareness.

Pharmacological Therapy That Blocks Angiotensin Provides Some Benefits

Pharmacological treatment of risk factors for cardiovascular disease such as dyslipidemia, hypertension, and diabetes ameliorate the effects of heart disease. Beyond that, evidence is accumulating that moderation of the renin–angiotensin system (a major hormonal regulatory system) with various inhibitory drugs may prevent premature age-associated changes in structure and function of the CV system. For example, use of a angiotensin II receptor antagonist (drug that blocks vasoconstriction and inflammation) is effective in improving vascular compliance and decreasing PWV in both healthy individuals and those with non-diabetic chronic renal disease, congestive heart failure, and CAD. Additionally, studies in rodents treated with inhibitors of the renin–angiotensin system throughout their life appear protected against CV age changes and live longer than untreated controls.

Caloric Restriction in Animal Models Retards CV Aging

Severe dietary reduction of caloric intake with adequate nutrition induces weight loss and exerts beneficial effects on the CV system in animal models of aging (Chapter 2). Reproducible benefits in animals subjected to caloric reduction (CR) for their lifespan include a reduction in the increase in IMT, improvement in left ventricular diastolic function, improvement in endothelial function in conduit (large) and resistance (small) arteries, reduction in whole body and regional sympathetic nerve activity, and enhancement of cardiac baroreceptor reflex activity. More importantly, the CR animals live longer than controls given unlimited access to food. Whether this regime will improve CV structure/function in man is not known, but a 2 year CR study in man is underway (Comprehensive Assessment of Long-Term Effects of Reducing Intake of Energy, CALORIE, phase 2). As reviewed by Lavie et al. (2014), there is a strong association between obesity and risk of CV disease as well as abnormalities of CV function. Although there exists a paradox in which obese CV patients have a better prognosis than lean ones, it has now been determined that that this applies to low fitness rather than weight. The paradox disappears in individuals with better fitness, although the reason for this is presently unknown.

Summary of Cardiovascular Age Changes

Heart
Structural
  Ventricular shape change
    Interventricular septum thickens
    Ventricle more spheroid
  Ventricles stiffen
    Fibroblast dysfunction
  Increase number of senescent cardiomyocytes
  Decrease in SA nodal cells; matrix change around conductive cells
  Atrial hypertrophy
  Fibrosis and calcification of aortic valve
Functional
  Diastolic dysfunction
    Increased systolic twist
    Delay in passive filling
  Systolic contractility of endocardial cells declines
  Decreased augmentation of heart rate, stroke volume, and ejection fraction during aerobic exercise
  Slowing of sino-atrial nodal activity
  Decreased sensitivity to sympathetic nervous system stimulation
Repair
  Time to replicative senescence of cardiomyocytes shortens
  Lower threshold to injury

Vasculature
Structural
  Artery wall stiffens: reduced compliance
  Occurrence of intimal–medial thickness
  Dysfunctional smooth muscle cells
  Evidence of inflammation, elastocalcinosis
Functional
  Endothelial cells senesce: become dysfunctional
  Loss of endothelium-dependent vasodilation
  Reduced sensitivity of the baroreflex to pressure changes

Consequences
  Decreased cardiac reserve
    Impaired response to physical activity above rest
  Reduced peripheral blood flow
  Reduce control of blood pressure fluctuations
  Reduced benefits from preconditioning
  Predisposition to systolic hypertension (arterial stiffness), atherosclerosis (endothelial dysfunction), diastolic heart failure (diastolic dysfunction), and arrhythmias (nodal, conductive changes)

SUMMARY

The heart pumps oxygenated, nutrient-fortified blood through the vasculature to the tissues of the body and returns with blood filled with carbon dioxide and waste products for exchange in the lungs or excretion by the kidneys. The pressure, blood flow, and resistance of the CV system are influenced by the structure of the heart and its vasculature, ANS, and local mediators. Age changes impact all components of this system and lead to a decrease in cardiac reserve during exercise (stress), reduced homeostatic blood pressure control, and a vulnerability to CV diseases.

The heart ages by becoming stiffer due to fibrosis, thickening (interventricular septum), and becoming more spheroidal rather than elliptical. Subendocardial myocytes contract less forcefully allowing a greater systolic twist and impairing diastolic filling and eventually inducing atrial hypertrophy. The aged heart accumulates senescent myocardial cell not because there is no renew but because newly replaced cells senesce faster.

The vasculature (large arteries) also becomes stiffer (collagen cross-linkage, elastocalcinosis) and thicker (IMT), changes that decrease arterial compliance, increase PWV, and facilitate systolic hypertension. Throughout there is evidence of endothelial dysfunction that not only reduces blood flow to the tissues but is permissive for the first steps to atherosclerosis.

Cardiovascular health can be optimized with a program of aerobic exercise. High-intensity aerobic exercise reduces/prevents large artery stiffness, endothelial dysfunction, increases VO_2max, and reduces established risk factors of CV disease. In animal models, the intervention of caloric reduction for a significant portion of the animal's life prevents aging of the heart and blood vessels.

CRITICAL THINKING

In what ways do senescent endothelial cells, fibroblasts, and cardiac myocytes influence CV aging?

How does aging of the CV system create a major risk factor for CV disease?

How might aerobic exercise ameliorate CV aging? Suggest ways to encourage the elderly to exercise.

In addition to aerobic exercise, what other interventions might prevent or reduce CV aging?

KEY TERMS

Adventitia outermost layer of the large conduit arteries; comprised of connective tissue, fibroblasts, nerves, and, sometimes, tiny blood vessels.

Aorta largest conduit artery in mammals; carries oxygenated blood from left ventricle to the peripheral organs.

Arterioles small blood vessels responsive to contractile and dilating substances from the endothelium. Activity of these vessels determines the total peripheral vascular resistance.

KEY TERMS

Artery large blood vessels that carry oxygenated blood away from the heart to the peripheral tissues. Carotid, femoral, and aortic artery are examples.

Atria upper chambers of the heart; contribution to overall pumping of the heart increases with age. Abnormal beating (arrhythmias, for example, atrial fibrillation) disrupts normal cardiac pumping and may lead to life-threatening ventricular fibrillation.

Baroreceptor reflex reflex to maintain blood pressure homeostasis. Example: sudden positional change (supine to standing), gravitational fall in blood pressure countered by reflex. Reflex is diminished with age and certain drugs.

Brachial artery large artery in arm; used in flow measurements.

Capillaries thin, one cell thick blood vessels; site of exchange of oxygen with carbon dioxide and nutrients with waste products.

Cardiac myocyte contractile cell of the heart.

Cardiac output amount of blood pumped from left side of the heart per minute. In humans, cardiac output is approximately ~5 L/min.

Compliance degree of flexibility or stretch. Loss of compliance indicates degenerative change in collagen and elastin components in the tissue. The compliance of large arteries and heart declines with age.

Elastocalcinosis accumulation of calcium in the large arteries to the extent that it damages the elastin fibers.

End diastolic volume (EDV) amount of blood in the ventricles at the end of diastole (relaxation); the heart adjusts to increases in EDV by a stronger contraction (Frank–Starling law).

End systolic volume (ESV) amount of blood in the ventricles at the end of systole (contraction).

Endothelial dysfunction repeated damage to endothelial cells leads to replicative senescence and loss of normal endothelial function, for example, release of the important vasodilator, nitric oxide.

Endothelin vasoconstrictor produced by endothelial cells; effects apparent in the presence of endothelial dysfunction.

Endothelium internal lining of blood vessels.

Homeostasis constancy of the internal environment; maintenance of normal functions within an optimal range.

Hypertrophy increase in size; enlargement.

Intima anatomically, the tunica intima or the inner portion of blood vessels contains the endothelium, intima, and internal elastic lamina. This area in arteries enlarges with age.

Intimal–medial thickening (IMT) thickening of the arterial wall. The degree of thickness positively relates to severity of CV disease.

Media the inner layer of blood vessels populated with smooth muscle cells.

Nitric oxide (NO) vasodilator released from healthy endothelial cells; also called endothelium-derived relaxing factor (EDRF).

Orthostatic hypotension fall in blood pressure on standing leading to light-headedness due to diminution of the baroreceptor reflex

p16^{INK4a} considered a mediator of cell senescence. It acts to inhibit cyclin-dependent kinases that usher the cell through mitosis.

Pulse wave velocity (PWV) the rate at which the pressure wave from each cardiac contraction cycle passes from the heart to the periphery. The faster the PWV, the less compliant (the stiffer) the arteries, suggesting aging or disease.

Systole/diastole names for contraction/relaxation period of each heart beat.

Systolic and diastolic blood pressure the blood pressure at the peak of systole (~120 mmHg) and the pressure during relaxation (~80 mmHg).

Vasoconstriction contraction of smooth muscle cells in the small arteries/arterioles to decrease the size of the lumen and reduce the amount of blood flow.

Vasorelaxation relaxation of smooth muscle cells in the arteries/arterioles to increase the size of the lumen and increase the amount of blood flow.

Vein blood vessels that carry blood back to the heart. Thin-walled and very compliant (stretch to contain and hold large amounts of blood).

Vena cava large vein that connects directly to the right atria and carries carbon dioxide-laden blood from the venules and veins.

Ventricle the lower large chambers of the heart; the left chamber pumps blood to the peripheral organs and the right chamber pumps blood to the lungs.

Venules small thin-walled blood vessels connected to the capillaries that return the blood with carbon dioxide and waste products to the larger veins and then the vena cava.

BIBLIOGRAPHY

Review

Atkinson J. 2008. Age-related medial elastocalcinosis in arteries: mechanism, animal models and physiological consequences. *J. Appl. Physiol.* **105**(5):1643–1651.

Dai D-F, Chen T, Johnson SC, Szeto H, Rabinovitch PS. 2012. Cardiac aging: from molecular mechanisms to significance in human health and disease. *Antioxid. Redox Signal.* **16**(12):1492–1526.

Fleg JL. 2012. Aerobic exercise in the elderly: a key to successful aging. *Discov. Med.* **13**(70):223–228.

Fletcher GF, Ades PA, Kligfield P, Arena R, Balady GJ, Bittner VA, Coke LA, Fleg JL, Forman DE, Gerber TC, Gulati M, Madan K, Rhodes J, Thompson PD, Williams MA. American Heart Association Exercise, Cardiac Rehabilitation, and Prevention Committee of the Council on Clinical Cardiology, Council on Nutrition, Physical Activity and Metabolism, Council on Cardiovascular and Stroke Nursing, and Council on Epidemiology and Prevention. 2013. 2013. Exercise standards for testing and training: a scientific statement from the American Heart Association. *Circulation* **128**(8):873–934.

Golbidi S, Laher I. 2013. Exercise and the aging endothelium. *J. Diabetes Res.* **2013**: 789607.

Guyton AC, Hall JE. 2006. *Textbook of Medical Physiology*, 11th ed. Philadelphia: Elsevier Saunders, p. 4.

Huang G, Gibson CA, Tran ZV, Osness WH. 2005. Controlled endurance exercise training and VO_2max changes in older adults: a meta-analysis. *Prev. Cardiol.* **8**(4):217–225.

Juhaszova M, Rabuel C, Zorov DB, Lakatta EG, Sollott SJ. 2005. Protection in the aged heart: preventing the heart-break of old age? *Cardiovasc. Res.* **66**(2):233–244.

Lakatta EG, Levy D. 2003. Arterial and cardiac aging: major shareholders in cardiovascular disease enterprises. Part II. The aging heart in health: links to heart disease. *Circulation* **107**(2):346–354.

Lakatta EG, Wang M, Najjar SS. 2009. Arterial aging and subclinical arterial disease are fundamentally intertwined at macroscopic and molecular levels. *Med. Clin. North Am.* **93**(3):583–604.

Lavie CJ, McAuley PA, Church TS, Milani RV, Blair SN. 2014. Obesity and cardiovascular diseases: implications regarding fitness, fatness, and severity in the obesity paradox. *J. Am. Coll. Cardiol.* **63**(14):1345–1354.

Lind L. 2006. Impact of ageing on the measurement of endothelium-dependent vasodilation. *Pharma. Rep.* **58** (Suppl.):41–46.

Moslehi J, DePinho RA, Sahin E. 2012. Telomeres and mitochondria in the aging heart. *Circ. Res.* **110**(9):1226–1237.

Okada Y, Galbreath MM, Shibata S, Jarvis SS, VanGundy TB, Meier RL, Vongpatanasin W, Levine BD, Fu Q. 2012. Relationship between sympathetic baroreflex sensitivity and arterial stiffness in elderly men and women. *Hypertension* **59**(1):98–104.

Strait JB, Lakatta EG. 2012. Aging-associated cardiovascular changes and their relationship to heart failure. *Heart Fail. Clin.* **8**(1):143–164.

Tanaka H, Monahan KD, Seals DR. 2001. Age-predicted maximal heart rate revisited. *J. Am. Coll. Cardiol.* **37**(1):153–156.

Wang M, Monticone RE, Lakatta EG. 2010. Arterial aging: a journey into subclinical arterial disease. *Curr. Opin. Nephrol. Hypertens.* **19**(2):201–207.

Wang M, Zhang J, Jiang L-Q, Spinetti G, Pintus G, Monticone R, Kolodgie FD, Virmani R, Lakatta EG. 2007. Proinflammatory profile within the grossly normal aged human aortic wall. *Hypertension* **50**(1):219–227.

Xin M, Olson EN, Bassel-Duby R. 2013. Mending broken hearts: cardiac development as a basis for adult heart regeneration and repair. *Nat. Rev. Mol. Cell Biol.* **14**(8):529–541.

Zouein FA, de Castro Brás LE, da Costa DV, Lindsey ML, Kurdi M, Booz GW. 2013. Heart failure with preserved ejection fraction: emerging drug strategies. *J. Cardiovasc. Pharmacol.* **62**(1):13–21.

Experimental

AlGhatrif M, Strait JB, Morrell CH, Canepa M, Wright J, Elango P, Scuteri A, Najjar SS, Ferrucci L, Lakatta EG. 2013. Longitudinal trajectories of arterial stiffness and the role of blood pressure: the Baltimore Longitudinal Study of Aging. *Hypertension* **62**(5):934–941.

Banks L, Sasson Z, Esfandiari S, Busato G-M, Goodman JM. 2011. Cardiac function following prolonged exercise: influence of age. *J. Appl. Physiol.* **110**(6):1541–1548.

Barasch E, Gottdiener JS, Larsen EK, Chaves PH, Newman AB, Manolio TA. 2006. Clinical significance of calcification of the fibrous skeleton of the heart and aortosclerosis in community dwelling elderly. The Cardiovascular Health Study (CHS). *Am. Heart J.* **151**(1):39–47.

Basso N, Cini R, Pietrelli A, Ferder L, Terragno NA, Inserra F. 2007. Protective effect of long-term angiotensin II inhibition. *Am. J. Physiol. Heart Circ. Physiol.* **293**(3):H1351–H1358.

Bergmann O, Bhardwaj RD, Bernard S, Zdunek S, Barnabé-Heider F, Walsh S, Zupicich J, Alkass K, Buchholz BA, Druid H, Jovinge S, Frisén J. 2009. Evidence for cardiomyocyte renewal in humans. *Science* **324**(5923):98–102.

Borlaug BA, Redfield MM, Melenovsky V, Kane GC, Karon BL, Jacobsen SJ, Rodeheffer RJ. 2013. Longitudinal changes in left ventricular stiffness: a community-based study. *Circ. Heart Fail.* **6**(5):944–952.

Calabia, J, Torguet P, Garcia M, Garcia I, Martin N, Guasch B, Faur D, Vallés M. 2011. Doppler ultrasound in the measurement of pulse wave velocity: agreement with the Complior method. *Cardiovasc. Ultrasound* **9**:13.

Christou DD, Seals DR. 2008. Decreased maximal heart rate with aging is related to reduced β-adrenergic responsiveness but is largely explained by a reduction in intrinsic heart rate. *J. Appl. Physiol.* **105**(1):24–29.

Correia LC, Lakatta EG, O'Connor FC, Becker LC, Clulow J, Townsend S, Gerstenblith G, Fleg JL. 2002. Attenuated cardiovascular reserve during prolonged submaximal cycle exercise in healthy older subjects. *J. Am. Coll. Cardiol.* **40**(7):1290–1297.

DeSouza CA, Shapiro LF, Clevenger CM, Dinenno FA, Monahan KD, Tanaka H, Seals DR. 2000. Regular aerobic exercise prevents and restores age-related declines in endothelium-dependent vasodilation in healthy men. *Circulation* **102**(12):1351–1357.

Ferlosio A, Arcuri G, Doldo E, Scioli MG, De Falco S, Spagnoli LG, Orlandi A. 2012. Age-related increase of stem marker expression influences vascular smooth muscle cell properties. *Atherosclerosis* **224**(1):51–57.

Fiechter M, Fuchs TA, Gebhard C, Stehli J, Klaeser B, Stähli BE, Manka R, Manes C, Tanner FC, Gaemperli O, Kaufmann PA. 2013. Age-related normal structural and functional ventricular values in cardiac function assessed by magnetic resonance. *BMC Med. Imaging* **13**:6.

Fleg JL, Morrell CH, Bos AG, Brant LJ, Talbot LA, Wright JG, Lakatta EG. 2005. Accelerated longitudinal decline of aerobic capacity in healthy older adults. *Circulation* **112**(5):674–682.

Fleg JL, O'Connor F, Gerstenblith G, Becker LC, Clulow J, Schulman SP, Lakatta EG. 1995. Impact of age on the cardiovascular response to dynamic upright exercise in healthy men and women. *J. Appl. Physiol.* **78**(3):890–900.

Fleg JL, Schulman S, O'Connor F, Becker LC, Gerstenblith G, Clulow JF, Renlund DG, Lakatta EG. 1994. Effects of acute beta-adrenergic receptor blockade on age-associated changes in cardiovascular performance during dynamic exercise. *Circulation* **90**(5):2333–2341.

Gando Y, Kawano H, Yamamoto K, Sanada K, Tanimoto M, Oh T, Ohmori Y, Miyatani M, Usui C, Takahashi E, Tabata I, Higuchi M, Miyachi M. 2010. Age and cardiorespiratory fitness are associated with arterial stiffening and left ventricular remodelling. *J. Hum. Hypertens.* **24**(3):197–206.

Gerhard M, Roddy M-A, Creager SJ, Creager MA. 1996. Aging progressively impairs endothelium-dependent vasodilation in forearm resistance vessels of humans. *Hypertension* **27**(4):849–853.

Hees PS, Fleg JL, Lakatta EG, Shapiro EP. 2002. Left ventricular remodeling with age in normal men versus women: novel insights using three-dimensional magnetic resonance imaging. *Am. J. Cardiol.* **90**(11):1231–1236.

James MA, Potter JF. 1999. Orthostatic blood pressure changes and arterial baroreflex sensitivity in elderly subjects. *Age Ageing* **28**(6):522–530.

Kajstura J, Gurusamy N, Ogórek B, Goichberg P, Clavo-Rondon C, Hosoda T, D'Amario D, Bardelli S, Beltrami AP, Cesselli D, Bussani R, del Monte F, Quaini F, Rota M, Beltrami CA, Buchholz BA, Leri A, Anversa P. 2010. Myocyte turnover in the aging human heart. *Circ. Res.* **107**(11):1374–1386.

Leibowitz D, Stessman J, Jacobs JM, Stessman-Lande I, Gilon D. 2013. Prevalence and prognosis of aortic valve disease in subjects older than 85 years of age. *Am. J. Cardiol.* **112**(3):395–399.

Lumens J, Delhaas T, Arts T, Cowan BR, Young AA. 2006. Impaired subendocardial contractile myofiber function in asymptomatic aged humans, as detected using MRI. *Am. J. Physiol. Heart Circ. Physiol.* **291**(4):H1573–H1579.

Mattace-Raso FU, van den Meiracker AH, Bos WJ, van der Cammen TJ, Westerhof BE, Elias-Smale S, Reneman RS, Hoeks AP, Hofman A, Witteman JC. 2007. Arterial stiffness, cardiovagal baroreflex sensitivity and postural blood pressure changes in older adults: the Rotterdam Study. *J. Hypertens.* **25**(7):1421–1426.

Monahan KD, Dinenno FA, Seals DR, Clevenger CM, Desouza CA, Tanaka H. 2001. Age-associated changes in cardiovagal baroreflex sensitivity are related to central arterial compliance. *Am. J. Physiol. Heart Circ. Physiol.* **281**(1):H284–H289.

Nagai Y, Metter EJ, Earley CJ, Kemper MK, Becker LC, Lakatta EG, Fleg JL. 1998. Increased carotid artery intimal–medial thickness in asymptomatic older subjects with exercise-induced myocardial ischemia. *Circulation* **98**(15):1504–1509.

Ogawa T, Spina RJ, Martin WH, 3rd, Kohrt WM, Schechtman KB, Holloszy JO, Ehsani AA. 1992. Effects of aging, sex, and physical training on cardiovascular responses to exercise. *Circulation* **86**(2):494–503.

Olivetti G, Giordano G, Corradi D, Melissari M, Lagrasta C, Gambert SR, Anversa P. 1995. Gender differences and aging: effects in the human heart. *J. Am. Coll. Cardiol.* **26**(4):1068–1079.

Olivetti G, Melissari M, Capasso JM, Anversa P. 1991. Cardiomyopathy of the aging human heart: myocyte loss and reactive cellular hypertrophy. *Circ. Res.* **68**(6):1560–1568.

Stratton JR, Cerqueira MD, Schwartz RS, Levy WC, Veith RC, Kahn SE, Abrass IB. 1992. Differences in cardiovascular responses to isoproterenol in relation to age and exercise training in healthy men. *Circulation* **86**(2):504–512.

Tanaka H, Seals DR. 2008. Endurance exercise performance in Masters athletes: age-associated changes and underlying physiological mechanisms. *J. Physiol.* **586**(1):55–63.

Tavakoli V, Sahba N. 2013. Assessment of age-related changes in left ventricular twist by 3-dimensional speckle-tracking echocardiography. *J. Ultrasound Med.* **32**(8):1435–1441.

Thorin E, Thorin-Trescases N. 2009. Vascular endothelial ageing, heartbeat after heartbeat. *Cardiovasc. Res.* **84**(1):24–32.

van Dalen BM, Soliman OI, Kauer F, Vletter WB, Zwaan HB, Cate FJ, Geleijnse ML. 2010. Alterations in left ventricular untwisting with ageing. *Circ. J.* **74**(1):101–108.

White M, Roden R, Minobe W, Khan MF, Larrabee P, Wollmering M, Port JD, Anderson F, Campbell D, Feldman AM, Bristow MR. 1994. Age-related changes in beta-adrenergic neuroeffector systems in the human heart. *Circulation* **90**(3):1225–1238.

10

AGING OF THE PULMONARY SYSTEM

OVERVIEW

Impact of the Environment More so than any other organ system, the pulmonary system is the one most intimately engaged with the environment. In facilitating gas exchange (its main function), the respiratory tract is subjected to pollutants, smoke, toxins, infectious agents, and industrial dust. Although many of these substances are known to cause lung disease, their role in aging of the pulmonary system is poorly defined. They, however, are considered prime suspects of accelerated aging. In this regard, chronic stimulation of the innate and adaptive immune cells of the respiratory tract by environmental hazards is postulated to instigate a proinflammatory state that promotes tissue damage and disease vulnerability (see Chapter 15). Biogerontologists are just beginning to understand the impact of the environment on the aging of the pulmonary system.

Despite the potentially adverse effects of the environment on the respiratory system, this system performs adequately for most of the human lifespan. However, in the presence of severe stress, for example, anesthesia and infections, age-associated pulmonary decrements become evident. An insufficient pulmonary reserve is the major reason for the higher morbidity and mortality of surgery, infectious diseases, and other chronic lung diseases experienced by the elderly. Results of interventional trials show that although chronic aerobic exercise training may not prevent the age-associated decline in respiratory reserve, the pulmonary system of athletic elderly operates at a higher physiological level, providing a significant advantage to overcome future severe stresses.

Human Biological Aging: From Macromolecules to Organ Systems, First Edition. Glenda Bilder.
© 2016 John Wiley & Sons, Inc. Published 2016 by John Wiley & Sons, Inc.

PULMONARY STRUCTURE/FUNCTION

Essential Components are Thoracic Cavity, Airways, and Lung Tissue The anatomical structures of the pulmonary system are the thoracic cavity, the airways, and the lung tissue. The thoracic cavity includes the bony spinal column, ribs, and connecting cartilage. It protectively encases the lungs and serves to create the pressure changes to move air in and out of the body. The airways start at the mouth and nose (oral/nasal cavities) than join to form the pharynx before becoming the trachea, the main airway passage. This latter structure branches into two bronchi that subsequently fork numerous times into smaller bronchioles and terminate in alveolar sacs. The bronchioles are comprised of smooth muscle cells (SMCs) whose contractile activity regulates the diameter of the bronchi and bronchioles and hence the amount of air exchange. The lung tissue includes the alveolar sacs, the pulmonary vasculature, and the supportive matrix. Figure 10.1 depicts these structures.

Respiration Requires Mechanical Effort from Skeletal Muscles and Regulation from Nerves Respiratory muscles include the diaphragm (base of chest cavity), internal and external intercostals (located between the ribs), and the abdominal muscles, used mainly in heavy breathing. Additionally, the autonomic nervous system (innervating the SMCs), somatic motoneurons (innervating the skeletal muscles), and coordination mediated by the central nervous system regulate involuntary (automatic) respiration that can be overridden by voluntary effort.

External respiration at rest is performed by the diaphragm. In quiet breathing, contraction of the diaphragm increases the size of the thoracic cavity, creates a negative pressure, and draws air in for inspiration. Relaxation of the diaphragm and recoil of the thoracic cavity and lung tissue push air out in expiration. With exercise or exertion, the intercostals, the diaphragm, and the abdominal muscles are necessary additions. In heavy breathing the contraction of the diaphragm and external intercostals enlarges the chest volume/size, and increases negative pressure to draw more air inward. During forced expiration, the external intercostals and diaphragm relax, and the internal intercostals and the abdominal muscles contract to reduce the chest cavity volume/size, and increase chest cavity pressure, which pushes a large volume of air outward (see Figure 10.2).

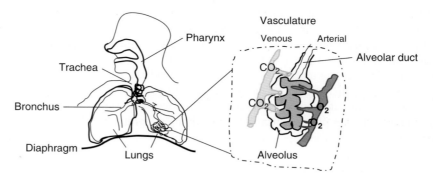

Figure 10.1. Anatomy of pulmonary system. Inset depicts terminal alveolar sac encased with blood vessels.

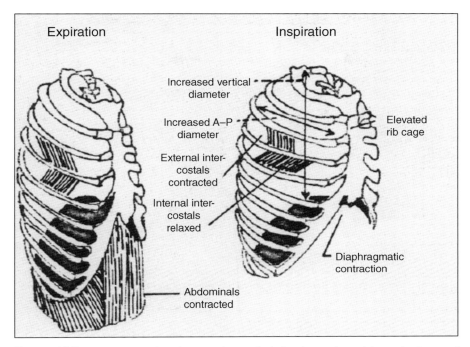

Figure 10.2. Changes in the chest cavity and skeletal muscles with expiration and inspiration to illustrate the mechanics of breathing. (Reprinted with permission from Guyton (1961).)

External Respiration Allows Exchange of Oxygen and Carbon Dioxide   *The function of the pulmonary system* is to exchange oxygen (from the air) for carbon dioxide (waste product from internal metabolism) in the process of external respiration (ventilation). Inspiration brings oxygen into the alveolar sacs, the site of gas diffusion and exchange. Oxygen diffuses into the pulmonary capillary blood for delivery to the peripheral tissues. In expiration, carbon dioxide from the peripheral tissues diffuses from the pulmonary capillaries into the alveolar sacs and moves out through the airways. Inspired oxygen, carried primarily by red blood cells, is utilized by tissue cells during cellular respiration (oxidative phosphorylation, Chapter 5). Oxidative phosphorylation generates carbon dioxide, a waste by-product that is expired.

Age Changes in the Thoracic Cavity

Thoracic Cavity Remodels with Hyperkyphosis and Becomes Stiffer   To perform the mechanics of respiration, the chest cavity must expand upward and downward during inspiration and expiration, respectively. *This movement becomes more difficult with age due to age-associated thoracic cavity remodeling.* Remodeling is attributed to a *decrease in thoracic cavity compliance (ease of distortion), largely dependent on the development of hyperkyphosis.*

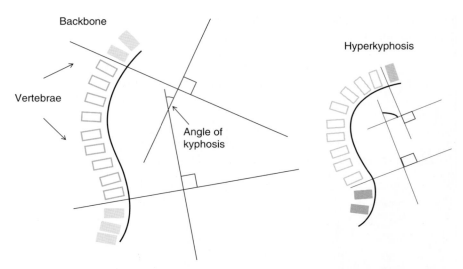

Figure 10.3. Measurement of angle of kyphosis using the Cobb technique. From X-rays, upper and lower vertebra defining the curve are selected and lines are drawn as shown.

Kyphosis refers to the forward curvature of the thoracic spine. Physiologically normal curvature values are between 20° and 40° (see Figure 10.3). With age (starting about age 40), the angle exceeds 40°, a state labeled *hyperkyphosis*. It is more severe in women than in men. The incidence and prevalence of hyperkyphosis in the elderly population is estimated between 20 and 40%. This forward arching of the spine is associated with reduced space between the ribs and a shortening of the intercostal muscles. Hyperkyphosis is not only associated with reduced ventilatory capacity but also with limitations of physical activity (stair-climbing speed), increased risk of future fracture (altered gait, increased positional sway), and increased mortality from pulmonary death.

Many factors contribute to the development of hyperkyphosis, chief among which are vertebral disc degeneration and vertebral fractures. Other factors are age-associated vertebral disc compression, dynapenia of the spinal extensors, and reduced sensory perceptions (visual, touch). Some hyperkyphosis is reversible as indicated by results of a program of spinal extensor exercises performed over a 2-year period that significantly reduces the kyphotic angle in women (45–69 years of age) especially those with substantial hyperkyphosis (Itoi and Sinaki, 1994).

Role of Cartilage Calcification in Loss of Chest Compliance

Based on chest radiographs, costal cartilage calcification increases with age from 6 to 45% from age 20 to 80 years. Those older than 70 years also exhibited calcification in the diaphragm and trachea. Although hyperkyphosis contributes significantly to reduced chest compliance, an age-associated increase in cartilage calcification (between ribs and sternum; between vertebrae) could conceivably enhance cavity stiffness. However, the extent to which cartilage calcification hinders thoracic compliance is not clear. Computer models of pulmonary dynamics suggest this to be minor.

Dynapenia of Respiratory Muscles Reduces Pulmonary Function
The diaphragm and the intercostals are postulated to weaken with age. Indirect data relate peripheral dynapenia (low grip strength) with poor performance on pulmonary function tests. Direct assessment of muscle strength derives from determination of maximal inspiratory pressure (MIP) and maximal expiratory pressure (MEP). In a large study of over 4000 participants >65 years of age (Cardiovascular Health Study), Enright et al. (1994) measured MIP (with a mechanical pressure gauge) and observed an age-associated decrease in MIP. Other factors, for example, current smoking, waist size, self-reported fair–poor health were confounding factors. A recent meta-analysis of 22 smaller studies (Sclauser Pessoa et al., 2014), designed to establish normative values for MIP, concluded that MIP was higher in men than in women and that over the span of 18–83 years of age, MIP declined with age in both sexes, beginning between the ages of 40 and 60 years. Others such as Tolep et al. (1995) and Polkey et al. (1997), using transdiaphramatic pressure measurements in young and elderly subjects, found that muscle strength was approximately 13–25% lower in elderly (65–75 years of age) subjects compared to young (19–28 years of age) subjects.

Motoneuron dysfunction is thought to account in part for respiratory muscle dynapenia. Action potential recordings from the nerves stimulating the diaphragm reveal distortions that worsen with age and have been associated with a degenerative appearance of the nerve terminals. Neuronal dysfunction exacerbates dynapenia.

Whether the thickness of the diaphragm changes with age is unknown. As assessed by computerized tomography scans in a study of 120 individuals (20–80 years of age), the thickness of the diaphragm remained constant with age. However, others have reported an increased thickness of the diaphragm (ultrasound determination) associated with a higher level of pulmonary function, observations evident only in elderly who habitually exercise.

Age Changes in the Airways

Airflow through the conduits (trachea, bronchi, and bronchioles) is largely regulated by the contraction and relaxation of the resident SMCs. SMC contraction produces constriction that creates a restricted airway of smaller diameter and reduced airflow; SMC relaxation produces dilation that opens up the airway and increases airflow.

Bronchodilators Are Less Effective in Elderly
Airflow decreases (resistance increases) normally in response to allergens (dust, dander, pollen, etc.), inflammatory mediators (leukotrienes), certain drugs (antihistamines), and the neurotransmitter of the parasympathetic nervous system (acetylcholine), all substances that stimulate SMC contraction. In contrast, airflow increases (SMCs relax and airway resistance decreases) in response to the sympathetic nervous system neurotransmitter (norepinephrine), bronchodilators (albuterol), anti-inflammatory drugs (hydrocortisone), and leukotriene antagonists (montelukast). *The response to bronchodilators lessens with age, favoring airway constriction*. Pharmacological study results show that the elderly require smaller amounts of constricting agents (e.g., methacholine that mimics acetylcholine) to induce bronchial constriction. This is not surprising in light of an age-associated decline in bronchodilator function. Additionally,

bronchodilation induced by a deep inspiration is less in older individuals compared to younger individuals. It is postulated that since bronchodilation requires activation of the β_2-adrenergic receptors that their number or affinity most likely decreases with age, in line with findings in several species (rat, rabbit, cow, guinea pig).

Airway Clearance Declines with Age The airway lining consists of a mix of epithelial mucus-producing cells with and without cilia and basal cells, a type of stem cell activated during injury. These cells provide the first line of defense against entry of pathogens and foreign material. Weakening of the airway barrier increases susceptibility to infections and exposure to inflammatory-inducing toxins (see Chapter 15). *Deterioration of the respiratory tract lining with age* has been reported. Specifically, *immediate and delayed clearance of foreign particles (examined with radiolabeled Teflon particles) decreases with age* and indicates that removal of debris by the upward beating of mucociliated cells of the large and small conduits becomes compromised with age. Additionally, the respiratory lining of elderly lungs compared to that of the young subjects contains more inflammatory cells (higher number of neutrophils, lower number of macrophages, and presence of activated immune cells) as determined from bronchoalveolar lavage of healthy individuals 20–78 years of age (Meyer et al., 1996). Indirect evidence of respiratory tract dysfunction is gleamed from reports that the elderly are also more sensitive to air pollution. Of relevance, results of a recent 4-year study with 11 million elderly (Dominici et al., 2006) showed that a modest elevation in fine particle number in the air (actually below EPA standards) significantly increases hospital admissions for respiratory and cardiovascular indications of individuals older than 75 years of age. Considering the role of antioxidative defense protection, for example, SOD, catalase, glutathione abundantly present in the respiratory lining of young individuals, and the increased sensitivity of the elderly to air pollution, notorious for its wealth of oxidative irritants, it has been proposed that the respiratory lining of the elderly lung is deficient in antioxidant defense mechanisms. This interesting hypothesis has been substantiated in aged animals and in elderly with lung disease, but needs confirmation in healthy elderly.

Age Changes in Lung Tissue

Dilated Alveolar Sacs and Decreased Number of Pulmonary Capillaries Limit Gas Diffusion Exchange of oxygen and carbon dioxide occurs across the alveolar basement membranes and the pulmonary capillaries surrounding the alveolar sacs. Transfer of gases depends on capillary blood volume/flow and available alveolar area and is measured by determining the uptake of tracer amounts of carbon monoxide (CO) (D_{LCO}). Although this measurement may be biased by numerous factors, for example, position, degree of ventilation during blood draws, and apparatus-induced artificial pressure changes, all *studies to date have consistently shown diffusion capacity to decrease with age.* In a study by Guénard and Marthan (1996) that compared 74 healthy subjects >68 years of age with 55 younger individuals (average age of 26 years), lung diffusion capacity decreased by as much as 50%. Histological studies support this functional change with evidence of *dilated alveolar sacs and increased alveolar space* sometimes termed "accelerated or senile emphysema." Additionally, there are *fewer capillaries per unit area*. These changes could explain the diffusion-limited flow observed with D_{LCO}. As *the area of functional gas exchange decreases (useful lung tissue), residual volume or dead space increases (dysfunctional lung tissue)*. In other words, a mismatch between ventilation (available gases) and perfusion (blood flow to pick up and drop off gases) develops over time.

Unlike the chest cavity that becomes less compliant with age, lung tissue becomes *more compliant with time*. Specifically, the ability of the lung tissue to "spring back," or *recoil after inflation, declines with age*. The reason for the loss of lung tissue recoil is not clear. One proposal suggests age-related structural deterioration in extracellular matrix proteins, especially elastin and collagen but there is no consensus regarding age-associated changes in these components. More likely, it is related to an age-dependent enlargement (dilation) of bronchiolar and alveolar ducts with and without wall thickening, mentioned above. Loss of lung recoil modifies the mechanics of breathing and reduces expiratory airflows and vital capacity discussed below.

Carotid Body Chemoreceptors Deteriorate with Age but the Significance Is Unclear A fall in blood oxygen tension (hypoxemia) or an elevation in blood carbon dioxide tension (hypercapnia) stimulates ventilation through a reflex mediated by the carotid body chemoreceptors. The elevated respiratory rate restores gas homeostasis. Since the *morphology of the carotid body deteriorates with age in man*, it has been hypothesized that the chemoreflex would also waiver allowing for acceptance of hypoxemia and hypercapnia, known cell stressors. However, Guénard and Marthan (1996) showed that blood oxygen tension is independent of age and more recently, albeit in a small study of 16 young women (average age 24) and 19 elderly women (average age 71), the chemoreflex to a hypoxemic test was similar in the two groups (Pokorski et al., 2004). Apparently, other chemoreceptors (centrally located) can compensate for carotid body degeneration and that tissue hypoxia comes about for other reasons not dependent on sensing of oxygen/carbon dioxide tension.

Mechanics of Breathing: Rest and During Exercise

Pulmonary Function Tests Measure Physical Fitness Pulmonary function tests capture the mechanics of respiration and demonstrate the degree of physical fitness of the system. The amount of air and its rate of flow into and out of the lungs at rest or during forced breathing are measured with an instrument called a spirometer (see Figure 10.4). Normal resting respiration rate is 12–14 breaths/min. Tidal volume (TV), the amount of air moving into and out of lungs in one cycle of quiet breathing, is about half a liter. The impact of age *per se* on *quiet breathing rate and TV is modest*. There is a slight decrease in TV and slight increase in rate and a possible contribution from abdominal muscles to facilitate quite breathing.

Maximal respiratory efforts are defined as the inspiratory reserve volume (IRV), the amount of air moving in during a maximal inspiration. It is approximately 3.3 L and conversely, expiratory reserve volume (ERV), the amount of air that moves out of the lungs during maximal expiration, is approximately 1 L. A hallmark measurement

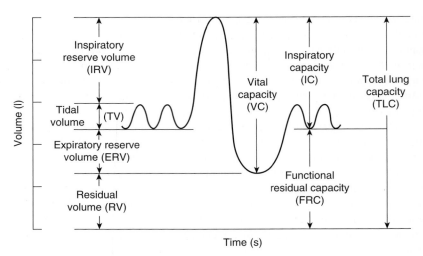

Figure 10.4. Lung volumes and capacity measurements. (Reprinted with permission from Barrett et al. (2012, p. 629).)

of physical fitness is forced vital capacity (FVC). *FVC equals the amount of air an individual can exhale after a forced inspiration.* The forced vital capacity of young females on average is ~3–4 L and of young males is slightly higher at 4.5–5.5 L. Even after a forced expiration, some air remains in the lungs and is termed the *dead space or residual volume (RV)*. RV and FVC sum to equal the total lung capacity (TLC) (see Figure 10.4, chart recording). In contrast to quiet breathing, *aerobic stress (forced breathing) reveals a decline in pulmonary function with age.*

Forced Expiration/Inspiratory Volumes and Forced Vital Capacity Decrease with Age
Pulmonary function tests affirm a decrease in maximal inspiratory/expiratory airflows, a decrease in FVC and indirectly an increase in RV. *Forced expiratory volume measured as a total (FVC) and quantity in 1 s (FEV1) is maximal at age 20 in females and age 27 in males, and declines as much as 75% thereafter.* This has been observed in cross-sectional as well as longitudinal studies, albeit with few individuals > 75 years of age.

The loss of lung recoil and decrease in chest compliance "trap" more air in the lungs leading to an increase in RV and a decrease in FVC. Because of these changes not only is the functional RC (more dead space) increased, but also *the elderly breath at higher lung volumes creating more work for respiratory muscles.* Even at rest, respiratory energy needs are higher in the 60 year old compared to the 20 year old. During forced expiration, the higher pressures and reduced alveolar support cause some airways to collapse. Thus, a vicious cycle develops that contributes to decreased diffusional capacity and oxygen uptake (described above).

Maximal Oxygen Consumption (VO_2max) Wanes with Age
An additional measurement, VO_2max, a measure of cardiopulmonary fitness, quantifies the elevation in oxygen consumption during a graded exercise. VO_2max peaks in

young adulthood (20–30 years of age) and declines thereafter. As shown by several large longitudinal studies, the decrease in VO_2max is nonlinear. This means that the decline in cardiopulmonary fitness accelerates with age. This observation was confirmed in a large study of cardiopulmonary fitness (VO_2max) evaluated in over 3000 women and 16,000 men (final age range 20–96 years) followed by serial health exams for 30 years (Aerobics Center Longitudinal Study) (Jackson et al., 2009). After the age of 45, there is an acceleration in the loss of cardiopulmonary fitness. There is also a clear gender difference such that women exhibit lower levels of VO_2max at all ages and a slower rate of fitness decline.

The effect of age on cardiopulmonary fitness is also discussed in Chapter 9. Recall that VO_2max depends on the cardiac output (stroke volume and heart rate) and peripheral oxygen extraction. This latter component relies on pulmonary activities of unobstructed gas diffusion, absence of dynapenia, and optimal chest/lung tissue compliance. Therefore, aging of the pulmonary system exerts a sizeable influence on VO_2max. The summary box reviews the age changes in the pulmonary system.

CONSEQUENCES

Severe Stresses Challenge Reduced Pulmonary Function Reserve of the Elderly
With age, there is a decrease in FVC and expiratory volumes/flow rates, an increase in RV (dead space), a mismatch between ventilation and blood perfusion, and reduction in innate and immunological defenses. For most elderly, gas exchange at rest is adequate and for the physically active (exercise trained) elderly, high levels of performance are possible. In contrast, exercise intolerance or rapid fatigue and shortness of breath (dyspnea) are common in elderly with a sedentary lifestyle. For all, situations of high stress seriously challenge the age-altered pulmonary system. The age-associated reduction in pulmonary functional reserve significantly *impacts the outcome of anesthetic-dependent surgeries and survival from infectious diseases*. In these states of high respiratory demand, the pulmonary system cannot relieve the state of hypoxemia and hypercapnia created by reduced function and elevated stress. The outcome is a high mortality rate. Pulmonary dysfunction accounts for 40% of the deaths following surgery in the elderly. Exposure to infectious agents, such as influenza and pneumonia, not only requires optimal innate and immunological defense mechanisms but also optimal lung function. Thus, infectious lung diseases are the fifth major cause of death in those >65 years of age.

In addition to infectious diseases, pulmonary age changes elevate the risk for other lung diseases, for example, chronic obstructive pulmonary disease (COPD), bronchitis, and asthma, and exacerbate existing lung disease. COPD is the fourth major cause of death in the elderly population. Additionally, pulmonary age changes are additive with cardiovascular disease, the major cause of death in the elderly population.

INTERVENTIONS FOR HEALTHY AGING

Exercise Training Improves Pulmonary Function
It was observed in the Aerobics Center Longitudinal Study (Jackson et al., 2009) discussed above that there

is a significant positive correlation between VO_2max and level of physical activity independent of age. Thus, although cardiopulmonary fitness may decline over time, higher levels of fitness allow one to perform well at higher levels of stress. See Chapter 9 for results of interventional programs of physical conditioning with sedentary elderly. In a 10-year study of competitive distance runners and sedentary controls >65 years of age, intensive exercise adequately maintained VO_2max up to the seventh decade. A decline thereafter appears concomitant with a reduction in exercise intensity level (Muster et al., 2010).

Another intervention seeks to reduce respiratory muscle dynapenia with inspiratory (IMST) and expiratory muscle strength training (EMST). Resistive or pressure threshold devices into which the trainee respires are used for respiratory muscle strength training (RMST) to increase maximal inspiratory and maximal expiratory pressure. As reviewed by Laciuga et al. (2014), EMST as short as several times a week for 4 weeks is effective in improving pulmonary function in several disease states, for example, COPD, Parkinson's disease, and multiple sclerosis. Thus far, only a handful of studies have investigated the possible benefits in health sedentary elderly. The results of these small studies are encouraging and show improvement in several parameters: MIP, MEP, walking performance, and cough reflex. Pulmonary interventionalists have emphasized the need for a large blinded randomized control trial (RCT) to assess the immediate and long-term benefits of RMST in healthy elderly.

Several studies have evaluated the effect of focal spinal extensor exercises to lessen the kyphotic angle and minimize thoracic cavity remodeling. Bansal et al. (2014) reviewed 13 similar studies and found that 8 presented convincing data that chronic resistance exercise improves posture and the preexisting hyperkyphosis in individual >45 years of age. As with RMST, a large RCT addressing the effect of spinal extensor exercise program on prevention and reversal of hyperkyphosis is required.

Summary of Age Changes of the Pulmonary System

Anatomical	Functional
Hyperkyphosis, remodeling of chest cavity	↓ Chest cavity compliance
Dynapenia of respiratory muscles	Airway constriction
Cartilage calcification	↓ Airway clearance of foreign debris
↓ Response to bronchodilators	↓ Gas diffusion
↓ Mucociliary function of respiratory lining	↓ Lung recoil
Dilated alveolar sacs,	↓ Mismatch between ventilation and perfusion
↓ Number of capillaries/sac	↓ $FEV1$, ↓ FEV, ↓ FVC
Carotid body chemoreceptor deterioration	↓ VO_2max

FEV1: forced expiratory volume in 1 s; FVC: forced vital capacity; VO_2max: maximal oxygen consumption to graded exercise.

Clearly, smoking is a risk factor for all lung diseases as well as for cardiovascular disease. It is also associated with reduced cardiopulmonary lung function in the absence of disease. Although it remains a lifestyle choice, it should be seriously considered as a habit without a single benefit.

SUMMARY

The main components of the pulmonary system are the thoracic chest, the airways, and the lung tissue. External respiration or ventilation facilitates the exchange of oxygen in the air for carbon dioxide generated through cell metabolism.

With time, the pulmonary system experiences the following changes: compliance of the thoracic cavity decreases due in part to hyperkyphosis and respiratory muscle dynapenia, airway resistance may increase due to a reduction in bronchodilator activity, and lung recoil declines due to dilation of the alveolar sacs. Additionally, the number of pulmonary capillaries decreases and adds to reduction of oxygen diffusion. Innate and immunological defense mechanisms of the respiratory tract are also compromised.

The loss of cavity compliance, lung recoil, and muscle strength contribute to recorded deficits in pulmonary function tests. With age, there is a decrease in $FEV1$, FEV, FVC, and VO_2max. Reduced pulmonary function becomes apparent during severe stresses of anesthesia-dependent surgery and pathogenesis of infectious diseases.

Chronic intense aerobic exercise programs and resistance exercises for respiratory and spinal extensor muscles show promise to improve pulmonary function in the elderly.

CRITICAL THINKING

How does aging affect the mechanics of breathing?

What recommendations would you give to the elderly to avoid exercise intolerance?

What in your opinion are the two most significant age changes in the pulmonary system? Support your choices.

Of what value is a spirometer?

What does FVC indicate about pulmonary function?

KEY TERMS

Allergen substance that activates the immune system. In relation to the pulmonary system, allergens such as dust, dust mites, dander, and pollens activate immune functions in the lungs that produce mediators of bronchial constriction and excessive mucus production.

Alveolar sac terminal structure in the lung and site of gas (oxygen/carbon dioxide) exchange.

Bronchus one of two main airway conduits that branch from the trachea.

Compliance term used to describe the degree of flexibility or ease of distortion of the thoracic cavity and the lungs.

Diaphragm major skeletal muscle of respiration; forms the base of the thoracic cavity.

Epithelial cell specific cell type found on epidermis of the skin, lining of the gastrointestinal tract, and the airways of the lungs.

External intercostals skeletal muscles located between the ribs that contract on inspiration.

Force expiratory volume1 (FEV1) volume of 1 s of forced expiration following forced inspiration. Predictive of pulmonary function.

Forced vital capacity (FVC) maximal amount of air that can be expired after a forced inspiration; gives estimate of thoracic compliance and respiratory muscle strength.

Internal intercostals skeletal muscles located between the ribs that contract on expiration.

Kyphosis curvature of the upper thoracic spine with vertebral compression attributed to osteoporosis, dynapenia and other factors.

Residual volume (RV) space in lungs that cannot be emptied of air and is considered nonphysiological.

Tidal volume (TV) volume of inspired and expired air that is exchanged at rest.

Trachea the major airway for respiration.

BIBLIOGRAPHY

Review

Bansal S, Katzman WB, Giangregorio LM. 2014. Exercise for improving age-related hyperkyphotic posture: a systematic review. *Arch. Phys. Med. Rehabil.* **95**(1):129–140.

Barrett KE, Barman SM, Boitano S, Brooks HL. 2012. *Ganong's Review of Medical Physiology*, 24th ed. New York: McGraw Hill.

Buchman AS, Boyle PA, Wilson RS, Gu L, Bienias JL, Bennett DA. 2008. Pulmonary function, muscle strength and mortality in old age. *Mech. Ageing Dev.* **129**(11):625–631.

Dominici F, Peng RD, Bell ML, Pham L, McDermott A, Zeger SL, Samet JM. 2006. Fine particulate air pollution and hospital admission for cardiovascular and respiratory disease. *JAMA* **295**(10):1127–1134.

Guyton AC. 1961. *Textbook of Medical Physiology* 2nd ed. Philadelphia: WB Saunders Company.

Kim J, Sapienza CM. 2005. Implications of expiratory muscle strength training for rehabilitation of the elderly: tutorial. *J. Rehabil. Res. Dev.* **42**(2):211–224.

Laciuga H, Rosenbek JC, Davenport PW, Sapienza CM. 2014. Functional outcomes associated with expiratory muscle strength training: narrative review. *J. Rehabil. Res. Dev.* **51**(4):535–546.

Lowery EM, Brubaker AL, Kuhlmann E, Kovacs EJ. 2013. The aging lung. *Clin. Interv. Aging* **8**:1489–1496.

Santulli G, Iaccarino G. 2013. Pinpointing beta adrenergic receptor in ageing pathophysiology: victim or executioner? Evidence from crime scenes. *Immun. Ageing* **10**(1):10.

Scichilone N, Messina M, Battaglia S, Catalano F, Bellia V. 2005. Airway hyperresponsiveness in the elderly: prevalence and clinical implications. *Eur. Respir. J.* **25**(2):364–375.

Sharma G, Goodwin J. 2006. Effect of aging on respiratory system physiology and immunology. *Clin. Intervent. Aging* **1**(3):253–260.

Sprung J, Gajic O, Warner DO. 2006. Review article: age related alterations in respiratory function—anesthetic considerations. *Can. J. Anaesth.* **53**(12):1244–1257.

Taylor BJ, Johnson BD. 2010. The pulmonary circulation and exercise responses in the elderly. *Semin. Respir. Crit. Care Med.* **31**(5):528–538.

Experimental

Aznar-Lain S, Webster AL, Cañete S, San Juan AF, López Mojares LM, Pérez M, Lucia A, Chicharro JL. 2007. Effects of inspiratory muscle training on exercise capacity and spontaneous physical activity in elderly subjects: a randomized controlled pilot trial. *Int. J. Sports Med.* **28**(12):1025–1029.

Ball JM, Cagle P, Johnson BE, Lucasey C, Lukert BP. 2009. Spinal extension exercises prevent natural progression of kyphosis. *Osteoporos. Int.* **20**(3):481–489.

Bartynski WS, Heller MT, Grahovac SZ, Rothfus WE, Kurs-Lasky M. 2005. Severe thoracic kyphosis in the older patient in the absence of vertebral fracture: association of extreme curve with age. *AJNR Am. J. Neuroradiol.* **26**(8):2077–2085.

Caskey CI, Zerhouni EA, Fishman EK, Rahmouni AD. 1989. Aging of the diaphragm: a CT study. *Radiology* **171**(2):385–389.

Edwards AM. 2013. Respiratory muscle training extends exercise tolerance without concomitant change to peak oxygen uptake: physiological, performance and perceptual responses derived from the same incremental exercise test. *Respirology* **18**(6):1022–1027.

Enright PL, Kronmal RA, Manolio TA, Schenker MB, Hyatt RE. 1994. Respiratory muscle strength in the elderly: correlates and reference values. Cardiovascular Health Study Research Group. *Am. J. Respir. Crit. Care Med.* **149** (2 Part 1):430–438.

Fleg JL, Morrell CH, Bos AG, Brant LJ, Talbot LA, Wright JG, Lakatta EG. 2005. Accelerated longitudinal decline of aerobic capacity in healthy older adults. *Circulation* **112**(5):674–682.

Fon GT, Pitt MJ, Thies AC, Jr. 1980. Thoracic kyphosis: range in normal subjects. *AJR Am. J. Roentgenol.* **134**(5):979–983.

Guénard H, Marthan R. 1996. Pulmonary gas exchange in elderly subjects. *Eur. Respir. J.* **9**(12):2573–2577.

Hurst G, Heath D, Smith P. 1985. Histological changes associated with ageing of the human carotid body. *J. Pathol.* **147**(3):181–187.

Imai T, Yuasa H, Kato Y, Matsumoto H. 2005. Aging of phrenic nerve conduction in the elderly. *Clin. Neurophysiol.* **116**(11):2560–2564.

Itoi E, Sinaki M. 1994. Effect of back-strengthening exercise on posture in healthy women 49 to 65 years of age. *Mayo Clin. Proc.* **69**(11):1054–1059.

Jackson AS, Sui X, Hébert JR, Church TS, Blair SN. 2009. Role of lifestyle and aging on the longitudinal change in cardiorespiratory fitness. *Arch. Intern. Med.* **169**(19):1781–1787.

Meyer KC, Ershler W, Rosenthal NS, Lu XG, Peterson K. 1996. Immune dysregulation in the aging human lung. *Am. J. Respir. Crit. Care Med.* **153**(3):1072–1079.

Muster AJ, Kim H, Kane B, McPherson DD. 2010. Ten-year echo/Doppler determination of the benefits of aerobic exercise after the age of 65 years. *Echocardiography* **27**(1):5–10.

Pokorski M, Walski M, Dymecka A, Marczak M. 2004. The aging carotid body. *J. Physiol. Pharmacol.* **55** (Suppl. 3):107–113.

Polkey MI, Harris ML, Hughes PD, Hamnegärd CH, Lyons D, Green M, Moxham J. 1997. The contractile properties of the elderly human diaphragm. *Am. J. Respir. Crit. Care Med.* **155**(5):1560–1564.

Schmidt CD, Dickman ML, Gardner RM, Brough FK. 1973. Spirometric standards for healthy elderly men and women: 532 subjects, ages 55 through 94 years. *Am. Rev. Respir. Dis.* **108**(4):933–939.

Sclauser Pessoa IM, Franco Parreira V, Fregonezi GA, Sheel AW, Chung F, Reid WD. 2014. Reference values for maximal inspiratory pressure: a systematic review. *Can. Respir. J.* **21**(1):43–50.

Simões RP, Deus AP, Auad MA, Dionísio J, Mazzonetto M, Borghi-Silva A. 2010. Maximal respiratory pressure in healthy 20 to 89-year-old sedentary individuals of central São Paulo State. *Rev. Bras. Fisioter.* **14**(1):60–67.

Svartengren M, Falk R, Philipson K. 2005. Long-term clearance from small airways decreases with age. *Eur. Respir. J.* **26**(4):609–615.

Teale C, Romaniuk C, Mulley G. 1989. Calcification on chest radiographs: the association with age. *Age Ageing* **18**(5):333–336.

Tolep K, Higgins N, Muza S, Criner G, Kelsen SG. 1995. Comparison of diaphragm strength between healthy adult elderly and young men. *Am. J. Respir. Crit. Care Med.* **152**(2):677–682.

Ware JH, Dockery DW, Louis TA, Xu XP, Ferris BG Jr., Speizer FE. 1990. Longitudinal and cross-sectional estimates of pulmonary function decline in never-smoking adults. *Am. J. Epidemiol.* **132**(4):685–700.

11

AGING OF THE GASTROINTESTINAL AND URINARY SYSTEMS

OVERVIEW

The gastrointestinal (GI) tract and the urinary system experience modest structural and functional changes with age. It is proposed that the redundancy (magnitude, duplication) of these tissues compensates somewhat for age-related losses. Redundancy is evident in the incredible length of the GI tract, ~20 ft. Additionally, the abundance and diversity of bacteria dwelling in the GI tract contribute greatly to the well-being of the GI tract and furthermore assist other systems, for example the immune system. Although geriatric physicians disagree on the best assessment of kidney function in the elderly, age-related changes in kidney function become apparent mainly during water and salt stresses such as dehydration or elimination of a water load.

Nocturia, chronic constipation, and urinary incontinence are age-associated GI and urinary system conditions that reduce quality of life (QOL). Entwined with sociological and psychological issues, management of these conditions improves QOL. Life-threatening conditions of dehydration and malnutrition are considered reversible, but in the elderly are frequently misdiagnosed, overlooked, and complicated by sensory deficits. Major age changes and related disorders are presented in this chapter.

GASTROINTESTINAL TRACT AND ASSOCIATED ORGANS: ROLE IN NUTRIENT ABSORPTION AND MICROBIOTA REGULATION

The gastrointestinal tract begins with the oral cavity (mouth, pharynx) and continues to the esophagus, stomach, small intestine, and large intestine (colon). With the exception of skeletal muscles of the oral cavity and external sphincter of the anus, the

---

Human Biological Aging: From Macromolecules to Organ Systems, First Edition. Glenda Bilder.
© 2016 John Wiley & Sons, Inc. Published 2016 by John Wiley & Sons, Inc.

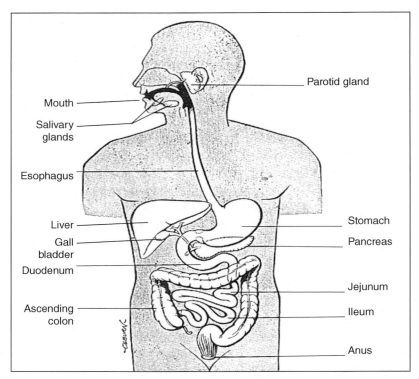

Figure 11.1. Anatomy of the GI system. (Reprinted with permission from Guyton (1961).)

GI tract is comprised of smooth muscles overlaid with an epithelial mucus layer. The liver, pancreas, and salivary glands are major accessory organs. Figure 11.1 displays the relative position of these glands, tissues and organs.

The GI tract performs two major functions: absorption of nutrients and oversight of bacterial colonies termed the microbiota. The well-known nutrient absorption function necessitates the breakdown of food (chewing, acid and enzymatic stomach activities) prior to absorption of small molecules, minerals, and water by the small intestine. Other than water, little absorption occurs in the colon. Rhythmic contractions of the GI tract are activated by the autonomic nervous system (ANS) that emanate from the vagus nerve (parasympathetic control) and intestinal enteric nervous system (complex neuronal system within the GI tract, autonomous but interacting with ANS). Counteracting influences originate from sympathetic nerve stimulation. Additionally, many hormones, for example, gastrin, ghrelin, cholecystokinin, motilin, gastric inhibitory peptide, and leptin participate in digestion through effects on hunger and satiety senses as well as regulation of GI motility and metabolism.

The accessory organs assist with digestive activities but perform additional distinct functions. The liver via a unique vascular network (hepatic portal system) receives sugars, amino acids, fats, and other small molecules directly from the intestine. The liver may process these substances further, for example, conversion of glucose to glycogen. The liver also synthesizes bile salts that make their way to the gall bladder and into the intestine to assist with fat breakdown. Fats and cholesterol are packaged with special proteins into structures called lipoproteins and secreted for use elsewhere and recycled back to the liver. The salivary glands produces saliva that is essential for oral health by aiding swallowing, speech, and providing protective effects on teeth and upper GI mucosal tissue (antibacterial and antifungal). Saliva also contains digestive enzymes for the catabolism of carbohydrates. The pancreas (hormonal functions discussed in Chapter 14) additionally produces an abundance of digestive enzymes (for breakdown of proteins, fats, and carbohydrates) and bicarbonate for neutralizing acid. The secretory process of the pancreas is closely regulated to assist with intestinal digestion.

GI Age Changes

Age Changes: Generally Minor with Exception of Reduced Colonic Transit Time
Aging of the oral cavity and esophagus is minor. Swallowing difficulties and reduced motility of the esophageal muscles may occur but they are related to disease (generally neurological) and not to aging *per se*.

Two different age-associated changes have been reported for the stomach: decreased compliance (reduced motility) of the upper (fundus) and lower (antral) sections of the stomach and a reduction in secretion of hydrochloric acid (hypochlorhydria). The former may contribute to an early decrease in satiety, considered insignificant; the latter is more important and may predispose to malabsorption and bacterial overgrowth. Hypochlorhydria is statistically associated with present and past infections with Helicobacter pylori.

Controversy surrounds the effect of age on nutrient absorption by the small intestine. Transit time (determined by an "eatable" camera) does not change with age but absorption of radiolabeled nutrients decreases especially at older ages (>80 years). As reviewed recently by Britton and McGlaughlin (2013), "there are no clinically significant changes in small intestinal function attributable to ageing." There is in contrast, a decrease with age in the propulsive activity of the colon associated with the disappearance of nerves and/or reduction in neurotransmitters. As a result, *colonic transit time increases, allowing for more water reabsorption and extended time for bacterial fermentation* of fiber. This predisposes to constipation and flatulence.

Gut Microbiota Exhibits Loss of Diversity
The gut microbiota are the resident bacteria in the intestine. They number approximately 10^{14} (that is, 10 followed by 14 zeros, outnumbering all of man's somatic and germline cells) and yet they coexist with and benefit man. Many (especially in the colon) are anaerobes, operating without a need for oxygen. Microbiota is invaluable in metabolizing nutrients (carbohydrates, proteins, and amino acids) that escape intrinsic enzymatic activity and hence add to energy metabolism. The microbiota is also essential for GI functions of cell renewal, mucus production, GI motility, and cooperation with immune cells in the maturation of dendritic cells (a kind of macrophage essential for normal immune function, see Chapter 15).

The diversity of the microbiota is established early in life with the introduction of a variety of foods and as expected is highly variable among individuals. Although the total number of microbes within the microbiota remains constant throughout the lifespan, *many factors affect the relative number of phylogenetically different bacteria within the microbiota*. Not surprisingly, the main factor is diet. For the elderly, additional influential factors include medications, immunosenescence, reduced GI motility, changes in dentition, and reduced salivary gland function.

Diversity of the Microbiota Decreases with Age and Diet Analyses of the microbiota among individuals of different ages are limited. Results of two studies found a *decrease in microbiota diversity* in those >65 years compared to younger subjects (20–40 years of age). These studies generated a "phylogenetic fingerprint" by sequencing DNA of fecal microbiota of 64 elderly (Biagi et al., 2010) and 178 elderly (Claesson et al., 2012), and smaller numbers of younger controls. Specifically, bifidobacteria, the bacteria prominent in childhood and young adulthood is present in lower amounts in the elderly. Similarly, the presence of other bacteria, the Firmicutes (some of which are anti-inflammatory) decrease and Bacteroidetes and Proteobacteria increase (see Box 11.1 for explanation of bacteria phyla). Interestingly, this *profile is most prominent in individuals in long-term care facilities, consuming a high fat/low fiber poor quality* (few fruits/vegetables) diet. This profile is also associated with markers of inflammation, for example, interleukin-6, interleukin-8, and increased incidence of frailty. These results suggest that a diet (low to moderate fat/high fiber, fruits and vegetables) has the potential to maintain a normal "ageless" microbiota and to additionally reduce the state of inflammation. Interestingly, elderly community dwellers

---

Box 11.1. More About Bacteria
Bifidobacteria: prominent gut bacteria; major bacteria in breast-fed babies (80% of cultivable fecal microbiota); anaerobic and functions in the absence of oxygen; also called lactic acid bacteria; component of probiotics to treat GI disorders.

Firmicutes: phylum of diverse bacteria; most are Gram-positive bacteria, spore-forming, with a cell wall but others are Gram-negative with no cell wall and are termed low G+C because the DNA is deficient in guanine (G) and cytosine (C). Examples include *Staphylococcus*, *Micrococcus*, *Streptococcus*, and *Lactobacillus*. *Clostridium difficile* and anti-inflammatory producing bacteria, for example, *Clostridium* cluster XIVa and *Faecalibacterium prausnitzii* belong to this phylum.

Bacteroidetes: Gram-negative, nonspore-forming, anaerobic, and rod-shaped; considered opportunistic bacteria; examples include four classes: Bacteroidia, Flavobacteria, Sphingobacteria, and Cytophagia. There are 7000 different species; function is degradation of protein and carbohydrates; some are pathogens when GI lining is breached (infections usually due to *Bacteroides fragilis* and *Bacteroides thetaiotaomicron*).

Proteobacteria: phylum of Gram-negative bacteria; mostly free moving; many are facultative (operate with and without oxygen); examples: *Escherichia*, *Salmonella*, *Vibrio*, *Helicobacter*, and *Yersinia*.

who adhered to this diet exhibit microbiota similar to that of young individuals (Claesson et al., 2012). However, the observation that centenarians also exhibit a reduction in microbiota in association with proinflammatory mediators creates a enigma and raises the possibility that important subtleties within the microbiota and the GI tract remain to be unraveled. Which aspects of the microbiota contribute to longevity are not known.

In addition to diet, *medications may adversely alter microbiota*. Chief on this list are the broad spectrum antibiotics with potential to shift the microbial balance in favor of pathogenic microbes. Other drugs include proton pump inhibitors, nonsteroidal anti-inflammatory drugs, and opioids that change the pH of the GI tract and enhance pathogen colonization, damage the GI tissue, or reduce transit time.

Age changes in the GI tract such as reduced transit time, irrespective of medication use, may affect microbiota diversity. Aging of the immune system (termed immunosenescence) also influences the intestinal flora since both innate and adaptive immunity are essential to oversee and keep in check this huge assortment of microbes.

Liver, Salivary Glands, and Pancreas Exhibit Modest Decline with Age

From a paucity of human data, it is concluded that *liver size decreases by 20–40% with age (24–91 years) and in one study that assessed dye (indocyanine green) clearance, hepatic blood flow also decreases with age*. Morphological studies from postmortem and surgical samples show a thickening of the endothelial lining within the liver, possibly accounting for decreased uptake of macromolecules, for example, cholesterol. Degradation of foreign compounds by specialized hepatic enzymes, termed P450 drug metabolizing enzymes, decreases significantly (30%) after the age of 70. This was observed in 226 subjects (20 to >70 years of age) by measurement of P450 enzyme content and clearance of a drug (in this case, antipyrine) normally metabolized by these enzymes. In consideration of the reduction in P450 content and activity it is recommended that medication for the elderly be prescribed at the lowest dose possible to avert an adverse drug reaction. There are some data to suggest that hepatobiliary function declines with age due to a reduction in bile salt synthesis, decreased contractility of the gall bladder, and reduced uptake of cholesterol. Other metabolic functions have not been formally assessed.

The *function of the salivary glands (production of saliva) is not impaired as a function of age*. Degenerative changes in the glands are evident but are insufficient to negatively impact saliva production. Nevertheless, *dry mouth due to hyposalivation (xerostomia) is a common complaint* in up to 30% of the elderly. This is *attributed not to aging but to polypharmacy*, primarily drugs with anticholinergic effects (opioids, tricyclic antidepressants, antihistamines, antiarrhythmics, and many others) and some diseases. Hyposalivation is serious as it predisposes to oral mucosal lesions (fungal, bacterial), difficulty in swallowing, and impaired speech.

The third accessory organ is the pancreas. Exocrine function is hardly perturbed by age. The *secretion of pancreatic enzymes and bicarbonate solution may decline up to 20–30%, but this change is considered clinically insignificant*. These alterations were assessed during exogenous stimulation, so their relevance to postprandial secretion (meal-induced pancreatic secretion) is unknown.

Age-Associated Disorders

Two conditions relevant to GI aging are malnutrition and anorectal disorders. Diseases such as gastroesophageal reflux and irritable bowel syndrome are beyond the scope of this book.

A Common Anorectal Disorder Is Chronic Constipation
Primary constipation is considered the most common GI disorder in the elderly. Although benign, persistent constipation may lead to more serious conditions of fecal incontinence, impaction, and perforation. Additionally, the sociopsychological associations decrease QOL.

Stool evacuation from the rectum is under both voluntary (motoneuron-skeletal muscle of the external anal sphincter) and involuntary (ANS-smooth muscle of the internal anal sphincter) reflex control. Continence is determined directly by sphincter activity. Although limited amount of data are available in humans, the *main age-associated change that contributes to constipation is enteric neurodegeneration that reduces colonic transit time*. The slowed transit allows for more water absorption and hard stools. Other factors such as diet (decreased fiber consumption), and decreased chewing, reduced stomach acid, medications and low levels of physical activity also contribute to constipation. Interestingly, both meals and walking increase intestinal motility. The mechanism of neural loss is unknown, although in animal models it is attributed to persistent oxidative damage.

Constipation may progress to fecal incontinence, accelerated by other age-associated alterations such as thinning of the internal anal sphincter (reported in women), reduced sphincter tone, and reduced rectal sensations. Treatment of chronic constipation proposed by Grassi et al. (2011) includes "adequate fluid intake, a diet rich in fiber (35–40 g/day), olive oil, physical activity, and laxative."

Malnutrition
Malnutrition in the elderly is complicated. It has many causes, and compared to this condition in the young, is more difficult to reverse, although there are strategies that help. It is often the cause of sarcopenia and frailty with increased morbidity and mortality. The initial issue is postulated as a reduction in food intake related mainly to a decrease in olfaction (smell) and less so to a decrease in taste sensations. Together, the enjoyment of eating is diminished, leading to decreased intake with little variation in food selection. The age-associated decrease in gastric compliance may add to reduced intake by accelerating the onset of satiety. Additionally, the "hunger-inducing hormones" appear to be less effective in the elderly possibly because some of these hormones rely on nitric oxide, a mediator in low abundance in the elderly. In addition to chronic diseases and associated medications, many psychological and social factors contribute to malnutrition such as depression, weakness, loneliness, fatigue, and physical disabilities.

Although multicausal, malnutrition is potentially reversible. The key is to identify those at risk, minimize weight loss and optimize nutritional status. One plan begins with an increased consumption of protein (exceeding RDA) spread out over the course of the day. Based on a meta-analysis of 36 randomized clinical trials of elderly consuming high protein oral nutritional supplement, this approach reduced disease complications and hospital readmissions, and improved strength and weight gain. Successful adjuncts to increased protein consumption include appealing food presentation and use of natural flavor enhancers.

URINARY SYSTEM

Components: Kidneys, Ureters, Bladder, and Urethra

Each kidney is comprised of nearly 1 million nephrons, substructures capable of filtering blood. Each nephron consists of a glomerulus and a series of tubules (Figure 11.2). Blood laden with waste products, for example, metabolic breakdown products, water-soluble drugs, enters the nephon's glomerular head via a small afferent arteriole. Within the glomerulus, liquid and dissolved substances, but not cells or particulate matter, are filtered and cleansed blood exits by way of the efferent arteriole. The glomerular filtrate traverses a series of tubules (descending limb, loop of Henle, ascending limb, and collecting duct) where the filtrate is modified by the process of reabsorption (ions, water taken back to the blood) and secretion (acids, bases added from the blood to the filtrate). The final product, the urine, passes through the ureters into the bladder and is excreted through the urethra.

The glomeruli and some of the tubules are located in the cortex (outermost section of the kidney); the remaining tubules (loop of Henle and collecting ducts) are situated in the medulla, the inner part of the kidney.

The kidney performs several functions. In addition to filtration of the blood and elimination of waste substances, the kidney assists with volume regulation (diluting and concentrating urine as needed to regulate blood volume), produces several important hormones, for example, renin (blood pressure regulation),

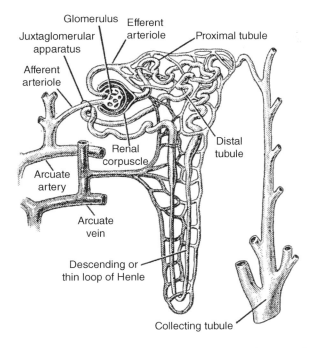

Figure 11.2. Anatomy of basic unit of the kidney, the nephron. (Reprinted with permission from Guyton (1961).)

1,25 dihydroxyvitamin D (calcium metabolism), and erythropoietin (EPO, blood cell formation (hematopoiesis)), and supports acid–base balance in the blood.

Age Changes in the Kidney

Structural Changes in Nephrons May Lead to Some Loss of Kidney Function
Structural changes observed with age are a *decrease in the number of active glomeruli and an alteration in their anatomy*. The basement membrane, essential for quality filtration, undergoes thickening. At present, this is related to the *dysfunctionality of podocytes*, cells that maintain this unique membrane "barrier" between blood and filtrate. This deteriorative change is labeled glomerulosclerosis. As glomeruli disappear, the remaining ones adjust with acceptance of additional pressure and filtration. *Tubules decrease in number, length, and volume, accompanied by interstitial fibrosis*, changes that account for a *decrease in renal mass with age* of about 1% per year after the age of 40. Similar to other tissues, renal vascular thickening occurs and afferent–efferent shunts that reduce filtration function are common.

A measurement of renal function is glomerular filtration rate (GFR) or how fast blood is filtered through the glomeruli. Cross-sectional, observational, and longitudinal studies conclude that *GFR decreases with age*. The BLSA reported a decrease of 0.75 ml/min/year (20–80 years) and other studies are within this range (0.4–2.6 ml/min/year). It is important to note that this is an averaged response. Interestingly, analysis of individual data complicate the picture and show that approximately 30% of elderly maintain a stable GFR and that another 30% exhibit an increase in GFR with time. Clearly many undefined factors (such as GFR measurement methods) contribute to this heterogeneity. In conjunction with a loss of glomeruli, renal blood flow decreases with age and blood is shunted from the cortex to the medulla.

A cautionary note is presented regarding the reliability of the age-associated change in GFR. The methods frequently used to determine GFR have been criticized. GFR is normally assessed by quantitation of serum creatinine, a muscle waste product. Since creatinine is neither reabsorbed nor secreted in the nephron its blood level reflects its renal clearance rate. In the elderly, *sarcopenia, weight loss, reduced protein intake, and hydration level all diminish the reliability of creatinine assessments*. Several formulas have been generated to compensate for age-related issues, but none have been convincingly validated in the elderly. A substitute for the measurement of creatinine is the assessment of the clearance of iohexol, a contrast dye, but this method requires intravenous infusion. Iohexol clearance appears useful in disease states in which a filtration value of 60 ml/min/1.73 m^2 of body surface area is set as the limit below which kidney function is considered impaired as confirmed by other criteria, for example, proteinuria. Its application to the elderly is complex because kidney filtration for many elderly fall within or below this bottom range and yet there is no corroborative evidence of renal dysfunction, that is, proteinuria or other deficiencies. Until better methods are developed for the elderly, effects of age regarding GFR may be misleading.

Structural changes at the tubule level predispose to aberrant regulation of blood levels of sodium and potassium, and water balance. Specifically, disturbed renal handling of sodium and potassium may lead to an *inability to retain sodium in elderly consuming a low sodium diet or conversely to excrete sodium in those ingesting large quantities of salt.* The former situation could cause volume loss and acute kidney damage and the latter would promote hypertension and cardiovascular damage. Additionally, perturbation of water management causes an *inability to concentrate urine* as in the case of nocturia (see below) or *to excrete a water load that results in critically low levels of blood sodium (hyponatremia).* The ability of the tubules to maintain the near alkaline pH may be impaired. This deficit is especially evident following consumption of a large protein meal or in the presence of stress.

Inhibition of RAS Promotes Longevity

In the inner wall of the glomerular afferent arteriole are specialized cells, called juxtaglomerular cells (modified epithelial cells), that produce renin, an important component of a blood pressure control system known as the renin–angiotensin system (RAS). The release of renin, is influenced by many factors, for example, decrease in arteriolar pressure, activation of sympathetic nerves, and low levels of sodium in the distal tubule. The release of renin initiates a cascade of events that culminates in formation of the final product of angiotensin II (Ang II), a powerful vasoconstrictor and effector of renal water and sodium reabsorption. Ang II binds to and activates Ang II receptors (AT1/AT2). RAS components reside not only in the kidney but have been found in other tissues: heart, brain, and immune cells. Chronic activation of RAS induces hypertension and understanding RAS is important for disease treatment. However, for the biogerontologist, Ang II is a harmful substance with life-shortening effects to include proinflammatory effects, induction of cellular senescence in endothelial cells, and promotion of ROS production.

Inhibition of the effects of Ang II results in amelioration of heart disease, reduction of kidney disease progression, and prevention of cognitive decline. In animal models, KO of AT1 receptors translates into a 26% increase in lifespan of mice, and pharmacological inhibition of Ang II production with lifelong enalapril treatment (AT1 inhibitor) increases the lifespan in rats. There is a role for Ang II and its receptor not only in disease but also in longevity that needs to be defined and explored in man. It is reasonable to avoid conditions that activate the release renin (hence, Ang II production) and some examples are use of diuretics, dehydration, sodium depletion, cardiac failure, and psychological stimuli of stress.

Age-Related Conditions/Issues: Dehydration, Medication Toxicity, Nocturia, and Urinary Incontinence

Dehydration is Common in the Elderly

One consequence of age-associated changes in body composition (see Chapters 6,7) is a reduced ability to compensate for dehydration. The *reduced fluid reserve from muscle and total body stores permits a more rapid onset of dehydration.* Additionally, polypharmacy (multiple drug use, especially laxatives and diuretics) and reduced kidney function both prevalent in the elderly promote dehydration.

Identification of dehydration in the elderly is challenging. Although measurement of osmolarity is the gold standard for dehydration, more apparent signs, such as lethargy, confusion, headache, and dizziness, may be more valuable. Unfortunately, these signs of dehydration, classic in young adulthood, are less commonly manifest in the elderly.

Current studies with the elderly have identified *potentially promising signs of dehydration* that correlate with serum osmolarity. They are *dry mouth, poor cognition, raised body temperature, low fluid intake, low urine volume and possibly change in urine color, auxiliary sweating, and foot vein filling* (Hooper et al., 2014).

There are two types of dehydration: water loss (insufficient fluid intake) and salt–water loss (loss of both water and salt as with diarrhea). Of these, the elderly are more sensitive to water loss dehydration because of the lower total water stores, reduced renal ability to concentrate the urine, and reduced thirst sensation. Water loss dehydration is associated with many chronic conditions related not only to kidney stones, drug toxicity, renal failure, and constipation but also fractures, falls, heat stress, and heart attack.

A number of organizations, for example, World Health Organization, European EFPA Panel on Dietetic Products, and Institute of Medicine, USA, have recommended daily total water intake (food and drinks) to be 2–2.8 L for women; 2.5–3.7 L for men and consumed as drinks alone to be 1–2.3 L for women; 1–3 L for men. Relevant studies show liquid intake in the elderly is on the low end of these values. To improve fluid intake in the elderly, education on the health benefits is essential. Reasons for avoiding liquids should be identified and overcome. Common ones, especially in nursing homes, include poor access to drinks and not enough help with toileting. A European trial, New Dietary Strategies Addressing the Specific Needs of Elderly Population for a Healthy Ageing in Europe, by FP7-funded NU-AGE seeks to address some of these issues. The effect of dietary and fluid intervention in elderly community dwellers will be studied. Important new information on adequate fluid intake and dehydration prevention in the elderly is expected.

Nocturia Affects Sleep but is Manageable Nocturia (nocturnal polyuria) is a condition that describes the change in the youthful pattern of urinary excretion. Young individuals excrete higher volumes of urine during the day than at night. In the elderly, this pattern is reversed. Nocturia interferes with sleep, increases the risk of falls, and reduces QOL. Its prevalence is high (~90% >80 years of age).

Many explanations for nocturia have been offered. The inability to concentrate urine, noted above, is a consequence of reduced levels of the pituitary hormone, vasopressin (antidiuretic hormone), and possibly loss of vasopressin receptors and associated water channels (latter two shown only in animal models). As vasopressin levels fall, less water is reabsorbed in the kidney tubules and more is excreted. A second explanation relates to the RAS that retains more sodium during the day (upright posture) and less at night (supine). Edema-producing diseases, for example, heart failure also contribute to nocturia. Edematous fluid of the legs is reabsorbed in the supine position and contributes to increased urine production at night.

Nocturia disrupts sleep and aggravates already distorted sleep patterns of the elderly. Nocturia creates fragmented sleep compounding the lack of deep sleep common in the elderly. The repeat voiding episodes further increase the chance of a fall and fracture.

Since the etiology of nocturia is varied, it is important to identify the main factor specific to each individual and treat it. Most of the conditions discussed above are amenable to pharmacotherapy, exercise, or behavioral therapy in the case of severe sleep disturbances.

Potential for Drug Toxicity The kidneys are the main site for elimination of most drugs. The hepatic cytochrome P450 enzymes biotransform drugs to a more water soluble entity that can be excreted by the kidneys. A reduction of renal function allows the drug to persist in the blood usually at a higher concentration. A drug concentration exceeding the therapeutic range induces an unwanted adverse drug reaction (ADR). Reduced kidney function is one of the many reasons for the high incidence of ADRs in the elderly population. Drugs of particular concern are antibiotics (specifically aminoglycosides, vancomycin), digoxin, lithium, histamine antagonists, and diuretics. Diuretics are especially troublesome in the presence of reduced kidney function since their site of action *is* the kidney and their persistence in a weakened kidney produces tissue toxicity that further compromises renal function. Since assessment of kidney function may be unreliable, *it is prudent to start all medications in the elderly at the lowest possible dose.*

Urinary Incontinence Has Many Causes Urine exits the renal medulla and flows through the ureters into the bladder, comprised of smooth muscle cells collectively called the detrusor muscle. Contraction of the detrusor expels the urine through the bladder neck or urethra. The neck is controlled by the involuntary smooth muscle internal sphincter and the voluntary skeletal muscle external sphincter. The bladder is innervated with sensory afferent fibers (sensing stretch) and efferent parasympathetic nerves, and some efferent sympathetic fibers regulating blood flow. Somatic motor fibers innervate the external sphincter. Basically, as the bladder fills, stretch sensations are generated and relayed back through the parasympathetic nerves to contract the detrusor. This occurs in phases of increase in pressure, sustained pressure, and relaxation until the bladder is maximally filled. These changes may be followed by another reflex that inhibits the external sphincter to allow urination (micturition).

A number of age changes have been observed in the bladder but the *distinction between disease and age per se is difficult to discern*. Results of clinical studies report age-associated reduction in bladder capacity, increased presence of unreliable bladder contractions, decrease in flow rates (due to reduced detrusor velocity but not strength), and increase in residual (post void) urine. These *changes are suggestive of fibrosis (collagen disposition) that diminish bladder functions*. Factors that initiate fibrosis are unknown, but at present they appear related to bladder ischemia. Neurotransmitter function may be disrupted with age, as supported by studies on detrusor biopsies that revealed a loss of muscarinic (parasympathetic) receptors with age (from males only) and a decrease in receptor mRNA in both sexes. Whether these changes are permissive for the most common and socially stressful conditions associated with the bladder, that of urinary incontinence (UI), is unknown.

UI is defined by the International Continence Society as "involuntary loss of urine sufficient in quantity and frequency to be a social or health issue." Although it may occur in young individuals, *its prevalence increases with age for both men and women, with gender preference for women*, in which UI prevalence is 30–50% after menopause compared to 3–11% in older men. Since UI is considered underreported, these values are low. One reason given for the underreporting is that symptoms are considered mild and simply part of aging. Treatment awareness will prevent worsening.

The types of UI include (i) stress urinary incontinence described as involuntary loss of urine during physical exertion or exercise; (ii) urge incontinence or involuntary loss of urine associated with sudden sense of urgency (sensory or motor dysfunction); (iii) mixed incontinence that combines stress and urge incontinence; and (iv) special forms such as overflow, fistula-dependent or neurogenic incontinence. *Stress incontinence is the most common type of incontinence (48%) in women while urge incontinence is the most common type in men.* Incidence of urge UI and mixed UI are 17 and 34%, respectively.

There are several factors that favor UI. Reduced muscle sphincter tone is responsible for stress incontinence. Structural alterations in the detrusor muscle that increase or decrease bladder contractility, that is, bladder dysfunction, is the basis for urge incontinence.

UI may also occur as a result of changes unrelated to bladder or sphincter function. These include the use of certain drugs, for example, diuretics, anticholinergics, or polypharmacy; restricted mobility; inflammation or infection; delirium; cognitive decline; comorbid diseases or disorders, for example, diabetes, congestive heart failure, degenerative joint disease, sleep apnea, and severe constipation. Treatment of these extraneous conditions alleviates UI.

UI is manageable but management is specific to the type of UI. The first step is a comprehensive physical exam, review of current mediations and fluid intake/micturition diary. Second, several diagnostic tests (from simple to complex) may be necessary to accurately identify the type of UI. In general, as summarized by a recent review (Goepel et al., 2010), UI may be treated with mediations (e.g., antimuscarinic drugs, alpha-receptor blockers, or serotonin/norepinephrine reuptake inhibitors), behavioral retraining (e.g., bladder management, changes in fluid intake), and pelvic floor exercises. The latter is particularly successful in the treatment of stress UI. Medications are beneficial for urge and mixed UI. Consideration of existing comorbidities and current use of anticholinergic drugs is necessary in prescribing antimuscarinic drugs for treatment of urge or mixed UI in the elderly. Drugs with selectivity for the bladder and not the brain or use of drugs other than antimuscarinic agents should be considered to prevent worsening of memory loss in cognitively impaired individuals.

Summary of Age Changes of the GI and Urinary Systems

Gastrointestinal system
  ↓ motility of stomach; ↓HCL secretion
  ↑ colonic transit time
  Loss of diversity of microbiota, partial influenced by diet
  ↓ hepatic P450 enzymes (?)
  ↓ salivary gland secretions (?)

Urinary system
  Deficits in glomerular basement membrane
  ↓ number of nephrons
  ↓ glomerular filtration rate
  ↓ renal blood flow
  Deficits in tubular functions of reabsorption/secretion

SUMMARY

The GI tract and associated organs (liver, pancreas, and salivary glands) are tasked with degrading food to forms useable by the cells of the body. The GI tract also oversees the extensive population of resident bacteria, microbiota, that assist with this function. Reduced colonic transit time due to enteric neurodegeneration is the main age change. This, along with other factors, contributes to chronic constipation, the most common GI condition of the elderly. Although the diversity of the microbiota lessens with age, this appears more related to diet than to age *per se*. Malnutrition is prevalent in the elderly population and fosters sarcopenia and frailty. Reversibility is possible but challenging.

The urinary system comprised of the kidneys, ureters, bladder, and urethra disposes of nitrogenous waste, regulates fluid volume and precious ions, and produces regulators such as renin, erythropoietin, and vitamin D. In the absence of disease, the observed deterioration of nephrons and their disappearance have variable effects on kidney function. A reliable measure of kidney function to replace creatinine measurement is sought.

In the presence of kidney dysfunction, nocturia (frequent voiding at night) and drug toxicity become serious issues. Nocturia is manageable with identification of one of its several causes; drug toxicity decreases with better drug selection and lower doses. The prevalence of urinary incontinence is high in the elderly. The two most common types in the elderly are stress (sphincter dysfunction) and urge (detrusor dysfunction) incontinence. Urinary incontinence is manageable following identification of the type and contributing factors.

CRITICAL THINKING

Is the measurement of GFR with creatinine reliable in the elderly? Support your answer.

Why is it important to accurately measure GFR in the elderly?

Why do you think urinary incontinence is under reported?

Of what value is the microbiota?

KEY TERMS

Angiotensin II a potent vasoconstrictor. It is the final protein in the RAS cascade the occurs following release and activation of renin, an enzyme from the kidney.

Creatinine a metabolic product of phosphorylcreatine in skeletal muscle. Considered a dependable catabolic compound to assess the filtration of blood by the kidney.

GFR glomerular filtration rate is the rate (ml/min) that blood is filtered in the glomerular substructures of the kidney.

Glomerulus a group of capillaries cells, and specialized basement membrane within a capsule (Bowman's) of the kidney. Multiples of hundreds of these substructures exist to filter blood.

Juxtaglomerular cells special cells located in the media of the afferent arteriole of each glomeruli. The cells secrete renin in response to neuronal, pressure, and local mediator effects.

Microbiota the multitude of bacteria that reside symbiotically in the GI tract.

Nephron the basic unit of the kidney of which there are millions. Each contains a glomerulus and a serious of tubules.

Nocturia a condition of excessive voiding at night rather than during the day.

Osmolarity measure of the quantity of solutes (particles, ions) in a specified amount (liter) of solution or plasma.

Proteinuria presence of significant amounts of protein in the urine, suggesting kidney damage and disease.

RAS renin–angiotensin system refers to the system whereby changes in the kidney set off a cascade of enzymatic changes that culminate in the formation of angiotensin II, a potent vasoconstrictor. The outcome is an increase in blood pressure. RAS components exist in many tissues in the body.

Urinary incontinence leakage of urine sufficient to be a health or personal embarrassment issue. Stress and urge are the main types evident in the elderly population.

Vasopressin also called antidiuretic hormone. Secreted by the posterior pituitary, it acts to reabsorb water in the renal collecting ducts to concentrate the urine. Decreased activity of vasopressin results in an increased volume of dilute urine.

BIBLIOGRAPHY

Review

Barrett KE, Barman SM, Biotano S, Brooks HL. 2012. *Ganong's Review of Medical Physiology*. New York: McGraw Medical.

Biagi E, Nylund L, Candela M, Ostan R, Bucci L, Pini E, Nikkïla J, Monti D, Satokari R, Franceschi C, Brigidi P, De Vos W. 2010. Through ageing, and beyond: gut microbiota and inflammatory status in seniors and centenarians. *PLoS One* **5**(5):e10667.

Bolignano D, Mattace-Raso F, Sijbrands EJ, Zoccali C. 2014. The aging kidney revisited: a systematic review. *Ageing Res. Rev.* **14**: 65–80.

Britton E, McLaughlin JT. 2013. Ageing and the gut. *Proc. Nutr. Soc.* **72**(1):173–177.

Cevenini E, Monti D, Franceschi C. 2013. Inflamm-ageing. *Curr. Opin. Clin. Nutr. Metab. Care* **16**(1):14–20.

Conti S, Cassis P, Benigni A. 2012. Aging and the renin–angiotensin system. *Hypertension* **60**(4):878–883.

Dannecker C, Friese K, Stief C, Bauer R. 2010. Urinary incontinence in women: part 1 of a series of articles on incontinence. *Dtsch. Arztebl. Int.* **107**(24):420–426.

Garrett WS, Gordon JI, Glimcher LH. 2010. Homeostasis and inflammation in the intestine. *Cell* **140**(6):859–870.

Gibson W, Wagg A. 2014. New horizons: urinary incontinence in older people. *Age Ageing* **43**(2):157–163.

Goepel M, Kirschner-Hermanns R, Welz-Barth A, Steinwachs KC, Rübben H. 2010. Urinary incontinence in the elderly: part 3 of a series of articles on incontinence. *Dtsch. Arztebl. Int.* **107**(30):531–536.

Grassi M, Petraccia L, Mennuni G, Fontana M, Scarno A, Sabetta S, Fraioli A. 2011. Changes, functional disorders, and diseases in the gastrointestinal tract of elderly. *Nutr. Hosp.* **26**(4):659–668.

Guyton AC. 1961. *Textbook of Medical Physiology*, 2nd ed. Philadelphia: WB Saunders Company, p. 819.

Guyton AC. 1961. *Textbook of Medical Physiology*, 2nd ed. Philadelphia: WB Saunders Company, p. 83.

Hooper L, Bunn D, Jimoh FO, Fairweather-Tait S. 2014. Water-loss dehydration and aging. *Mech. Ageing Dev.* **136–137**: 50–58.

Martone AM, Onder G, Vetrano DL, Ortolani E, Tosato M, Marzetti E, Landi F. 2013. Anorexia of aging: a modifiable risk factor for frailty. *Nutrients* **5**(10):4126–4133.

Nitti VW. 2001. The prevalence of urinary incontinence. *Rev. Urol.* **3** (Suppl. 1):S2–S6.

Putignani L, Del Chierico F, Petrucca A, Vernocchi P, Dallapiccola B. 2014. The human gut microbiota: a dynamic interplay with the host from birth to senescence settled during childhood. *Pediatr. Res.* **76**(1):2–10.

Salles N. 2007. Basic mechanisms of the aging gastrointestinal tract. *Dig. Dis.* **25**(2):112–117.

Schmucker DL. 2005. Age-related changes in liver structure and function: implications for disease? *Exp. Gerontol.* **40** (8–9):650–659.

Siroky MB. 2004. The aging bladder. *Rev. Urol.* **6** (Suppl. 1):S3–S7.

Wiggins JE. 2012. Aging in the glomerulus. *J. Gerontol. A Biol. Sci. Med. Sci.* **67**(12):1358–1364.

Yu SW, Rao SS. 2014. Anorectal physiology and pathophysiology in the elderly. *Clin. Geriatr. Med.* **30**(1):95–106.

Experimental

Claesson MJ, Jeffery IB, Conde S, Power SE, O'Connor EM, Cusack S, Harris HM, Coakley M, Lakshminarayanan B, O'Sullivan O, Fitzgerald GF, Deane J, O'Connor M, Harnedy N, O'Connor K, O'Mahony D, van Sinderen D, Wallace M, Brennan L, Stanton C, Marchesi JR, Fitzgerald AP, Shanahan F, Hill C, Ross RP, O'Toole PW. 2012. Gut microbiota composition correlates with diet and health in the elderly. *Nature* **488**(7410):178–184.

Mansfield KJ, Liu L, Mitchelson FJ, Moore KH, Millard RJ, Burcher E. 2005. Muscarinic receptor subtypes in human bladder detrusor and mucosa, studied by radioligand binding and quantitative competitive RT-PCR: changes in ageing. *Br. J. Pharmacol.* **144**(8):1089–1099.

SECTION V

REGULATORY ORGAN SYSTEMS: CENTRAL NERVOUS SYSTEM, SENSORY, ENDOCRINE, AND IMMUNE SYSTEMS

INTRODUCTION

Regulatory organ systems are those that exert widespread influence over all other organ systems in an organism and additionally coordinate successful responses to both external (environmental) and internal stresses. The central nervous system (CNS) and sensory systems process multiple sensory modalities that assist countless cognitive activities of humans and manage movement and balance. The endocrine system with its diversity of hormones ensures a constant milieu for cells and tissues. Its homeostatic expertise guarantees normal levels of essential ions, nutients, oxygen, and mediators, to name a few. It is master of growth and reproduction. As watchdog over all organ systems, the immune system produces an unlimited variety of responses that thwart damage from pathogens, toxins, and cell mutations. Undoubtedly, age-associated deterioration within these systems produces profound losses that affect quality of life issues, and increases liability to accidents, disease, and death. Major age-related changes in these systems are discussed in this section.

It is noted that present-day neuroimaging techniques have figured prominently in unraveling some aspects of the aging brain and additionally have created new avenues of research. Progress has been achieved in distinguishing age changes from those wrought by disease. Small advances have also been achieved in the use of brain exercise training to minimize some cognitive loss, for example, visual processing speed. Other improvements relate to awareness (and removal) of confounding factors in CNS and immunological studies. As data are lacking, debate continues on the use of hormone replacement therapy for the numerous age-associated hormonal deficits. Since poorly vetted antiaging hormones abound, cllinical trials with select hormones are sorely needed.

12

AGING OF THE CENTRAL NERVOUS SYSTEM

OVERVIEW

The Central Nervous System (CNS), Responsible for Cognitive Behaviors, Motor Coordination, and Regulatory Activities, Consists of the Brain and the Spinal Cord

The brain receives sensory information from the environment and internal organs through specialized cells called neurons (see Figure 12.1). Neurons sort, process, and direct responses or outputs in the form of cognitive behaviors, for example, memory, reasoning, decision making, emotions, creativity; skeletal motor activity, for example, movement, balance, coordination; and oversight of internal hormonal and system functions, for example, wakefulness, sleep, satiety, thirst, heart rate, blood pressure, respiration, and temperature.

Many Issues Hamper Progress in Understanding CNS Aging Unlike most other organ systems, the brain remains an enigma. The lack of knowledge of basic neuronal circuitry poses a challenge to scientists of all disciplines and confounds studies on complex behaviors such as creativity and reasoning. Second, specific to aging, there is no consensus on exclusion criteria for CNS studies in the elderly. Few studies reject those with covert disease for example, pathology discovered with neuroimaging scans and sadly many studies even fail to bar subjects with established risk factors for disease, such as hypertension. Cognitive function tests also harbor serious problems such as the practice effect, absence of baseline data prior to aging measurements, and failure to equalize technological and educational levels among cohorts of different ages.

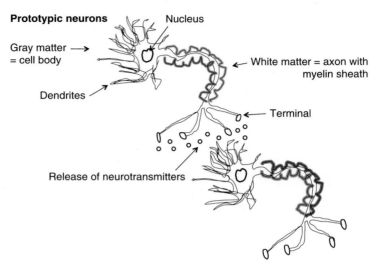

Figure 12.1. Schematic of two communicating neurons illustrating dendrites, cell body, and axon with terminals; synapse is the space between the terminus of one neuron and receptors on dendrites of adjacent neuron.

Although a boon to this field, neuroimaging techniques (discussed next) are not without limitations. Brain function is measured indirectly and interpretation requires expertise that is difficult to critique. These and other issues are enumerated in Table 12.1.

Neuroimaging Technology Stirs Debate on the Aging Brain A number of imaging tools are now available to observe the human brain at work and assess changes over time. The available instruments include the magnetic resonance imaging (MRI), functional MRI (fMRI), positron emission tomography (PET), and diffusion tensor imaging (DTI). Another technique, computerized axial tomography (CAT), also provides quality brain scans but its use of a large number of X-rays limits its utility in healthy volunteers. Neuroimaging is far superior to postmortem brain analyses that are fraught with artifacts due to unavoidable random tissue swelling and shrinkage.

MRI scans provide baseline information on brain structure and in this respect has been valuable in the assessment of age-associated changes in brain volume. *Brain functionality is measured with the fMRI or PET neuroimaging.* FMRI measures aspects of blood flow that are considered representative of neuronal activity and PET uses radiolabeled tracers to characterize metabolic changes or neurotransmitter release indicative of function. Both techniques reveal age-related physiological changes in response to specific cognitive tasks. Whether neuronal conductivity (connectedness or circuitry among neurons) deteriorates with age is probed with the DTI. DTI detects components of water motion in neurons (and surrounding myelin) and interruptions in motion are construed as impedance to nerve impulse traffic. Although scan interpretation is intensely debated, neuroimaging has produced fascinating hypotheses on brain aging, providing a foundation for future investigations. More information on neuroimaging is found in Box 12.1.

TABLE 12.1. Challenges to Investigations on CNS Aging

Brain physiology
- Basic understanding of neuronal connections, functions, repair, and maintenance minimally understood in animals and scarcely at all in man

Study design
- No consensus on exclusion criteria; no cause–effect data in man; paucity of study subjects; many studies with cross-sectional design (see Chapter 2)

Covert disease (detected in neuroimaging scans)
- Significance not clear, but possible early cardiovascular disease; presence confounds age change

Cognitive function tests
- Factors that *temporarily* decrease cognitive function need identification; level of test difficulty must be comparable among age groups and the practice effect minimized; baseline measurements in adulthood generally lacking

Paucity of relevant animal models
- Majority of studies with rodents; constraints with use of nonhuman primates; conserved functions need validation

Imaging limitations
- Interpretation of neuroimaging scans requires exceptional technical expertise; many assumptions inherent in methods, for example, ability of instruments to detect structure/function is independent of subject's age

Variability; heterogeneity
- Individuals exhibit loss of cognitive function at different rates; factors such as genetics, environment, and lifestyle choices play a complex role

---

Box 12.1. More About Neuroimaging Techniques

The MRI determines brain structure and allows for measurement of brain volume. This unique noninvasive technology employs magnetic forces and radio frequency pulses to briefly vibrate subatomic particles. The particle vibrations create rapid "signature" changes over time that are complied and viewed as multiple internal views of the brain. Scans are interpreted in relation to known anatomical markers.

Whereas the MRI determines anatomical alterations, fMRI noninvasively measures functional or physiological events in the brain. The technology relies on the physiological increase in neuronal metabolism and subsequent increased blood flow that reproducibly occurs on activation of neuronal pathways or circuits. FMRI detects the hemoglobin of the red blood cell after it has unloaded its reserve of oxygen. This indirect signal of neuronal activation is termed the blood oxygen level dependent (BOLD) signal.

FMRI has been successful in identifying regions of the brain responsive to specific cognitive, sensory, and motor tasks. Brain mapping data has been obtained from both cross-sectional and longitudinal study designs. However, overinterpretation of fMRI data is cautioned due to (i) the indirect nature of the measurement, (ii) the rapidity of the measured response exceeds the physiological response and thus narrowly evaluates brain function, (iii) the fMRI measured response is correlatively and not causally related to the task, (iv) the complexity of the method generates numerous analytically issues, and (v) age changes in cerebral blood flow may alter the BOLD signal in potentially complex and unknown ways.

PET imaging combines the use of a radioactive tracer with multiple X-rays to produce images of various organs. The PET scanner detects the radioactive decay given off by the tracer. Frequently, the radioactive analogue of glucose, for example, fluorodeoxyglucose, is injected to detect regions that are metabolically active (and presumably neuronally active) and hence utilizing sugar. Radioactive analogues of neurotransmitters are also used to assess brain function. To anatomically identify areas of metabolism or neurotransmitter activity, MRI scans are generally obtained simultaneously with PET scan.

DTI or diffusion tensor imaging is a technique that detects thermal motion of the water found in neurons. Deviations in water flow permit sensitive detection of axon substructure within the white matter (axons). DTI scans are employed to detect connectivity between brain regions. Age changes in connectivity have been interpreted as relating in part to deficiencies in the myelin. Similar to PET and fMRI, anatomical maps generated by MRI are obtained simultaneously with DTI scans.

AGING OF THE BRAIN

The CNS Is Comprised of Billions of Neurons and Assisted by Several Other Cell Types

Neurons and supporting cells microglia, astrocytes, oligodendrocytes, and Schwann are the resident cells of the brain. The *prototypic neuron has a cell body (containing the nucleus), dendrites, or extensions that receive information and an axon that relays information to the adjacent neuron* (Figure 12.1). *Neurons send information through a nerve impulse* generated by movement of ions (sodium, potassium, and calcium). Nerve impulses pass from dendrites to the axon and onto the neighboring neuron(s). Many *axons are encircled with a fatty sheath (myelin) that accelerates transmission of nerve impulses along its length.* Myelin is produced and maintained by Schwann cells and oligodendrocytes. The microglia handle infections and injury repair; astrocytes fortify the brain capillaries to create the blood–brain barrier.

The CNS collaborates with another system, the peripheral nervous system (PNS), which includes (i) sensory neurons (afferent neurons) of pain, light, temperature, sound, and vibration, which directly sense the environmental and relay information to the brain (discussed in Chapter 13), (ii) motor or somatic neurons (efferent neurons) that send information from the brain and spinal cord to the skeletal muscle (discussed in Chapter 7), and (iii) autonomic nerves (efferent neurons) that send information from the brain and spinal cord to the internal organs such as blood vessels, heart, airways, and gastrointestinal tract (discussed in Chapters 9, 10 and 11). Figure 12.2 depicts the relationship between the CNS and the PNS.

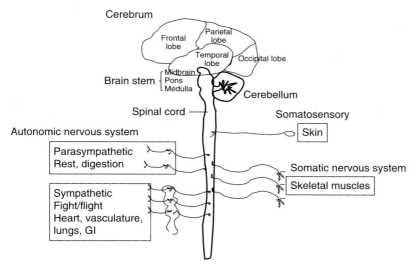

Figure 12.2. Overview of the central nervous system and the peripheral nervous system (somatic nerves; autonomic nervous system).

Structural Changes

The Major Anatomical Divisions of the Brain Are the Cerebrum, Cerebellum, and the Brain Stem
The cerebrum is divided into two hemispheres connected by nerves collectively designated the corpus callosum. The uniquely undulating surface of the cerebrum creates convolutions (gyri), small indentations (sulci), or large indentations (fissures). The *cerebrum is subdivided into distinct regions or lobes*: the *frontal lobe, the parietal lobe* (behind the frontal lobe), the *temporal lobe* (below the frontal and parietal lobe on both sides), and the *occipital lobe* (posterior portion of cerebrum) (see Figure 12.2). Each lobe undertakes specific functions although reciprocal interactions between the lobes are evident and essential. In brief and concluded largely from studies of brain lesions arising from surgery or stroke, the frontal lobe is the locus of intelligence, reasoning, creativity, and motor function. The parietal lobe is implicated in speech interpretation. The temporal lobe is involved with hearing and the occipital lobe processes visual input. Located within the temporal lobes are the hippocampus and entorhinal cortex, discrete regions central to formation, processing, and maintenance of memory. Three cavities termed ventricles containing spinal fluid are located within the cerebrum; one cavity is found in the brain stem (Figure 12.3). More information on the CNS is found in Box 12.2.

Age-Associated Brain Shrinkage Remains Controversial
Some of the first studies on the aging brain dealt with measurement of postmortem specimens of brains from individuals of different ages. The observed differential tissue shrinkage led to the conclusion that *loss of brain mass (neurons) was time*

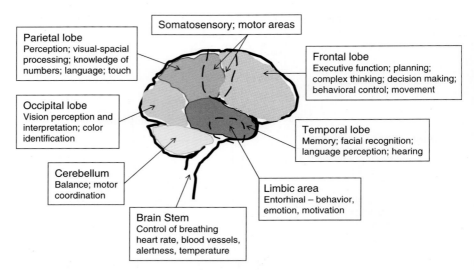

Figure 12.3. Regions of brain associated with different cognitive behaviors and motor and sensory processing.

> **Box 12.2. More About the CNS**
> The brain and spinal cord are surrounded by fluid (cerebral spinal fluid), membranes (meninges), and bone (cranium and vertebra). Under normal conditions, the brain receives approximately 20% of the cardiac output (blood pumped by the heart).
>
> The brain stem contains several subdivisions to include the thalamus, a relay center of sensory impulses to cerebrum, and the hypothalamus, locus of control neurons that regulate pituitary (hormone) function, body temperature, water balance, weight, and hunger (see Chapter 14 Aging of the Neuroendocrine System). Below these structures are the midbrain, pons, and medulla oblongata, which are involved in mediating reflexes, regulation of cardiovascular function, and respiration and relay of information upward to the thalamus and hypothalamus. An important nexus of neurons called the limbic system, located around the upper brain stem, regulates emotions, the reward system, and indirectly personality. The prevalence of white matter hyperintensities (WMHs) on MRI scans of the elderly is higher than that of younger individuals. WMHs are associated with a rise in risk of vascular disease and a decline in cognitive function. Located primarily in the regions around the ventricles and in frontal regions, WMHs are *strongly associated* with age, but their relationship to normal aging is not clear. WMHs are the products of chronic ischemia that kill brain cells and/or damage the myelin sheath. Hypoxia or toxicity from homocysteine or amyloid protein products are possible initiators of cell damage and death. Changes in regional cerebral blood flow (rCBF) are also associated with increases in WMHs. In a longitudinal study of 8 years, elderly individuals whose MRI showed an increasing number of WMIs also showed measureable changes in rCBF (PET scans). However, the relation between changes in rCBF and the onset of WMIs remains unclear.

dependent and cognitive deficits would follow. Results from MRI studies have moderated this view and hold promise to explain the presently unknown mechanism(s) for the age-associated changes in brain mass and their relation to cognitive function.

An extensive analysis by Hedman et al. (2012) of data from 56 longitudinal MRI studies of brain volume of healthy individuals, 4–88 years of age (combined subject number >2000), concluded that *brain volume changes significantly over the lifespan*. Brain growth occurs during childhood and adolescent phases (~1%/year) up to about 13 years of age. Between 18 and 35 years of age, additional growth is possible. *After age 35, brain volume begins to decline at a rate of 0.2% per year with additional loss after age 60 (0.5% per year)*. Individual cross-sectional and longitudinal MRI studies describe the brain volume diminution as attributable to cortex thinning (possibly due to reduced brain convolutions and furrows), subcortical size reductions, and purportedly associated ventriculomegaly (enlargement of the ventricles). The underlying assumption is that neurons are disappearing.

Differential regional shrinkage as well as individual variations in shrinkage rates are evident. Brain regions *vulnerable* to shrinkage include the hippocampus, entorhinal cortex (medial temporal lobe), orbital-frontal cortex, and cerebellum. Other regions of the caudate nucleus (part of limbic region), prefrontal subcortical white matter, and the corpus callosum also show signs of volume loss but at later times. Areas of the lateral prefrontal and primary visual cortices (outer regions), putamen (located at the base of cerebrum), and pons (brain stem) maintain their volume or appear to be more resistant to volume loss.

An unsolved mystery is why brain aging in healthy cognitively competent individuals tends to overlap with that of Alzheimer's disease (AD) pathology. The common thread is the observation that the entorhinal cortex, the so-called AD-prone region that shows marked atrophy in AD (Fjell et al., 2014), also thins with age in individuals without dementia or AD. Why some individuals transition from normal aging to AD pathology is unknown and is under intense study.

The above findings are tempered with *other studies that show no change in brain volume with age*. In studies that *exclude* individuals with *covert brain disease* or with *cardiovascular risk factors, brain structure changes little or not at all*. In particular, the rejection of MRI scans from individuals who developed dementia 6 years after an MRI study, Burgmans et al. (2009) demonstrated that gray matter volume (neuron cell body) in seven brain regions (including regions of hippocampus and prefrontal cortex) do not change with age in the remaining healthy subjects (52–82 years old). Whereas most studies exclude individuals with neurological and psychiatric disorders, *few* studies, as the one just described, reassess subjects after study completion to detect onset of late cognitive decline and thus confirm the prior existence of covert brain disease. Other studies fail to eliminate subjects with risk factors for cardiovascular disease, prime initiator of covert brain damage and firmly associated with brain shrinkage.

Serotonin and Dopamine Pathways Experience Reduced Activity

Neurons are separated from one another by an intervening space called the synaptic cleft or synapse. Since the nerve impulse cannot bridge the cleft, small molecules termed neurotransmitters are released from each terminus, diffuse across the synapse,

TABLE 12.2. Neurotransmitters Involved in Brain Aging

Acetylcholine (ACh)
- Memory and executive function
- Deficit in AD producing severe memory loss

Dopamine (DA)
- Motor coordination; reward-related adaptability; episodic memory; age-related cognitive loss associated with reduced receptor binding and altered dopamine biotransformation
- Deficit in Parkinson's Disease producing severe motor disorder

Serotonin (5-HT)
- Reward-related adaptability; episodic memory; cognitive loss related to altered receptor binding

Gamma-aminobutyric acid (GABA)
- Visuospatial performance; implicated in neuroplasticity in animals; loss contributes to cognitive deficiencies in animals
- Deficit in Huntington's disease producing rapid, uncoordinated movements

Glutamic acid (glutamate)
- Spatial memory; age-related alteration in rodents

Norepinephrine (NE)
- Alertness and cortical function?
- Decrease in depression?

Modulators (neuropeptides; nitric oxide; ATP and adenosine; brain-derived neurotropic factor (BDNF))
- Involved in neuroplasticity, neurogenesis, and vasculogenesis in animal models

and stimulate receptors on neighboring dendrites, thereby maintaining the flow of information by the generation of a nerve impulse in the adjoining neuron. Abnormalities at the synapse produce dysfunction in nerve transmission. Table 12.2 describes some of the known neurotransmitters and assisting compounds termed neuromodulators in the CNS. With the exception of dopamine, serotonin, and acetylcholine, few have been studied with regard to a possible role in age-related changes in brain function in man. Neurons producing dopamine (termed dopaminergic) and acetylcholine (termed cholinergic) figure prominently in Parkinson's disease (PD) and AD, respectively. In these diseases, level of the neurotransmitters declines as the neurons disappear.

Age-related changes in dopaminergic (dopamine) and serotonergic neurons (serotonin) have received the most attention and in general *functional contribution by these two systems declines with age*. It is noted that all studies thus far (except for genotyping) evaluated a limited number of participants (less than 100) and used the cross-sectional study design.

Dopaminergic neurons are best known for their involvement in PD, *the most common movement disorder of the elderly*. Disease progression leads to *abnormal and disabling skeletal muscle activity* characterized as shaking tremors, muscle rigidity, bradykinesia (slowed gait) with difficulty initiating and completing movement, reduced facial expressiveness, and slowed and monotonal speech.

In addition to skeletal muscle control and coordination, *dopamine additionally (in conjunction with serotonin) modulates reward-based learning and decision making*. Results of cognitive testing indicate that the elderly experience reduced adaptive behavior in which they are unable to modify their future goal-oriented actions when faced with partially unreliable or probabilistic conditions and rewards. This "learning reluctance" has been related to various decrements in the dopaminergic and serotonergic systems. Specifically, results of PET scans using radiolabeled selective neurotransmitter ligands (substitute compounds for the endogenous neurotransmitter) *reveal that dopamine and serotonin receptor binding decrease as much as 10–11% per decade beginning from early to late adulthood*. The decreased availability of dopamine and serotonin receptors correlates with a reduction in cognitive performance. *Low receptor number predicts poor cognitive behavior.* Receptor deficits appear in relevant cortical–striatal–midbrain loops that mediate reward-based behavior. More on the dopaminergic neurons is found in Box 12.3.

A role for cholinergic (acetylcholine, ACh) neurons in memory has been suspected for some time and reinforced with the loss of cholinergic neurons in memory-robbing AD and the ameliorating effect of anti-cholinesterase inhibitors, for example, donepezil, that block the degradation of ACh in AD therapy. Additionally, a cholinergic hypothesis of geriatric memory dysfunction has existed for some time (Bartus et al., 1982), albeit with limited support. With the advent of imaging techniques, renewed interest in the cholinergic hypothesis has emerged. However, aside from an involvement in memory, the newer view, as reviewed by Dumas and Newhouse (2011), concludes that cholinergic neurons are essential to tasks necessitating serious attention (either difficult tasks or tasks of sorting relevant from irrelevant information). This, coupled with observations that pharmacological manipulation of the cholinergic system produces activation (fMRI scans) in the same regions (prefrontal cortex) triggered by specific attentional memory tasks, supports the cholinergic functional compensation model of cognitive dysfunction. This *hypothesis proposes that prefrontal activation in the elderly is associated with better*

Box 12.3. More About Aging of Dopamine
Specific genotypes have been correlated with enhanced cognitive difficulties relative to reward-based adaptability. In particular, *genetic variants* (polymorphisms) of genes that influence dopamine functions, for example, transport uptake protein, dopamine degradative enzyme called catecholamine-*o*-methyltransferase, receptors, and auxiliary proteins, are associated with changes in measurable components of reward-related behavior, episodic memory, and brain glucose metabolism. Variants that reduce dopamine function magnify age changes in these cognitive areas. Finally, in healthy elderly subjects, low midbrain dopamine synthesis (radiolabeled dopamine precursor and PET analysis) is associated with a brain pattern (measured fMRI in the prefrontal lobe during reward-related processing) *opposite* to that observed in young subjects. Interestingly, administration of levodopa, the precursor to dopamine, to elevate the level of brain dopamine, improves task-based learning rate and task performance in the older participants only. These are provocative findings that require further investigation.

attentional performance because of a requisite activation of cholinergic neurons. Those that cannot activate this pathway, or do so weakly, perform poorly on attentional tasks.

Connectivity between Neurons May Decrease with Age
The gray matter (neuronal cell bodies) forms the outer layer of the brain and of some regions of the spinal cord. In contrast, white matter contains neuronal axons; the whiteness is due to the presence of myelin. Myelin is susceptible to damage by inflammatory, infectious, and immunological processes and is a likely target of oxidative events, abundant in aging.

DTI measures changes in connectivity of neuronal circuits and has potential to identify reduced communications between brain regions. Cross-sectional studies probing 20–100 subjects consistently report *DTI changes indicative of reduced connectivity with age in white matter fibers in anterior brain regions*. Specifically, connectivity deficits occurs in (i) fibers of the corpus callosum, presumed essential in working memory (WM), (ii) fibers participating in spatial memory, and (iii) fibers connecting the limbic system (emotion/motivation center) with the frontal lobe. These changes are correlated with functional changes in problem solving and WM. Furthermore, reduced white matter connectivity in specific anatomical tracks projecting from the medial lobe has been correlated with decreased performance on executive function, processing speed, and memory. Others have found a correlation between changes in connectivity in the prefrontal region with perceptual speed in memory retrieval in healthy elderly.

A rare longitudinal study using DTI by Barrick et al. (2010) showed that a minimum of 2 years is sufficient to detect changes in white matter integrity (using three different methods of DTI analysis) in the group of 50- and 90-years old subjects. Significantly, the *extent of connectivity decline exceeds that estimated from cross-sectional studies*, and *affects all brain pathways not just frontal or temporal lobe circuitry*. These findings emphasize a serious age-related detriment in impulse trafficking.

Auxiliary Cells Senesce and Fail to Maintain Homeostasis
The microglial cell, an important auxiliary brain cell, is a type of macrophage, with innate and adaptive immunity capabilities (see Chapter 15). Microglia protect the brain in several ways: handle microbial invasions, clean up toxic events, and monitor neuronal health. Microglia are masters of neuronal maintenance. Their functions include (i) phagocytosis of apoptotic neuronal remnants, pathogens, and abnormal proteins, and (ii) modulation of neuronal circuitry by trimming collaterals and synapses.

The *homeostatic function of the microglia wanes with age*. This has been observed primarily in animal models and in AD. Based on morphological changes, it is postulated that microglia eventually senesce and thereby express a new phenotype (SASP) consisting of reduced surveillance, slowed phagocytosis or clearance, and production of proinflammatory mediators, changes that promote neurotoxicity and a vulnerability to disease, that is, dementias. Initiators of microglia senescence are likely oxidative events. This is based on findings of aged microglia with dramatic telomere shortening, DNA damage, and autophagy deficits.

FUNCTIONAL CHANGES

Cognitive tests seek to define the loss of brain function To assess cognitive competency, a plethora of tests have been developed. Many of these tests are described in Table 12.3 appropriate for age-related domains of interest and are some of those proposed by Alexander et al. (2012) as part of "a set of measures to characterize cognitive profiles." According to Alexander et al. (2012), a proposed list of *validated* tests should facilitate a collection of reliable data for comparisons across the lifespan, among and within different aging populations, between normal and pathological aging and where conserved functions exist, between man and animal models (monkey, rat, and mouse).

Some tests in Table 12.3 are subcategories found within more comprehensive and established tests such as the Wechsler Adult Intelligence Scale (WAIS), known as an IQ test and now in its fourth edition and the Rey Auditory Verbal Learning test (RAVLT), a popular verbal learning and memory test. One test absent from

TABLE 12.3. Cognitive Tests[a]

Encoding/retrieval of verbal memory
- number of learned words recalled with and without delay (Rey Auditory Verbal learning test); number of ideas recalled from auditory stories (subtest from Wechsler Memory Scale (WMS))

Encoding/retrieval of visual spatial memory
- recall of series of faces or geometric figures (asked to reproduce) (WMS-III); recall of series of colored doors; copy and later recall from memory a complicated geometric figure

Spatial navigation
- no standard test—physical or computerized tests using virtual environments to find hidden objects

Associative memory—verbal; visual
- early and delayed recall of learned word pairs; recall test includes novel pairs (WMS-III); computer program to locate position of pattern contained in a box; face-name associations.

Source memory
- not standardized—recall lists of words/sentences/objects in relation to the context or sentence-voice recall of which one of two speakers previously read multiple sentences

Prospective memory
- not standardized—engage in primary task that is interrupted by another "must do or should do" task such as press a key if a certain word appears on screen during primary task

Processing speed
- timed test with low cognitive demand is the Coding and Symbol Search of the Wechsler Adult Intelligence Scales-IV. Test for number of correct symbols drawn for digits from code list of digit symbols
- others: (i) finger tapping (taps/10 s) as measure of simple motor processing speed; (ii) press button within 1–3 s interval when zero appears on computer screen or press one of four keys to match numbers 1–4 on screen

(continued)

TABLE 12.3. (*Continued*)

Language
- vocabulary rules for words and sounds/symbols
- vocabulary comprehension and conversational discourse
- confrontation naming (Boston naming test—60 line items to be named; vocabulary knowledge test; verbal fluency test)

Visuospatial memory includes appreciating, integrating, and using objects in space context; complex
- Rey-Osterrieth complex figures test—copy complex geometric design, immediate recall, delay recall; complete fragmented picture as it is being presented

Executive function
- task or set shifting; Wisconsin Card Sorting Test—Test of global shift is a plus–minus task of three lists of 30, two-digit numbers; one must add 3 to digits in list1, subtract 3 from those in list 2 and alternate (add and subtract) from those in list 3. No cues, for example, + or − signs. Variations: A number–letter set appears in one of four quadrants, if set appears in top two quadrants, one determines if number is odd or even, if set appears in bottom two quadrants, one determines if letter is a consonant or vowel
- updating and monitoring: Updating tests include consonant updating in which letters appear on a screen at 2 s intervals and one repeats the last four letters with each update and. At end, recalls last four consonants
Classic working memory test is the operation span in which one verifies a math solution and at the same time retains a word in memory, repeated for 2–5 problem-word pairs
- inhibition of prepotent responses: (a) inhibition of dominate factor/distracters, (b) resistance to proactive interference (major issue in WM tasks); Tests of inhibition; Stroop task—color–word task in which color of incongruent word is named compared to naming color of something neutral (symbol); pressing right key for right pointing arrow on right side of screen and left key for left pointing on left screen; incongruent is left pointing arrow on right screen

[a]Summarized from Alexander et al. (2012).

Table 12.3 is the Mini Mental State Exam (MMSE). Although used extensively, its purpose is to screen for mild cognitive impairment, a risk factor for dementia. The MMSE test measures a range of cognitive behaviors that include orientation memory (date and location), attention (following directions), skill application (math subtraction), verbal recall (word list immediate and delayed), language (writing, naming), and motor/spatial skills (drawing object). It can be found at www.health.gov.

Cognitive Behavior Exhibits Selective Decline with Age Substantial interest exists in understanding age-associated cognitive behavior changes in man. Specifically, biogerontologists seek to know what domains are vulnerable to aging, the time of onset of deteriorative changes, the possibility of improvement with cognitive training and other interventions, and the mechanisms that deter pathological aging. At present, the field offers only a few reliable observations.

Using a multiplicity of neuropsychological tests (as noted in Table 12.3), results of cross-sectional and longitudinal studies have shown that (i) cognitive skills of information processing speed, executive function (as in WM), and memory, including specialties of encoding and retrieval processes (episodic), associative processes, source finding, and prospective retention, decline with age, as well as reward-based behavior noted above, and (ii) functions such as semantic memory (facts and knowledge), most aspects of language, autobiographical memory (personal history of self), emotional processing, and automatic memory processes remain fairly stable throughout life. *It is concluded that the aging brain experiences a selective decline in some cognitive domains (episodic and WM) but stability in others (semantic, autobiographical, and automatic skills).*

A *decrease in information processing speed is a consistent and prominent age-associated change* and is considered *the major factor* that influences performance by the elderly on tests of other cognitive activities. As is clear from test results, reduced information processing speed negatively impacts visual and motor responses and causes an elevated risk of errors, for example, driving and decision making. *This plus other mental deficits,* for example, *in executive function, has been identified as contributing to a rise in the incidence of falls with age.*

Some elderly experience a *reduction in one or more aspects of executive function*. Neuropsychological tests of this function subcategorize executive function into (i) task (set)-shifting, (ii) updating and monitoring, and (iii) blocking/disengaging from inhibitory (disruptive) interference. The *ability to shift from one task to another or in general not to persevere in a fruitless action tends to decline with age*. Generally, low performance on these tests correlates with poor performance of instrumental activities of daily living (IADLs, for example, money management and essential housekeeping). Updating, monitoring, and blocking/disengaging are significant components of WM. *According to cognitive testing, these aspects of WM decline with age.*

Similar to evaluations of executive function, specific types of memory have been probed with a battery of neuropsychological tests. *Poor recall, especially of words, is perhaps the most frequent subjective complaint of the elderly that relates in part to a deficiency in encoding and retrieval of memory*. Inability to establish new episodic memories (events and places around an individual) is a related outcome. Age-associated deficits in associative, source, and prospective memory have been reported and impact memory clarity and accuracy (associative memory), ability to remember the origin of an event (source memory), and recall of "to be done" future events (prospective memory). *Spatial and navigational memory may also diminish with age* and contribute to reduced ability to find ones way geographically or mentally on a complicated trail.

While age-dependent changes in cognition are highly variable and *not obligatory*, they may explain some additional observations in the elderly such as (i) reduced memory in the presence of distracters or during multitasking (reduced inhibitory processing in the presence of high memory load), (ii) decreases in flexibility to shift strategies in a particular task (task-shifting process), and (iii) story repetition (reduced destination recognition).

Semantic memory, autobiographical memory, and automatic skill memory appear to sidestep the effects of aging. Controversy swirls around which components of language (vocabulary, grammar, and pronunciation/symbols) might decline with age. Language comprehension is a complex function that is affected by changes in processing speed and WM capacity especially challenged during rapid speech and in the presence of hearing loss. Although older individuals may not vary their selection of words, use simple sentence constructs, and pause frequently, these behaviors are not necessarily due to age-related deficits in vocabulary and grammar application but more likely a result of reduced WM and information processing speed.

There Is No Consensus On Time of Onset of Cognitive Decline
Although the onset of cognitive decline is debated, few would disagree as to the importance of pinpointing its commencement. An exact start point would enable researchers to uncover causal effects and additionally ensure optimal success of interventions. According to Finch (2009), maximal cognitive function peaks in the early twenties and begins to decline after age 26 as noted in a study that repeatedly and comprehensively tested 1000 subjects 2–95 years of age. Results of most cross-sectional studies agree and show that age-related decline in memory, spatial visualization, executive function, and speed of information processing begins at about 20–30 years of age. However, *results of several longitudinal studies indicate a delay in cognitive decline that does not occur until midlife or about the sixth decade*. The discrepancy between onset of cognitive decline has been attributed to inherent bias in the study designs, generally the cohort effect in the case of the cross-sectional study design and the practice effect in the longitudinal study design. To add to the complexity, a large study on the practice effect that involved assessment of cognitive abilities (reason, speed, memory, vocabulary, and space) of >1500 subjects, 18–80 years of age over a 2.5-year period, observed that the practice effect "varied in magnitude across neurocognitive abilities and as a function of age" and in adjusting for this effect, longitudinal changes in cognitive abilities are less positive in the young and less negative in the older subjects (Salthouse, 2010). The controversy remains unresolved. However, removal of a cohort effect, for example, education from cross-sectional studies and the practice effect from longitudinal studies, favors a later onset of cognitive decline, closer to 60 years of age.

There exist *conditions of cognitive decline that are temporary and hence potentially reversible*. Their existence is highlighted by the results of a 6-year longitudinal study of >1800 subjects (55–85 years of age) in which a small percentage (~18%) of subjects exhibited cognitive decline (assessed by the MMSE). Of note, half of these individuals regained normal cognition on retest after the study closed. A *transient reduction in cognitive performance may be caused by social stress, anxiety, head trauma, a recent fall, severe bacterial infection, polypharmacy (multiple drug use), or an endocrine imbalance*, for example, hypothyroidism. On the other hand, irreversible cognitive decline is statistically associated with persistent memory complaints, poor memory performance, and progression of cardiovascular disease.

Brain Patterns from Neuroimaging Reveal Differences in Elderly Compared to Young Subjects
FMRI and PET neuroimaging produce brain images that purport to capture neurophysiological activity during cognitive testing. Animated neurons (neuron circuitry) are detected indirectly by an increase in cerebral blood flow (BOLD signal) or by presence of radioactive metabolic tracer. The association of a particular "activated" brain region with performance on a cognitive test allows a *correlation* to be established between *in vivo* brain activity and cognitive behavior.

In general, fMRI brain scans of elderly engaged in objective tests of memory display *activation patterns that differ from younger cohorts (cross-sectional) or change over time (longitudinal) irrespective of the performance level*. In the elderly, the regional location of the brain activation and the degree of primary and secondary activation differ from that of young individuals. For example, during tests for episodic memory and WM, the elderly show a *decrease in brain activity in the medial temporal lobe* (parahippocampal region) and an enhanced or *overactivity in the prefrontal cortex* (ventral, middle, and dorsal prefrontal cortex) and parietal regions (depending on the task) compared to the response of young subjects. Additionally, *use of both hemispheres (bilateral activation) is prevalent in fMRI scans of the elderly* compared to the unilateral activation characteristic of fMRI scans obtained from young subjects.

A consistent fMRI finding is prefrontal overactivation in the elderly. This is gleaned from results of cross-sectional studies and a few longitudinal studies. In cases where *cognitive performance in the elderly is preserved (i.e., same as in young or no decline over time), the novel brain scans have been attributed to age-related compensation*, also called *cognitive reserve*. It is proposed that the cognitive success of so-called high-performing elderly results from a *reorganization of the brain circuitry*. Recruitment of other neuronal pathways (hence overactivation) yields a better performance outcome. In the absence of compensation, reduced performance is predicted. It is speculated that compensation arises from a lifelong maturational accumulation of experiences that beneficially "rewires" the brain.

Cognitive reserve, a concept proposed by Stern (2012), allows for structural changes that include brain volume reorganization in addition to circuitry remodeling. It is proposed that these changes together preserve cognition in the face of pathology or excessive cognitive load. It accounts for the ability of some individuals to maintain cognitive skills in the face of age changes and significant pathology (amyloid plaques, WMH, Lewy bodies). Thus, *cognitive reserve minimizes cognitive behavioral loss accelerated by aging and disease*. Cognitive reserve embraces brain plasticity or malleability of brain structure/function through stimulating interventions (see plasticity below).

The apparent failure of compensation to aid cognitive performance in some elderly, for example, those with prefrontal overactivation associated with poor performance scores, poses the need for a different explanation. *One possibility is dedifferentiation, a condition that renders the brain less able to select appropriate regions for activation and/or to selectively inhibit unnecessary neuronal pathways*. Dedifferentiation is illustrated in studies that compare a battery of test outcomes in relation to fMRI scans. Scans of some elderly show no selective neuronal activation during the various cognitive tasks, that is, the same regions "light up" for all the tasks, while discrete task-related regional neuronal activation is evident on scans of the young subjects.

A third explanation of fMRI scans of the elderly proposes that cognitive performance is related to *preservation of brain structure and function* such that age-resistant cognitive performance in the elderly requires a "youthful" brain volume, connectivity, and functional activity, not regional over- or underactivation. Hence, fMRI scans of the older cohort need to mimic that of the young cohort and can do so through *brain maintenance not compensation*. This recent explanation originated from a comparison of cognitive performance and fMRI scans from a cross-sectional *and* longitudinal study design using participants of the BETULA prospective cohort study on memory, health, and aging, a Swedish study of community dwellers (Nyberg et al., 2010). In essence, cross-sectional and 6-year longitudinal data of fMRI scans

TABLE 12.4. Interpretations of fMRI Scans in Elderly

fMRI Outcome	Cognitive Performance	Explanation
Overactivation	High	Compensation
Over-, under-, or nonselective activation	Low	Dedifferentiation
Same as in youth	No change with age	Maintenance

during cognitive testing (word categorizing tests) of 60 individuals (49–79) were viewed together. The fMRI *overactivity so prevalent on this and other cross-sectional studies was absent in the longitudinal study that instead revealed underactivation.* The discrepancy is explained by the presence of a cohort *subgroup effect of high-performing elderly* whose fMRI overactivity (supposedly high performers always increase their brain activity) eventually lessened with time in conjunction with reduced performance. It was concluded by Nyberg et al. (2010, 2012) that age-related reduction of structure (brain volume) and function (fMRI) are predictive of memory deficits. Less brain volume (cells, synapses, and axons) and less brain function (underactivation) predict memory decline while preserved brain structure and function in the *absence of disease is the main determinant of cognitive competency*. Data in the elderly that show a preservation of brain volume, for example, in hippocampal volume and white matter connectivity (DTI results), consistently report better cognitive performance. As expected, disease produces brain shrinkage, loss of cells, and connectivity and poor cognition.

In summary (see Table 12.4), changes in brain activity associated with preserved memory performance have been explained by the phenomenon of (i) cognitive reserve/compensation or (ii) brain structure/function maintenance with little or no change with age. Dedifferentiation on the other hand contributes to poor cognitive performance. Distinguishing among these pathways is key to the selection of effective cognitive interventions.

Cognitive Reserve/Maintenance Is Improved by Physical, Social, and Mental Enrichment

Whether through development of cognitive reserve or through maintenance mechanisms, considerable effort has been directed to optimizing brain function in the elderly. Results from epidemiological studies, interventional (randomized controlled trials (RCT)) studies, and experimental animal models have identified cognitive enrichment, including physical (aerobic and resistance exercise), social (networks), and mental (complex stimulation and skill learning) interventions as potential avenues to preserving cognitive skills over time. However, this field is young and thus far temporal associations, not cause and effect results prevail in human studies.

Epidemiological/Cohort Studies Show Cognitive Stimulation Associated with Higher Cognitive Test Scores

As reviewed by Daffner (2010), epidemiological findings indicate "that as people age, participation in intellectually stimulating activities may sustain cognitive function, create a buffer against mental decline and even promote longevity." Engagement in intellectually stimulating activities is *associated* with preserved cognition and lower risk for mild cognitive impairment (MCI) and AD. One representative study (Wilson et al., 2003) of >4000 participants over 65 years of age evaluated the impact of participation in cognitively stimulating activities (characterized as seeking and processing information with the least amount of physical and social requirements). On a scale of 1–5, subjects rated their participation in the following activities: viewing TV; listening to radio; reading newspapers; reading magazines; reading books; playing games of cards, checkers, crosswords, or other puzzles; and going to a museum). A composite average was compared to results from a battery of cognitive assessments taken several times over a 5-year period. After controlling for chronic medical conditions and depression, the greater the frequency of participation in the sum of cognitively stimulating activities, the smaller the decline in cognitive function. Epidemiological studies cannot establish cause and effect, but the study of large numbers of individuals over long time periods favors the identification of useful factors appropriate for future probing.

Mental Training through Brain Exercises Improves Some Aspects of Cognitive Behavior Cognitive training programs although fairly new, are evolving to achieve effects that translate in improved competency in everyday life not just to a better test score. The first study on the effect of cognitive training in the elderly showed that a 5-week training program that used the method of loci (MOL) technique (ancient Greek way of tethering items to be remembered to a mental map of a familiar area) improves spatial and episodic memory scores and produces significant metabolic changes in the hippocampus. Results of two large RCT, ACTIVE (Advanced Cognitive Training for Independent and Vital Elderly) and the IMPACT (Improvement in Memory with Plasticity-based Adaptive Cognitive Training), have provided further insights into the effect of cognitive training on the aging brain.

Briefly, in the ACTIVE trial, over 2500 participants (65–94) were divided into four groups that received 10 sessions of training in one of the following: memory (verbal episodic memory); reasoning (ability to solve problems that follow a serial pattern); speed of processing (visual search and identification); and no-contact control group. Additional "booster" training was available 11 and 35 months later for the first three groups. Cognitive function improved significantly in all treatment groups with additional improvement seen with booster sessions. When measured after the second booster session 5 years from start of study, gains in cognitive performance were maintained in areas of training compared to the controls. However, improvement in day to day *function (IADLs) was evident only in the group receiving reasoning training* (but not in the memory or speed training groups). When assessed 10 years after training, reasoning and speed of processing performance but not memory performance was maintained and significantly more individuals in the three training groups reported less difficulty with IADLS compared to controls (60% of training group and 50% of controls) suggesting some contribution to daily life, albeit small. Interestingly, individuals initially identified as memory impaired did not benefit from memory training but *did* benefit from reasoning and speed training.

In the IMPACT, trial participants (>450 participants) were divided into two groups, one of which received a computerized cognitive training program and the other a general cognitive stimulation program modeling treatment (active, control for 1 h/day, 5 days/week, for 8 weeks, for a total of 40 h). The novel training protocol included mental exercises in time ordered judgment tasks, discrimination tasks, spatial match tasks, forward span tasks, instruction following tasks, and narrative-memory tasks. Cognitive performance was determined before and after training with six subtests of the Repeatable Battery for the Assessment of Neuropsychological Status (RBANS). Those taking the experimental cognitive training improved on the RBANS as well as other memory and attention tests compared to the controls (Smith et al., 2009). Improvement in IADLs was not studied.

In addition to training-induced improvements in test scores, learning-specific changes in electroencephalography (EEG) measurements and connectivity (DTI) have been associated with improved WM.

Clearly, cognitive training improves test scores and possibly brain functionality, but what role it will play in significantly improving performance of IADLs remains to be demonstrated.

Physical Exercise Is Associated with Improved Cognitive Function
Physical exercise increases cerebral blood flow and glucose metabolism, thereby providing at least two major benefits to the brain. Other pluses such as maintenance of brain structure, connectivity, and neuronal health through activation of hormones, growth factors, and production of anti-inflammatory mediators have been consistently observed in animal models. Confirmation of these exercise-induced positive cellular effects in man is eagerly awaited.

Data from epidemiological, longitudinal cohort, cross-sectional, and interventional studies provide "evidence that links regular physical activity or exercise to higher cognitive function, decreased cognitive decline and reduced risk of AD and dementia" (Brown et al., 2013). A number of supportive studies are summarized in Table 12.5. Some are highlighted below.

Epidemiological studies, for example, Nurses' Health Study, Honolulu-Asia Aging Study, Health and Retirement Study, and longitudinal cohort-based studies (Treviso Longeva Study; Glostrup 1914 cohort; Rush Memory and Aging Project) (Wendell et al., 2013) analyzed the relation between exercise and cognition in the elderly and observed (i) a positive association between the level of physical activity, for example, walking >1.5 h/week, and preservation of scores on objective measures of cognition or (ii) a positive link between higher levels of physical activity, for example, walking 2 miles/day and reduced risk of dementia. Generally, these studies lasted 2–26 years with one or more measurements of physical activity and cognition and included participants within a narrow range of ages as well as over the lifespan. Physical activities were judged mostly by self-reported levels of frequency of activity (physical activity, formal exercise, and leisure activities). Rarely was the intensity of physical activity quantified and only one study determined the level of fitness with measurement of VO_2 max. Several studies assessed cognition with the MMSE but the majority of studies determined cognitive performance with three or more validated executive function and memory tests.

In the Glostrup 1914 cohort study that assessed participants at 50, 60, 70, and 80 years of age, baseline cognition was set at 50 years of age. When performance at later ages is compared with the "baseline," the association of higher levels of physical

TABLE 12.5. Studies of Physical Activity and Cognition in the Elderly

Nurses' health study (Weuve et al., 2004)
- >18,000 women, aged 70–81, tested every 2 years for 6 years; higher levels of physical activity (walking 1.5 mile/week) associated with better performance (general cognition, verbal memory, category fluency, and attention)

Honolulu-Asia aging study (Abbott et al., 2004)
- >2000 men, ages 71–93, 3 years duration; walking >2 miles/day associated with lower risk of dementia

Karolinska Institute (Rovio et al., 2005)
- >1400 men/women, ages 65–79, 21 years duration; vigorous physical activity (20–30 min, twice weekly) associated with lower risk of dementia and AD

Health and retirement study (Infurna and Gerstorf, 2013)
- 4177 (59% women), aged 30–97, 2 years duration; higher levels of continuous physical activity associated with better pulmonary function, lower systolic blood pressure, lower HbA1c, higher HDL-C, and less memory decline

Prospective study (Wendell et al., 2013)
- 1400 men/women, aged 19–94, tested 6 times over 18 years; higher baseline VO_2 max associated with better trajectory of performance (verbal and visual memory)

Danish cohort comparison (Christensen et al., 2013)
- 2262 men/women born 1905; 1584 born 1915, tested at ages 93, 95; cohort born 1915 scores better on MMSE, 5 other cognitive test and ADLs; no difference in physical fitness tests between cohorts

Treviso Longeva Study (Gallucci et al., 2013)
- 120 men/189 women, aged 77+, 7 year duration; low score on physical performance battery associated with poor MMSE score

Glostrup 1914 cohort study (Gow et al., 2012)
- 802 men/women, born 1914 and tested at ages 50, 60, 70, and 80. Greater activity (self-reported physical and leisure) associated with higher level of cognition (scores on four tests); preserved differentiation

Australian imaging, biomarkers and lifestyle study of ageing (Brown et al., 2013)
- 546 men/women, ages 60–95, tested once; higher levels of self-reported physical activity level associated with higher HDL-c and lower levels of insulin, triglycerides, and amyloid Aβ1 (PET)

Risk of AD study (Woodard et al., 2012)
- 27 participants; 18 months duration; higher self-reported levels of physical and cognitive activities associated with greater hippocampal volume (MRI) and task-activated fMRI, but benefit only in those at high risk of AD (APOEε4)

Prospective cohort (Iwasa et al., 2012)
- 567 men/women, ages 70+, 5 years duration; hobby participation but not self-reported social or physical activity preserves cognition (MMSE)

(continued)

TABLE 12.5. (*Continued*)

Cardiovascular health study (Erickson et al., 2010)
- 299 men/women, average age 78, 9 years duration; physical activity (walking 72 blocks/week) predicts increase in gray matter (MRI) and reduction in cognitive loss

Age gene/environment susceptibility: Reykjavik Study (Chang et al., 2010)
- 4761 men/women, baseline at 50 years of age; 26 years duration; self-reported physical activity (greater than 0 but less or equal to 5 h or more) predicts faster processing speed in executive function and lower risk of dementia

MMSE: mini mental state exam; PET: positron emission tomography; MRI: magnetic resonance imaging; fMRI: functionalMRI; HDL-C: high density lipoprotein cholesterol; HbA1c: glycated hemoglobin; ADLs: activities of daily living; AD: Alzheimer's disease.

activity with higher levels of cognitive performance is reduced but not obliterated. The authors attribute this to a confounding factor of "preserved differentiation," which suggests that inherent cognition at midlife may bias the effect of exercise on cognitive changes.

A review of interventional RCTs gleamed from 7 data bases concluded that 8 of the 11 merit trials show an association between improvements (up to 14%) in cardiorespiratory fitness (measured as VO_2 max) and cognitive performance especially for motor function and auditory attention and less so with information processing speed and visual attention (Angevaren et al., 2008). Several yearlong RCTs reported that (i) twice weekly resistance training improves cognitive performance relating to response inhibition processes at the test level and improves blood flow in regions of the cortex mediating these processes and that (ii) chronic endurance exercises in 65–74 year olds (supervised gym exercises, 3 h/week) preserved cognition (assessed by the MMSE) during the exercise period. Shorter trials of community dwellers (65–75 years of age) participating for 3 months biweekly in multicomponent training of neuromuscular coordination, balance, agility, cognitive executive function, or progressive resistance training and additionally tested before and after training with tests for executive function and functional mobility found that *at least several months of physical training benefit executive function*. Both types of training, multicomponent and resistance, improve functional mobility. Several studies found that acute exercise (20 min of walking) improves speed of processing but not function in other cognitive areas (interference/inhibition tests).

Many studies have measured the effect of physical activity and fitness on brain volume and brain gray matter (MRI) (reviewed in Erickson et al. (2014)). Despite the variation in study designs, analytical methods, assessment of fitness, and differences in exclusion criteria, there is agreement that higher levels of physical activity and fitness are associated with larger volumes of the prefrontal lobe and the hippocampus, the same regions necessary for memory and executive function. Moderate intensity exercise for 6 months to 1 year is sufficient for brain size enhancement and where determined, preserved cognitive function. These finding reinforce the concept of brain plasticity.

Occasionally, functional brain (fMRI) or connectivity (DTI) measurements are made in studies comparing physical activity levels with cognitive performance. One important report that completed two studies with elderly subjects (cross-sectional study of high- and low-fit elderly subjects and a 6-month longitudinal aerobic training program of average-fit elderly) showed that the greater the fitness level, the better the cognitive test results (flanking arrow test that measures attentional circuitry), and the more "plastic" or flexible the fMRI results.

Neuroplasticity

Neuroplasticity has been defined as *the changes that occur in the functional and anatomical organization of the brain as a result of experience* (Spolidoro et al., 2009). Although evident at any age, in cognitive aging neuroplasticity represents the final common pathway of neurobiological processes, including structural, functional, or molecular mechanisms, that result in stability or compensation for age- or disease-related changes. It is the *means to achieving cognitive reserve*.

In the developing organism, neuroplasticity happens in a major way at certain time periods and as a result, verbal, motor, and mental experiences achieve their most intense effect on neuronal reorganization. Select examples that lend themselves to study include the extensive cortical mapping that develops in individuals with intense practice for example, of a musical instrument at a very early age. Cortical connections are established with training at later ages (preteen and older) but never to the same extent and necessitating considerably more practice. Although initially deemed unlikely, neuroplasticity appears possible in the adult and senior. Reports of favorable changes in brain structure, connectivity and cognitive performance associated with exercise and cognitive training in the older individual support this view.

Although molecular mechanisms of neuroplasticity in man are lacking, results of animal studies in rodents, canines, and monkeys show that the addition of an enriched environment, for example, new toys, additional space for exploration, exercise, and balanced diet, prevents or reduces the decline in measured cognitive function, for example, spatial memory examined with the water maze test in rats or finding the hidden toy in canine studies. Furthermore, *examination of brain tissue reveals the presence of supplementary neuronal brain connections that include new dendritic spines, additional synapses, and birth of new neurons* specific to regions involved in learning, for example, hippocampus. Others demonstrate the beneficial effect of a 4 month enrichment on amelioration of pathology in a mouse model of AD. The amount of the hallmark amyloid β protein is reduced by exposure of the AD mouse model to enriched spatial surroundings that include exercise wheels, tunnels, and toys.

Neurogenesis or the generation of new neurons from neural stem cells is an established mechanism to support cognitive resiliency in animals. This phenomenon has been convincingly established in mice, primates, and fetal human brain but whether it occurs in the adult brain is debatable. Using a ^{14}C tracking technique to date cells in human postmortem brains, Spalding et al. (2013) reported that "700 neurons are added to each hippocampus per day, corresponding to an annual turnover of 1.75% of the neurons within the renewing fraction, with a modest decline during aging." In the mouse, neurogenesis is essential for cognitive adaptability and since the estimated rates of neurogenesis in the mouse and man are similar, it is suggested that neurogenesis in man is also of critical importance, but *at present in vivo neurogenesis has not been demonstrate*d. Confirmation of neurogenesis with exercise and cognitive training in man awaits the advent of more sensitive noninvasive techniques.

Summary of Age Changes of the CNS
Structural changes
- Volume (cells)—two views
  1. ↓ brain volume (0.2%)/year after age 35; 0.5% loss after age 60, cortex thins, subcortical region atrophies, ↑ ventricle size
     vulnerable: hippocampus, entorhinal cortex, orbital-frontal cortex/cerebellum
  2. No change with exclusion of covert/overt disease; normal gray matter volume even in vulnerable regions
- Neurons/neurotransmitters (synapses)
  ↓ Dopamine/serotonin receptor binding (10%/decade from adulthood) associated with reduced reward-based behavior
- Connectivity (axons)—two views
  1. ↓ connectivity in white matter fibers in specific fiber tracks (corpus callosum, limbic system with the frontal lobe, prefrontal, and medial projection)
  2. ↓ connectivity diffuse, more widespread
- Microglia
  Cell senescence—↓ phagocytosis, ↓ neuronal modulation, ↑ proinflammatory mediators (results in animal models)

Functional changes
- Onset debated: 35–60 years
  - ↓ Select cognitive behaviors include
    ↓ information processing speed (impacts visual/motor responses)
    ↓ executive function (reduced working memory for updating, monitoring, set-shifting, and reducing inhibitory influences)
    ↓ memory for encoding/retrieval processes (episodic), associative processes, source finding, and prospective retention
    ↓ reward-based behavior
  - No change in semantic memory (facts/knowledge), language, autobiographical memory (personal history of self), emotional processing, automatic memory processes
  - ↓ cognitive behavior due to ↓ cognitive reserve, ↓ maintenance or dedifferentiation

SUMMARY

Neurons, the major cell type of the CNS, receive, interpret, and generate responses to external and internal stimuli, thereby facilitating sensory, somatic, and internal organ functions as well as cognitive behaviors of intelligence, reasoning, creativity, emotions, and memory. Neurons distribute information through means of the nerve impulse and communicate with one another via release of neurotransmitters. Auxiliary cells of microglia, oligodendrocytes, astrocytes, and Schwann cells ensure neuronal health.

The advent of sophisticated neuroimaging techniques (MRI, fMRI, PET, and DTI) has advanced understanding of the aging brain. Study results are tempered with many issues: incomplete knowledge of brain physiology, inherent bias of study designs, effects of covert brain disease, relevance of cognitive behavior tests, imaging interpretation limitations, heterogeneity of response, and correlative nature of studies.

A modest age-related reduction in brain volume may occur, although this is debated. A loss of brain connectivity (axonal deficiencies) has been observed and synaptic transmission in select neuronal systems (dopaminergic/serotonergic) slow with age. Microglia appear to senesce with age, a change that disrupts brain homeostasis. A decline in cognitive performance with age is highly variable. The most common age change is a slowing of information processing speed. Deficits in executive function, memory, and reward-based behavior may occur with age; generally semantic memory, language, autobiographical memory, emotional processing, and automatic memory processes remain constant throughout the lifespan.

Brain patterns in the elderly as measured by fMRI or PET scans during various cognitive tests differ from those of the young. Main differences reported by some are prefrontal overactivation and bilateral hemispheric activation. These changes in relation to the level of cognitive performance are explained by the phenomena of compensation, dedifferentiation, or maintenance.

Environmental enrichment, cognitive skill training, and exercise are under intense investigation as interventions to preserve cognition. The molecular mechanisms for preservation of cognitive performance or reduction of risk for MCI or AD with these interventions are poorly understood in man, although animal studies suggest these interventions stimulate neurogenesis and the growth of dendritic spines and new synapses.

CRITICAL THINKING

Explain the reasons for the existing controversy in the following areas of brain aging: brain volume changes, cognitive function changes, brain exercises, and neuroimaging scans.

In relation to fMRI scans, what is meant by compensation, dedifferentiation, and maintenance?

How is cognitive behavior assessed in the elderly? Is this approach convincing?

Describe interventions that seem to preserve or improve cognitive performance with age? What are some reservations regarding them?

KEY TERMS

ADL activities of daily living related to basic grooming such as bathing, oral hygiene, toileting, and eating.

Amyloid Aβ protein protein fragment that accumulates in the amyloid plaques, characteristic of AD.

Axon extension from the neuron's cell body that relays information outward to another neuron. Many are wrapped in a fatty sheath of myelin.

Brain stem includes multiple structures: the thalamus, the hypothalamus, the midbrain, pons and medulla oblongata, and the limbic system. These regions are involved in involuntary reflex and hormonal control of all systems of the body. The limbic system is responsible for emotions and rewards.

Cerebellum separate posterior region of the brain involved in balance and motor coordination.

Cerebrum the largest portion of the brain containing distinct lobes: frontal, parietal, occipital, and temporal.

Dendrite extension from the neuron's cell body that receives input from another neuron.

Entorhinal cortex area of relay from the hippocampus to the cerebrum; important in memory and thinking.

Gray matter neuron cell body.

Hippocampus region involved in various types of memory and thought.

IADLs instrumental activities of daily living include handling finances, shopping, using the phone, and housecleaning.

Imaging techniques variety of techniques using magnets, X-rays, radioactivity, and computers; determine brain activity during cognition such as fMRI, PET, DTI, or brain structures such as MRI and CAT.

Lewy bodies protein aggregates that accumulate in neurons. Excessive amount present in PD and dementias.

Limbic system a collection of nerve pathways that include the hippocampus and amygdala and others that integrate and relay information (emotional) to other regions, for example, hypothalamus and cortex.

Neuron basic cell unit of the brain, central, and peripheral nervous systems. This specialized postmitotic cell possesses a cell body, dendrites, and an axon; relays electrical activity throughout the nervous system; neurons have many dendrites and hence receive a complexity of incoming information; terminals of the axon release neurotransmitters to activate adjacent neurons.

Neuroplasticity ability of neurons to change in response to external stimuli or "learning"; neuroplasticity influenced by environmental enrichment.

Prefrontal cortex the front most (anterior) part of the frontal cortex. Area associated with executive function.

PNS peripheral nervous system that includes the parasympathetic and sympathetic components. Nerves traverse and exit the spinal cord.

Small vessel disease (SVD) insidious and covert cardiovascular disease in the brain that may account for reduction in cognition; identified on brain scans.

Ventricles cavities in the cerebrum that contain cerebrospinal fluid and are thought to enlarge with age.

Ventriculomegaly enlargement of cerebral ventricles due to weakening of supporting structures.

White matter axons of neurons; surrounded by myelin sheath and appear white.

White matter hyperintensities (WMH) area on brain scans that appear super white and indicate areas of poor blood flow; significance not fully understood but potentially related to cognitive decline.

BIBLIOGRAPHY

Review

Alexander GE, Ryan L, Bowers D, Foster TC, Bizon JL, Geldmacher DS, Glisky EL. 2012. Characterizing cognitive aging in humans with links to animal models. *Front. Aging Neurosci.* **4**:21–39.

Angevaren M, Aufdemkampe G, Verhaar HJ, Aleman A, Vanhees L. 2008. Physical activity and enhanced fitness to improve cognitive function in older people without known cognitive impairment. *Cochrane Database Syst. Rev.* **16**(2):CD005381.

Bartus RT, Dean RL, 3rd, Beer B, Lippa AS. 1982. The cholinergic hypothesis of geriatric memory dysfunction. *Science* **217**(4558):408–414.

Chowdhury R, Guitart-Masip M, Lambert C, Dayan P, Huys Q, Düzel E, Dolan RJ. 2013. Dopamine restores reward prediction errors in old age. *Nat. Neurosci.* **16**(5):648–653.

Colcombe SJ, Kramer AF, Erickson KI, Scalf P, McAuley E, Cohen NJ, Webb A, Jerome GJ, Marquez DX, Elavsky S. 2004. Cardiovascular fitness, cortical plasticity, and aging. *Proc. Natl. Acad. Sci. USA* **101**(9):3316–3321.

Daffner KR. 2010. Promoting successful cognitive aging: a comprehensive review. *J. Alzheimers Dis.* **19**(4):1101–1122.

Dumas JA, Newhouse PA. 2011. The cholinergic hypothesis of cognitive aging revisited again: cholinergic functional compensation. *Pharmacol. Biochem. Behav.* **99**(2):254–261.

Erickson KI, Leckie RL, Weinstein AM. 2014. Physical activity, fitness, and gray matter volume. *Neurobiol. Aging* **35 suppl 2**:S20–S28.

Eriksson PS, Perfilieva E, Björk-Eriksson T, Alborn A-M, Nordborg C, Peterson DA, Gage FH. 1998. Neurogenesis in the adult human hippocampus. *Nat. Med.* **4**(11):1313–1317.

Finch CE. 2009. The neurobiology of middle-age has arrived. *Neurobiol. Aging* **30**(4):515–520.

Grady C. 2012. The cognitive neuroscience of ageing. *Nat. Rev. Neurosci.* **13**(7):491–505.

Han SD, Bangen KJ, Bondi MW. 2009. Functional magnetic resonance imaging of compensatory neural recruitment in aging and risk for Alzheimer's disease: review and recommendations. *Dement. Geriatr. Cogn. Disord.* **27**(1):1–10.

Harry GJ. 2013. Microglia during development and aging. *Pharmacol. Ther.* **139**(3):313–326.

Hedden T, Gabrieli JD. 2004. Insights into the ageing mind: a view from cognitive neuroscience. *Nat. Rev. Neurosci.* **5**(2):87–96.

Hedman AM, van Haren NE, Schnack HG, Kahn RS, Hulshoff Pol HE. 2012. Human brain changes across the life span: a review of 56 longitudinal magnetic resonance imaging studies. *Hum. Brain Mapp.* **33**(8):1987–2002.

Ho NF, Hooker JM, Sahay A, Holt DJ, Roffman JL. 2013. In vivo imaging of adult human hippocampal neurogenesis: progress, pitfalls and promise. *Mol. Psychiatry* **18**(4):404–416.

Madden DJ, Bennett IJ, Burzynska A, Potter GG, Chen NK, Song AW. 2012. Diffusion tensor imaging of cerebral white matter integrity in cognitive aging. *Biochim. Biophys. Acta* **1822**(3):386–400.

Morrison JH, Baxter MG. 2012. The ageing cortical synapse: hallmarks and implications for cognitive decline. *Nat. Rev. Neurosci.* **13**(4):240–250.

Nilsson L-G, Sternäng O, Rönnlund M, Nyberg L. 2009. Challenging the notion of an early-onset of cognitive decline. *Neurobiol. Aging* **30**(4):521–524.

Nyberg L, Lövdén M, Riklund K, Lindenberger U, Bäckman L. 2012. Memory aging and brain maintenance. *Trends Cogn. Sci.* **16**(5):292–305.

Salthouse TA. 2011. Neuroanatomical substrates of age-related cognitive decline. *Psychol. Bull.* **137**(5):753–784.

Sambataro F, Safrin M, Lemaitre HS, Steele SU, Das SB, Callicott JH, Weinberger DR, Mattay VS. 2012. Normal aging modulates prefrontoparietal networks underlying multiple memory processes. *Eur. J. Neurosci.* **36**(11):3559–3567.

Shohamy D, Wimmer GE. 2013. Dopamine and the cost of aging. *Nat. Neurosci.* **16**(5):519–521.

Smith GS. 2013. Aging and neuroplasticity. *Dialogues Clin. Neurosci.* **15**(1):3–5.

Spolidoro M, Sale A, Berardi N, Maffei L. 2009. Plasticity in the adult brain: lessons from the visual system. *Exp. Brain Res.* **192**(3):335–341.

Stern Y. 2012. Cognitive reserve in ageing and Alzheimer's disease. *Lancet Neurol.* **11**(11):1006–1012.

Szczepanski SM, Knight RT. 2014. Insights into human behavior from lesions to the prefrontal cortex. *Neuron* **83**(5):1002–1018.

Wechsler D. 1987. *Wechsler Memory Scale: Revised.* New York: Psychological Corporation.

Wechsler D. 2008. *Administration and scoring manual for the Wechsler Adult Intelligence Scale-IV.* San Antonio: NCS Pearson Inc.

Experimental

Abbott RD, White LR, Ross GW, Masaki KH, Curb JD, Petrovitch H. 2004. Walking and dementia in physically capable elderly men. *JAMA* **292**(12):1447–1453.

Barrick TR, Charlton RA, Clark CA, Markus HS. 2010. White matter structural decline in normal ageing: a prospective longitudinal study using tract-based spatial statistics. *Neuroimage* **51**(2):565–577.

Beason-Held LL, Kraut MA, Resnick SM. 2008a. I. Longitudinal changes in aging brain function. *Neurobiol. Aging* **29**(4):483–496.

Beason-Held LL, Kraut MA, Resnick SM. 2008b. II. Temporal patterns of longitudinal change in aging brain function. *Neurobiol. Aging* **29**(4):497–513.

Brown BM, Peiffer JJ, Taddei K, Lui JK, Laws SM, Gupta VB, Taddei T, Ward VK, Rodrigues MA, Burnham S, Rainey-Smith SR, Villemagne VL, Bush A, Ellis KA, Masters CL, Ames

D, Macaulay SL, Szoeke C, Rowe CC, Martins RN. 2013. Physical activity and amyloid-β plasma and brain levels: results from the Australian Imaging, Biomarkers and Lifestyle Study of Ageing. *Mol. Psychiatry* **18**(8):875–881.

Bucur B, Madden DJ, Spaniol J, Provenzale JM, Cabeza R, White LE, Huettel SA. 2008. Age-related slowing of memory retrieval: contributions of perceptual speed and cerebral white matter integrity. *Neurobiol. Aging* **29**(7):1070–1079.

Burgmans S, van Boxtel MP, Vuurman EF, Smeets F, Gronenschild EH, Uylings HB, Jolles J. 2009. The prevalence of cortical gray matter atrophy may be overestimated in the healthy aging brain. *Neuropsychology* **23**(5):541–550.

Chang M, Jonsson PV, Snaedal J, Bjornsson S, Saczynski JS, Aspelund T, Eiriksdottir G, Jonsdottir MK, Lopez OL, Harris TB, Gudnason V, Launer LJ. 2010. The effect of midlife physical activity on cognitive function among older adults: AGES—Reykjavik Study. *J. Gerontol. A Biol. Sci. Med. Sci.* **65**(12):1369–1374.

Christensen K, Thinggaard M, Oksuzyan A, Steenstrup T, Andersen-Ranberg K, Jeune B, McGue M, Vaupel JW. 2013. Physical and cognitive functioning of people older than 90 years: a comparison of two Danish cohorts born 10 years apart. *Lancet* **382**(9903):1507–1513.

Comijs HC, Dik MG, Deeg DJ, Jonker C. 2004. The course of cognitive decline in older persons: results from the longitudinal aging study amsterdam. *Dement. Geriatr. Cogn. Disord.* **17**(3):136–142.

Dixon RA, de Frias CM. 2014. Cognitively elite, cognitively normal, and cognitively impaired aging: neurocognitive status and stability moderate memory performance. *J. Clin. Exp. Neuropsychol.* **36**(4):418–430.

Dreher J-C, Meyer-Lindenberg A, Kohn P, Berman KF. 2008. Age-related changes in midbrain dopaminergic regulation of the human reward system. *Proc. Natl. Acad. Sci. USA* **105**(39):15106–15111.

Erickson KI, Raji CA, Lopez OL, Becker JT, Rosano C, Newman AB, Gach HM, Thompson PM, Ho AJ, Kuller LH. 2010. Physical activity predicts gray matter volume in late adulthood: the Cardiovascular Health Study. *Neurology* **75**(16):1415–1422.

Erickson KI, Voss MW, Prakash RS, Basak C, Szabo A, Chaddock L, Kim JS, Heo S, Alves H, White SM, Wojcicki TR, Mailey E, Vieira VJ, Martin SA, Pence BD, Woods JA, McAuley E, Kramer AF. 2011. Exercise training increases size of hippocampus and improves memory. *Proc. Natl. Acad. Sci. USA* **108**(7):3017–3022.

Erickson KI, Voss MW, Prakash RS, Basak C, Szabo A, Chaddock L, Kim JS, Heo S, Alves H, White SM, Wojcicki TR Mailey E, Vieira VJ, Martin SA, Pence BD, Woods JA, McAuley E, Kramer AF. 2011. Exercise training increases size of hippocampus and improves memory. *Proc. Natl. Acad. Sci. USA* **108**(7):3017–3022.

Fjell AM, Westlye LT, Grydeland H, Amlien I, Espeseth T, Reinvang I, Raz N, Dale AM, Walhovd KB, Alzheimer Disease Neuroimaging Initiative. 2014. Accelerating cortical thinning: unique to dementia or universal in aging? *Cereb. Cortex* **24**(4):919–934.

Forte R, Boreham CA, Leite JC, De Vito G, Brennan L, Gibney ER, Pesce C. 2013. Enhancing cognitive functioning in the elderly: multicomponent vs resistance training. *Clin. Interv. Aging* **8**:19–27.

Gallucci M, Mazzuco S, Ongaro F, Di Giorgi E, Mecocci P, Cesari M, Albani D, Forloni GL, Durante E, Gajo GB, Zanardo A, Siculi M, Caberlotto L, Regini C. 2013. Body mass index, lifestyles, physical performance and cognitive decline: the "Treviso Longeva (TRELONG)" study. *J. Nutr. Health Aging* **17**(4):378–384.

Gow AJ, Mortensen EL, Avlund K. 2012. Activity participation and cognitive aging from age 50 to 80 in the Glostrup 1914 cohort. *J. Am. Geriatr. Soc.* **60**(10):1831–1838.

Infurna FJ, Gerstorf D. 2013. Linking perceived control, physical activity, and biological health to memory change. *Psychol. Aging* **28**(4):1147–1163.

Iwasa H, Yoshida Y, Kai I, Suzuki T, Kim H, Yoshida H. 2012. Leisure activities and cognitive function in elderly community-dwelling individuals in Japan: a 5-year prospective cohort study. *J. Psychosom. Res.* **72**(2):159–164.

Lazarov O, Robinson J, Tang Y-P, Hairston IS, Korade-Mirnics Z, Lee VM, Hersh LB, Sapolsky RM, Mirnics K, Sisodia SS. 2005. Environmental enrichment reduces Abeta levels and amyloid deposition in transgenic mice. *Cell* **120**(5):701–713.

Nyberg L, Salami A, Andersson M, Eriksson J, Kalpouzos G, Kauppi K, Lind J, Pudas S, Persson J, Nilsson LG. 2010. Longitudinal evidence for diminished frontal cortex function in aging. *Proc. Natl. Acad. Sci. USA* **107**(52):22682–22686.

Peich M-C, Husain M, Bays PM. 2013. Age-related decline of precision and binding in visual working memory. *Psychol. Aging* **28**(3):729–743.

Rebok GW, Ball K, Guey LT, Jones RN, Kim H-Y, King JW, Marsiske M, Morris JN, Tennstedt SL, Unverzagt FW, Willis SL. 2014. Ten-year effects of the advanced cognitive training for independent and vital elderly cognitive training trial on cognition and everyday functioning in older adults. *J. Am. Geriatr. Soc.* **62**(1):16–24.

Rovio S, Kåreholt I, Helkala E-L, Viitanen M, Winblad B, Tuomilehto J, Soininen H, Nissinen A, Kivipelto M. 2005. Leisure-time physical activity at midlife and the risk of dementia and Alzheimer's disease. *Lancet Neurol.* **4**(11):705–711.

Salthouse TA. 2009. When does age-related cognitive decline begin? *Neurobiol. Aging* **30**(4):507–514.

Salthouse TA. 2010. Influence of age on practice effects in longitudinal neurocognitive change. *Neuropsychology* **24**(5):563–572.

Smith GE, Housen P, Yaffe K, Ruff R, Kennison RF, Mahncke HW, Zelinski EM. 2009. A cognitive training program based on principles of brain plasticity: results from the Improvement in Memory with Plasticity-Based Adaptive Cognitive Training (IMPACT) study. *J. Am. Geriatr. Soc.* **57**(4):594–603.

Spalding KL, Bergmann O, Alkass K, Bernard S, Salehpour M, Huttner HB, Boström E, Westerlund I, Vial C, Buchholz BA, Possnert G, Mash DC, Druid H, Frisén J. 2013. Dynamics of hippocampal neurogenesis in adult humans. *Cell* **153**(6):1219–1227.

Unverzagt FW, Kasten L, Johnson KE, Rebok GW, Marsiske M, Koepke KM, Elias JW, Morris JN, Willis SL, Ball K, Rexroth DF, Smith DM, Wolinsky FD, Tennstedt SL. 2007. Effect of memory impairment on training outcomes in ACTIVE. *J. Int. Neuropsychol. Soc.* **13**(6):953–960.

Wang M, Gamo NJ, Yang Y, Jin LE, Wang X-J, Laubach M, Mazer JA, Lee D, Arnsten AFT. 2011. Neuronal basis of age-related working memory decline. *Nature* **476**(7359):210–213.

Wendell CR, Gunstad J, Waldstein SR, Wright JG, Ferrucci L, Wright JG, Ferrucci L, Zonderman AB. 2013. Cardiorespiratory fitness and accelerated cognitive decline with aging. *J. Gerontol. A Biol. Sci. Med. Sci.* **69**(4):455–462.

Weuve J, Kang JH, Manson JE, Breteler MM, Ware JH, Grodstein F. 2004. Physical activity, including walking, and cognitive function in older women. *JAMA* **292**(12):1454–1461.

Willis SL, Tennstedt SL, Marsiske M, Ball K, Elias J, Koepke KM, Morris JN, Rebok GW, Unverzagt FW, Stoddard AM, Wright E, ACTIVE Study Group. 2006. Long-term effects of cognitive training on everyday functional outcomes in older adults. *JAMA* 20: **296**(23):2805–2814.

Wilson RS, Bennett DA, Bienias JL, Mendes de Leon CF, Morris MC, Evans DA. 2003. Cognitive activity and cognitive decline in a biracial community population. *Neurology* **61**(6):812–816.

Woodard JL, Sugarman MA, Nielson KA, Smith JC, Seidenberg M, Durgerian S, Butts A, Hantke N, Lancaster M, Matthews, MA, Rao S. 2012. Lifestyle and genetic contributions to cognitive decline and hippocampal integrity in healthy aging. *Curr. Alzheimer. Res.* **9**(4):436–446.

Wong DF, Young D, Wilson PD, Meltzer CC, Gjedde A. 1997. Quantification of neuroreceptors in the living human brain: III. D_2-like dopamine receptors: theory, validation, and changes during normal aging. *J. Cereb. Blood Flow Metab.* **17**(3):316–330.

Zahr NM, Rohlfing T, Pfefferbaum A, Sullivan EV. 2009. Problem solving, working memory, and motor correlates of association and commissural fiber bundles in normal aging: a quantitative fiber tracking study. *Neuroimage* **44**(3):1050–1062.

13

AGING OF THE SENSORY SYSTEM

GENERAL PRINCIPLES

Age Increases Threshold, Decreases Discrimination, and Slows Reaction Time

Our perceptual senses of sight, hearing, smell, taste, position, movement, touch, pain, and temperature are essential for an active life. A reduction or distortion in these senses negatively impacts the quality of life (QOL), contributes to accidents, and in some cases accounts for fatalities.

The basic component of each sense is the specialized cell/receptor or nerve ending that responds to a specific environmental stimulus. Photoreceptor cells of the eye convert light waves to vision. Hair cells of the inner ear transfer sound waves to audition-mediated information. Nasal mucosal receptors and receptors within the taste buds detect hundreds of chemicals to supply the sensation of smell and taste, respectively. Proprioceptors within the muscles and tendons sense position and movement providing information to the brain on balance. Multiple cutaneous somatosensory organs and nerve endings relay information on touch, different types of pressure and vibration, pain, and temperature.

Detection of environmental stimuli by sensory receptors obeys general principles. There exists a minimal amount of stimulus that will elicit a response defined as the *threshold*. If the stimulus intensity is below the threshold, the stimulus is not detected. *With age, the threshold tends to increase such that more and more stimuli go undetected.* Therefore, the stimulus intensity must *increase* in order to elicit a response (reach the threshold). For example, this translates into (i) use of higher wattage or (ii) elevation in

---

Human Biological Aging: From Macromolecules to Organ Systems, First Edition. Glenda Bilder.
© 2016 John Wiley & Sons, Inc. Published 2016 by John Wiley & Sons, Inc.

loudness (decibels, e.g., turning up the volume). These principles apply to all senses—sight, hearing, pressure, smell, taste, temperature, and so on.

The ability to distinguish between two sensory stimuli is called *discrimination*. *With age, discriminatory faculty declines*, a loss that requires one stimulus to exceed the other in some quality to emphasize the distinction. For example, with reduction of discrimination, two adjacent gray objects may appear as one. A distinction between them would be facilitated by a change in color intensity (dark gray aside of light gray).

Sensory receptors process environmental stimuli in a time-dependent manner. Most responses are rapid, if not instantaneous. *With age, sensory "reaction" time tends to decrease*. Therefore, the *time needed to process a sensory input increases*. Since the processing of sensory information requires relays to the brain, it is unclear where in the complex pathway speed reduction occurs.

In cases where sensory loss includes multiple deficiencies, for example, reduced discrimination, increased reaction time, and elevated thresholds in multiple senses, a *phenomenon of sensory overload* is likely. *Difficulty in concentration, confusion, and withdrawal are characteristic manifestations of this state*.

SENSORY ORGANS: AGE CHANGES IN STRUCTURE/FUNCTION

Sense of Vision

Reception of Light Waves Modified by Age at Multiple Loci: Cornea, Iris, Lens, Retinal Receptors, Optic Nerve, and Visual Cortex The main structural components of the eye are the cornea, the iris, the lens (and ciliary body), the chambers (anterior/posterior), the retina, and the optic nerve. The eye is encased in the sclera, part of which forms the transparent cornea. The anterior chamber containing liquid termed aqueous humor, is enclosed between the cornea and the lens, a flexible clear capsule whose ends are attached via ligaments to a small muscle, the ciliary body. The posterior chamber of vitreous humor fills the bulk of the eye, cushioning the surrounding retinal layer that contains the specialized photoreceptors and supporting nerve cells. The retinal cells communicate with the optic nerve fibers that exit the eye collectively as the optic nerve. Neuronal information traverses the optic nerve to reach the primary visual cortex (occipital lobe) of the brain. Another important structure is the iris (pupil), a compound muscle situated in front of the lens. Figure 13.1 displays an anatomical schematic of the eye.

The function of the eyes is visual perception. Light waves are refracted (bent) on their way through the cornea. The lens, behind the cornea, further refracts the light in adjustment to distance. The light continues through the posterior chamber where individual *photons activate the photoreceptors of the outer retinal layer*. This activation causes neurotransmitter release (chemical mediators aiding communication between nerve cells), stimulation of auxiliary cells and relay of nerve impulses through the optic nerve to the visual cortex (occipital lobe) for interpretation.

The iris, a composite of two sets of smooth muscles, the dilator (radial) and the constrictor (circular), reflexly dilate or constrict the pupil in response to a change in light intensity (called dark and light adaptation). The ciliary body is in part muscular and when contracted allows a shape change in the lens capsule that enables near

SENSORY ORGANS: AGE CHANGES IN STRUCTURE/FUNCTION

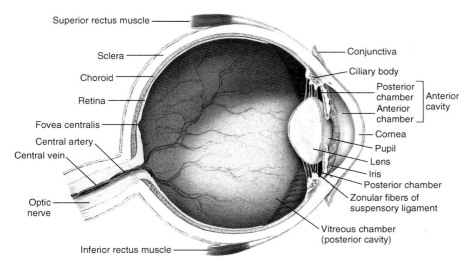

Figure 13.1. Anatomy of the eye. Fovea centralis is the central portion of eye and functions to provide clear central vision. (Reprinted with permission from Barett et al. (2012).) (See plate section for color version.)

vision. All of these structures are altered with age and are discussed in the following section.

Presbyopia Is the Loss of Accommodation: Near Vision

To see near objects (closer than 20 ft), the eyes rely on the process of *accommodation*. To focus light waves of near objects on the retina (photoreceptors), the lens needs to be more convex (round outward). This is achieved by contraction of the ciliary body that allows the ligaments to relax and the capsule and lens to seek a new shape (see Figure 13.2). In far vision, the lens is elongated, pulled by the ligaments. *Ciliary body contraction removes ligament tension, so the lens shape changes to refract the light to "accommodate" near vision.*

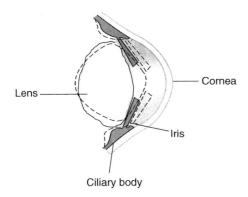

Figure 13.2. Accommodation. Dotted line illustrates changes in lens shape with accommodation; iris also constricts. (Reprinted with permission from Barett et al. (2012).) (See plate section for color version.)

Accommodation for near vision worsens with age, noticeable in the fifth decade and seriously declining thereafter. The *loss of accommodation is labeled presbyopia*. A 10 year old sees objects clearly within 3.5 inches. Thirty-five years later, clarity occurs at ~8 inches and by 85 years of age, at an unrealistic distance of 33 inches.

The explanation for presbyopia is simply that the *lens becomes less flexible with age*. Because there is no effective way to eliminate the transparent proteins of the lens, their continual production and secretion by the outer rim of epithelial cells leads to detrimental accumulation. This eventually reduces the "elasticity" of the lens such that with time, it fails to bulge freely. In addition to presbyopia, protein accumulation decreases lens transparency, reduces light refraction for both near and far vision, so objects at all distances are less sharp and alters color perception. Specifically the thickened *lens elevates the threshold for short wave lengths* thereby making it *more difficult to see objects in colors of blue, green, and violet*. Greater contrast (texture, intensity, and shape) among these colors is required to distinguish one from the other. Unfortunately, *the short wave lengths (which are reduced by the fatter lens) are also the most effective wave lengths for stimulation of circadian-melatonin sleep–wake cycle*. Age-related lens distortions are postulated to disturb normal sleep patterns (see Chapter 14).

Reduced Diameter of the Iris Causes Senile Miosis

The iris controls the size of the pupil by a reflex response to light (pupillary reflex). In low light or in the dark, the pupil dilates (pupil diameter widens as radial muscles contract) to allow in more light. In bright sunlight, the pupil constricts (pupil diameter decreases as the circular muscles contract) to minimize light entrance. *With age, there is a gradual reduction in pupil size under all conditions of illumination. This is termed senile miosis*. As calculated by Winn et al. (1994), there is a linear decrease in pupil size with age in subjects 17–82 years of age. For example, in the dark, the pupillary diameter of a child is approximately 8 mm. By age 80, the diameter is about 5 mm, a loss of some 37%. Similarly, over the same age range (10–80 years), pupillary diameter in low light conditions declines about 30% (~4 –~2.8 mm).

Senile miosis makes it difficult to (i) see in the dark, (ii) see in low light conditions, and (iii) restricts the visual field by as much as 12% by age 65. Possible benefits of senile miosis are a reduction in the light scatter created by the thickened lens and protection against phototoxicity.

Several mechanisms to explain senile miosis have been proposed. They include (i) a relative weakness of the opposing radial muscle (atrophy due to apoptosis) compared to the circular muscle favoring a smaller diameter, (ii) a relative reduction in sympathetic tone (innervating the radial muscle so that the muscle does not contract as forcefully to dilate the pupil), and (iii) less inhibition of parasympathetic tone (innervating the circular muscle permitting more circular muscle contraction). Additional research is needed to distinguish among these hypotheses.

Photoreceptors and Neuronal Tissue Degenerate Over Time

Age-associated changes in photoreceptors (rods) and neuronal tissue of the eye have been observed. These changes are thought to contribute to elevated thresholds of light sensitivity, slowed visual processing speed, and reduced motion detection. In studies that control for aging of the lens and iris, an elevation in threshold sensitivity could be explained by age changes in photoreceptors (rods). Exactly how the rods deteriorate is not clear but their loss adds to the slowed dark adaptation response and reduced spatial contrast sensitivity experienced by the elderly in low light.

Visual processing speed may slow with age. The variation from individual to individual is very large. However, slowed visual processing necessitates more time to complete daily activities, even simple tasks of positional change, for example, sitting to standing. Additionally, this deficit may contribute to accidents, for example, motor vehicle, falls. Multitasking exacerbates slowed visual processing time.

Motion perception is also at risk of aberration, primarily the result of neuronal disruptions in the visual cortex. Studies suggest that in individuals >70 years of age, there is an increased risk of reduced direction detection and sensitivity to motion. This creates a tendency to ignore small movements in lieu of a background focus and to misinterpret three-dimensional and spatial cues, self-motion, and collision paths, all of which favor an accident, motor or otherwise.

The Cornea Thickens and Changes Curvature, Resulting in Decreased Refraction
A *thickened cornea with an altered curvature decreases light refraction*, diminishing a major function of the cornea. These structural changes are a postulated consequence of slowed repair that over time allows the cornea to thicken. Various substances tend to precipitate in the cornea but are generally harmless. Substances such as cholesterol and calcium appear as yellow-white rings along the outer edge (arcus senilis) and iron precipitates appear as horizontal brown lines (Hudson-Stahli lines).

In summary, age changes in the lens, iris, and the retinal tissues contribute to alterations in visual threshold, discrimination, and reaction time in the elderly. Aging in the visual cortex are also possible but less well understood. Minor structural changes have been recorded in the cornea but the contribution to vision is modest. Severe loss in vision is a consequence of diseases such as glaucoma and macular degeneration.

Audition: Hearing Function

Sound Waves Travel through the Three Chambers of the Ear
The ear receives and processes sound waves through three specialized chambers. The outer chamber is the external cartilaginous structure containing the ear canal up to the membranous partition, the tympani membrane (eardrum). The sensitive tympani membrane responds to the pressure of the sound waves and passes vibrations to the middle chamber. Here three small bones, the malleus, the incus, and the stapes convert the sound waves for use by the inner chamber. The inner chamber, a coiled cavernous structure or cochlear, houses the auditory sensory fibers (hair cells) bathed in fluid. The modified sound pressures disturb the fluid surrounding the sensory hair cells, generating electrochemical information that travels as nerve impulses via the emanating auditory nerve to the auditory cortex for interpretation. The inner chamber connects with another structure, the semicircular canals of the vestibular apparatus, a system essential for positional balance. These structures are shown in Figure 13.3.

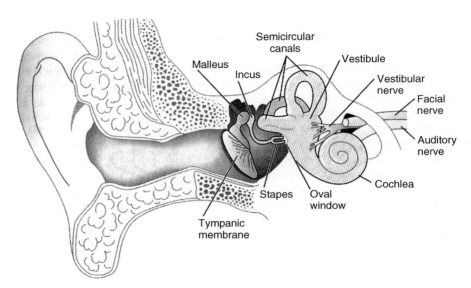

Figure 13.3. Anatomy of the ear. (Reprinted with permission from Costanzo (2006).) (See plate section for color version.)

Sound waves are characterized by amplitude and frequency. In general, *wave amplitude determines loudness*, the intensity of which is measured by the decibel (dB) scale (dB of various human activities are given in Table 13.1). *Frequency (number of waves per unit time given in cycles per second, cps, or hertz, Hz) represents the pitch of sound*. Humans can discriminate between about 2000 pitches in an optimal range of 1000–4000 Hz.

Presbycusis Is Age-Related Hearing Loss

The prevalence of hearing impairment (HI) also termed presbycusis or age-related hearing loss (ARHL) in the elderly is high. The prevalence of ARHL has been assessed by several studies over the past 10 years: Beaver Dam Offspring study of 3285 subjects 14–82 years of age (55% males; 33% females with ARHL in a subgroup of 65–84 years of age); BLSA (~40% ARHL in 347 individuals age 55+); Blue Mountain Study (~40% ARHL in individuals 65+), and Centers for Disease Control

TABLE 13.1. Significance of Decibels

Decibels (dB)	Activity	Comments
30	Whisper	Barely audible
40–50	Average classroom, moderate rainfall	Moderate
60–80	Alarm clock, busy traffic	Very loud
>85 prolonged exposure		Hearing loss
90–110	Subway, bass drum, chain saw	Extremely loud
120–160	Firearms, jackhammer, jet taking off	Painful

and Prevention and the National Health Interview Survey (47% males and 30% women have trouble hearing, 65 years and over). ARHL is not only widespread in the elderly but unfortunately, underdiagnosed and undertreated. Many individuals for social and psychological reasons persist in denial of HI.

Results of studies using animal models (genetically modified mice) clearly confirm the detrimental effect of noise exposure on hearing loss. *Chronic noise exposure causes an incsrease in hearing thresholds and accelerates presbycusis.* An increase in reactive oxygen species (see Chapter 4) has been implicated in hearing loss. Mice with a reduced ability to repair mitochondrial DNA experience accelerated hearing loss and conversely, mice on a caloric restricted diet exhibit a delay in onset of hearing loss. Although not as well studied, other pathways such as calcium metabolism, sex hormone levels, and glucocorticoid levels appear to affect hearing loss. These findings have not been confirmed in man.

Based on histological analyses, there are four subtypes of presbycusis. *They are sensory, neural, metabolic (striatal), and cochlear conductive.* Sensory presbycusis is the most common type. It reults from apoptosis of cochlear hair cells and causes *ARHL at high frequencies*. Hearing of consonants is diminished. Despite the clarity of vowels, inability to hear consonants prevents normal conversation. Neural presbycusis associated with reduced function of the spiral ganglion neurons (primary auditory neurons) *prohibits discrimination between two pitches of the same intensity*. Metabolic presbycusis related to decreased blood flow to the cochlea *produces an elevation of hearing thresholds*. Sound volume must increase to compensate ARHL. Thickening and stiffening of the basilar membrane produces *cochlear conductive presbycusis with modest ARHL over all frequencies*. Although defined as separate entities, in reality, the *elderly usually exhibit a mixture of the 4 subtypes*.

Often individuals with ARHL have difficulty admitting to this loss. Behaviors that suggest denial of ARHL are (i) difficulty in understanding the speaker, (ii) frequent requests for repetition, (iii) a slow hesitant response, (iv) no facial response, (v) cocking head to listen, (vi) failure to respond, and/or (vii) reply in monotone voice either too loud or too soft.

Olfaction: The Sense of Smell

An Abundance of Olfactory Receptors Reside in the Mucosa of the Upper Nasal Cavity
The olfactory receptor cells are located in the olfactory mucosa lining of the upper third of the nasal cavity. The receptors respond to odorants and relay the neuronal information through the olfactory bulb to the olfactory nerve and then to several brain areas. Olfactory receptor cells constantly renew themselves from the basal layer and can detect nearly 400 different chemicals.

Olfaction Atrophy Diminishes Sense of Smell and Depresses Sense of Taste
Although, *olfaction function is highly variable among the elderly, there is, nevertheless, a measurable decline after the age of 60 with earlier and greater decline in men.* Major olfactory impairment is present in 75% of those older than 80 years of age. This was established by Doty et al. (1984) in a study that administered a validated olfactory test, the University of Pennsylvania Smell Identification Test, to

1955 volunteers ranging in age from 5 to 99 years. In another study, the Beaver Dam Epidemiological Study (Murphy et al., 2002), ~63% of 80–97 year old exhibit loss of olfaction, although self-reported incidence in this age group is considerably smaller (12–18%). Generally, olfactory impairment in threshold and perceived intensity occurs across a range of odorants.

Histological evidence suggests that the olfactory mucosal undergoes decline with age as olfactory receptors disappear (apoptosis) and are poorly replaced. Reduced CNS processing may also contribute to a loss of smell.

In addition to age, other conditions contribute to olfactory impairment: head trauma, viral infections, rhinosinusitis (inflammation of the nasal cavity and surrounding sinuses), and neurodegenerative diseases, for example, idiopathic Parkinson's disease (IPD) and Alzheimer's disease (AD). Treatment of the first three conditions facilitates return of some olfactory function. Loss of olfaction is one of the first signs of IPD and AD and is not reversible. In addition to specific diseases and injury, medications belonging to numerous drug classes (e.g., antibiotics, anticonvulsants, antidepressants, antihistamines, antihypertensives, anti-inflammatory agents, antipsychotics, etc. (Bromley, 2000) depress olfaction.

Olfactory impairment causes several problems. There is an inability to detect chemicals that may lead to accidental gas exposure or food poisoning. Inability to detect body odor may alter relationships. Finally, the sense of smell contributes to the sense of taste. Taste impairment lessens interest in food, may alter food selection, reduces QOL, and promotes malnutrition.

Gustation: Sense of Taste

Decreased Taste Perception by the Elderly May Relate to Changes Other than Aging of Taste Buds The gustatory receptors (taste buds) located on the tongue (and throughout the pharynx, larynx, and into the first third of the esophagus) sense five different tastes: sweet, sour, bitter, salt, and umami (savory). The afferents from the gustation receptors are numerous and travel in different pathways to the nucleus of the tractus solitarius tract, and then onto the gustatory cortex of the brain.

Although there are subjective complaints of decreased taste perception by the elderly, *specific data on receptor sensitivity is scant*. In a small study of 12 young (20–29 years of age) and 12 elderly (70–79 years of age) subjects, Matsuda and Doty (1995) found that the response to increasing concentrations of salt (NaCl) in two tongue regions was markedly depressed in the older age group.

Taste perception in the elderly may be depressed for secondary reasons, including (i) *a decrease in saliva* that prevents access of savory chemicals to the taste buds, and (ii) *oral diseases*, for example, gingivitis, which produces interfering offensive chemicals. Additionally, several medications, especially anticholinergics and angiotensin-converting enzyme inhibitors, and some diseases, for example, diabetes, repress saliva production and produce a "dry" mouth that reduces taste sensation. Also, there are data to show that age-related changes in chewing and swallowing may understimulate olfactory receptors in the retronasal area and reduce their contribution to taste.

Somatosensory Afferents: Multiple Modalities of Touch, Vibration, Pressure, Temperature, Pain, and Proprioception

Somatosensory input comes from modalities of proprioception, touch, vibratory sense, pain, and temperature. Special receptors for these modalities relay neuronal information through ascending tracts to the cerebral cortex. The first three use the dorsal column medial lemniscus pathway and the last two use the ventrolateral spinothalamic pathway. These modalities are lost or compromised with spinal injury, sensory neuropathologies, and aging.

Proprioception Decreases with Age; Elevates Risk of Falls Proprioception is defined as the "senses of position and movement of our limbs and trunk, the sense of effort, the sense of force, and the sense of heaviness" (Proske and Gandevia, 2012). The uniqueness of this sensory avenue is that the individual generates these perceptions but is usually unaware of them. For example, one does not need to see the limb to know where it is. There are several receptors that facilitate proprioception. They are, in order of degree of contribution to proprioception, the muscle spindles (found in skeletal muscles), the golgi tendon organ (located within the tendons), and joint and skin receptors (mechanoreceptors discussed below). As the *main sensory component of proprioception, the muscle spindle* mediates both conscious sense of position, motion, and dynamics (position during movement) and the unconscious reflexes. Normal muscle spindle proprioception can be perturbed with vibrational forces that produce "noise" in the system and inhibit muscle spindle function. This manipulation induces an increase in errors relating to position and movement sense, sharing similarities with aging of the muscle spindles.

Studies on the effect of age on proprioception are characterized as small in size, cross-sectional in design, and with a focus primarily on positional sense of the lower limbs. As reviewed by Goble et al. (2012), *most of these studies have demonstrated a reduction in positional proprioception with age under a variety of test conditions*. In contrast, the *effect of age on joint movement also called kinesthesia is not as clear*. The threshold for detection of movement increases with age, but discrimination of the degree of movement varies with the muscle group under study. Although there are limited data, dynamic sense (combining motion and position) in which, for example, one monitors the passive movement of the ankle at different velocities and signals when a specific joint displacement angle has been achieved also lessens with age. However, dynamic acuity appears to worsen only with increase in velocity of rotation.

A number of studies have related deficits in proprioception acuity with instability of upright balance and incidence of falls. Lord et al. (1999) determined stance stability with a simple task whereby individuals stand with one foot in front of the other, slightly to the side, for 20 s, without and then with eyes closed. Interestingly, poor performance on this balance test is correlated with a higher incidence of self-reported falls of the preceding year. Only among the oldest age group is there a relation between reduced proprioception acuity and total time needed to perform ADLs (walk 50.8 ft, get out of a chair and walk 50.8 ft, ascend 11 stairs and descend 11 stairs).

Recent studies employ interventions such as muscle tendon vibration, surfaces changes, for example, compliant versus supportive, and the platform-based sway referencing to perturb proprioception acuity in the elderly in the presence of different stressors. Examples of these stressors are (i) muscle fatigue (Bisson et al., 2014), (ii) chronic physical exercise (Maitre et al., 2013), and (iii) reintegration challenge (Doumas and Krampe, 2010). Muscle fatigue decreases proprioception acuity at all ages, but it is especially harmful when proprioception information is less reliable as in the elderly. The *combination of muscle fatigue with reduced proprioception elevates the risk of falls. Chronic physical exercise is correlated with an increased ability to counteract declinations in proprioception*, and is a positive influence. The *ability to reintegrate positional adaptation information is age-sensitive*. This means that neural information used to adapt to one positional challenge is more likely to be repeated without modification in a different positional challenge and so is more likely to cause a mistake and a possible fall. Others report that *visual motion adds to positional instability* to a greater extent in older (70–79 years) individuals compared to other age groups (60–69 years and 18–20 years). Addition of a static reference point within the visual field produces less positional movement and more stability.

Loss of proprioception acuity is attributed to several factors: changes in muscle spindles (structure/function), sarcopenia, reduced neuronal conduction, and altered central cortex processing. There are specks of evidence for each. One early histological examination of human muscle spindles found a thickened capsule, decreased spindle diameter, decreased number of intrafusal muscles, and degenerated endplates. Sarcopenia is common in the elderly and a loss in muscle mass and change in composition appear related to reduced proprioception but the mechanism is unknown. Some evidence suggests that afferents of the ascending tracts disappear with age and recent MRI studies, probing brain areas related to proprioception, note a relation between proprioception acuity and reduced structure-dependent neuronal activity in a specific brain region (right putamen, areas similarly affected by Parkinson's and Huntington's diseases) in the elderly. These interesting findings need additional investigation.

Cutaneous Mechanoreceptors: Sense of Touch, Vibration, and Pressure

Mechanoreceptors are the ends (dendritic) of afferent fibers intertwined with other cell types to form a sense organ. They include Meissner's corpuscles, Merkel cells, Ruffini corpuscles, and Pacinian corpuscles (see Figure 13.4). Mechanoreceptors respond to modalities of touch/slow vibrations, pressure/touch, sustained pressure, and deep pressure/fast vibration, respectively. Since mechanoreceptors are activated during joint movement (skin stretch), they, especially Pacinian corpuscles, factor in joint kinesthesia proprioception, as discussed above.

Goble et al. (2012) reported that the density and absolute number of Meissner's and Pacinian corpuscles decrease with age. Functionality of Meissner's corpuscles (nerve recording of postmechanical vibrations and electrical stimulation) in 12 individuals 18–64 years of age shows loss of Meissner's corpuscles mechanical responsiveness. Bruce (1980) reported a twofold elevation in touch threshold of the hand of older subjects compared to data obtained in an earlier comparable study of young subjects. Gescheider et al. (1994) reported that the threshold for detection of vibrations increases with age (10–89 years, onset near 65 years of age) and is most

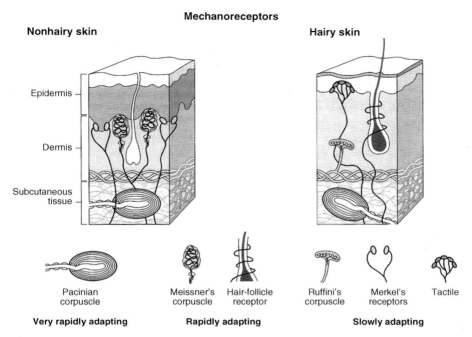

Figure 13.4. Skin mechanoreceptors. (Reprinted with permission from Costanzo (2006).) (See plate section for color version.)

evident in measurement of function of Pacinian corpuscles. Subjectively determined impression of the magnitude of vibrations delivered to the skin of the hand is lower for elderly (~68 years, five subjects) compared to younger subjects (~23 years, five subjects) at all intensities (Verrillo et al., 2002).

With changes in touch afferents, two-point discrimination declines. Several studies indicate the loss is greater in distal extremities (feet and hands) and is associated with increased risk of falls as well as clumsiness in handling small objects, for example, table utensils. Clearly, larger studies are needed to assess the role of cutaneous mechanoreceptors in ADLs.

Nociceptors Sense Pain from Chemical, Mechanical, and Temperature Modalities; Role in Aging Unknown

Nociceptors, basically bare nerve endings, are widely distributed in skin and deep tissue. They sense reactive chemicals, for example, histamine, irritants, strong acids, heavy pressure, and temperature extremes. Microelectrode recordings of electrophysiological activity of C-fibers (mediating dull, intense pain) of the elderly indicate the presence of unusual properties (mix of hyper- and hypofunction), similar to recordings from patients with neuritic pain. However, relevance of these findings to pain perception in health elderly was not determined.

It is postulated that a decrease in pain perception leads to untreated tissue trauma. Loss of sensations for deep pressure may manifest itself as bed sores or bruises without discomfort. This is a plausible explanation but at present there are no supportive data.

Thermoreceptors The various cold thermoreceptors differ in their responses to thresholds of coolness, ranging from pleasant cooling to painful cold (<10 °C). Warm thermoreceptors sense warmth from 30 to 40 °C. The sensors are afferent fibers and recent findings have identified specific ion channels that detect discrete temperatures.

The effect of age on thermoreceptors is poorly defined. Reports indicate that the number and conduction velocity of thermoreceptors remains constant with age. Despite this, thermal sensitivity decreases with age and a number of studies report increased cutaneous threshold to warm stimuli, although some parts of the skin exhibit increased threshold to cold stimuli. Considerable variability in responsiveness with age is also evident.

A decline in temperature sensitivity may result in tissue damage and burns that go unrecognized, and misjudgment in temperature assessment. Additionally, it is postulated that altered thermoreception is one of several factors that contribute to disturbances in the circadian rhythm of core body temperature, typically highest in the late afternoon and lowest in the early morning. With age, the amplitude (high versus low) deceases and the onset of the trough appears later in the morning (advanced phase). While there is expected variability in circadian core temperature rhythm, it is emphasized that core temperature is absolutely essential for optimal cell, tissue, and organ homeostasis (affecting other body rhythms, enzyme activities, general metabolism, etc.). It is predicted that distortions in the circadian temperature pattern will exert a domino effect throughout the body.

CONSEQUENCES AND COMPENSATION

Visual Deficits Minimized with Refractive Extra/Intraocular Lens, Common Sense Practices, and Possible Exercises The most recognizable sensory age change is presbyopia (loss of accommodation), affecting most if not all elderly. There is no known way to avoid presbyopia except perhaps with cataract surgery and implantation of the multifocal intraocular lens. Another procedure, conductive keratoplasty (corneal remodeling with heat from radio waves), provides a temporary fix (several months) relief. In lieu of these procedures, the inability to see small print or surface details necessitates "reading" glasses, monovision with contact lenses, or use of a magnifying glass.

The thickened yellowed lens reduces color perception, causing a failure to distinguish among objects of blue, green, and violet. Depth perception also decreases and distorts judgment, for example, regarding descending stairs. These visual changes promote errors, accidents, and falls. Practical compensation suggests use of hand rails on stairs, increase in light intensity (sunlight preferable), and alteration of the texture, design, and/or brightness between similar objects. Finally, corneal, lens, and retinal cell loss decrease the size and acuity of the visual field and necessitate frequent updates of corrective lenses.

Due to senile miosis, objects in dimly lit rooms are more difficult to discern and night vision is poor, additional changes that add to judgment errors and accidents.

TABLE 13.2. Interventions for Sensory Aging

Presbyopia
- Magnifying glasses for reading and close workSenile miosis

Senile miosis
- Improve lighting day (sunlight, ↑ wattage) and night (nightlight)

Reduced acuity/decreased processing of visual information
- Frequent change of corrective lenses; visual search and object recognition exercises; vitamin A?

Presbycusis
- Audiometric testing and if needed, use of hearing aid and/or assistive devices; learning other means of communication
- Noise abuse prevention
- Avoid ototoxic drugs

Positional sway and increased risk of falls
- Chronic physical exercise
- Yoga, tai chi, and balance training

Olfaction
- Avoid environmental toxins
- Suppress sinonasal infections/inflammation
- Retraining?
- Determine food safety by date not smell

Thermoreception
- Use a thermometer to gauge temperature

Dimly lit rooms can be accommodated with lights of increased wattage at altered angles, and/or increased exposure to sunlight. Night vision is assisted with use of night lights. Of interest are the results of a small RCT of 104 elderly subjects, half of which received a high daily dose of vitamin A (50,000 IU oral) for 30 days (Owsley et al., 2006) that showed improvement in night vision (increased rate of rod-mediated dark adaptation) in the group taking vitamin A. According to Owsley et al. (2006), this meant that those taking vitamin A compared to placebo could detect an object in an environment that was twice as dim. Whether long-term use of oral vitamin A in the elderly continues to improve/maintain dark adaptation would be important to know.

Analyses of several interventional studies show that visual search and object recognition exercises improve visual processing time and the benefit of such exercises could last up to 2 years or longer with booster sessions. This improvement profits daily visual tasks (driving, finding items, counting change, and reading directions). Availability of these exercises for interested elderly would be the logical next step.

ARHL Receives Partial Compensation from Hearing Devices but the Best Treatment Is Prevention of Noise Abuse

ARHL (presbycusis) is largely due to chronic exposure to environmental noise. Several regulations (Noise-Induced Hearing Loss Regulation, 2003; Environmental Regulations for Workplaces, 1987) have put in place requirements for safe levels of workplace noise. However, many home tools, audio devices, restaurants/bars, and Rock/Pop concerts exceed normal levels and are unregulated. Thus, their avoidance is prudent. Even short exposure to high decibels is considered toxic to hair cells. In addition, there are several medications (e.g., aminoglycoside antibiotics) that are ototoxic and should be avoided.

Considering the elevated prevalence of hearing loss among the elderly and the resultant restraints: communication handicap with family and friends in all settings, interference with leisure activities (listening to TV, radio, and music), and gathering information (directions, alerts), population screening for ARHL would seem reasonable. However, as noted by Walling and Dickson (2012), an assessment of the success of such a screen, a prerequisite as justification, for its implementation is impossible. This is because only a handful of screened individuals that are found positive for ARHL actually follow-up with audiometric testing and of those, only a few implement the recommendations of an audiologist. The consensus opinion from three prominent organizations (American Academy of Family Physicians, American Speech-Language-Hearing Association, and Institute for Clinical Systems Improvement) recommend physicians discuss ARHL with their patients, assess communication difficulties and possibly administer the whisper test (details available at http://webmedia.unmc.edu/intmed/geriatrics/reynolds/pearlcards/functionaldisability/whisper_test.htm). Audiometric testing should be proposed based on the patient's own hearing handicap or failure on the whisper test. Alternatively, testing is recommended every 3 years after age of 50 if there is evidence of risk factors, for example, noisy workplace environment.

A physical exam will reveal the presence of obstruction of the tympani membrane with cerumen (ear wax) that augments hearing loss. Cerumen is removed by "curetting, nonprescription solutions (hydrogen peroxide-based), warm water irrigation, and prescription cerumenolytics" (Walling and Dickson, 2012). A history of the onset of hearing loss is a valuable means to distinguish ARHL (bilateral, slow, and progressive) from age-independent unilateral and abrupt (perforated or obstructed tympani membrane, trauma) hearing loss.

Hearing aids are of two types: analogue, which basically amplifies sound, and digital, which improves sound sensitivity and can more accurately target the specific hearing problem. Regardless of the style (behind the ear, in the ear, or in the canal), *less than a quarter of elderly that need hearing aids purchase them and of those only a third actually use them*. McCormack and Fortnum (2013) reviewed 10 articles that examined the explanations for nonuse of fitted hearing aids. The main reason was lack of benefit. *Present-day hearing aids cannot replace normal hearing and fail to adequately suppress background noise*. Other reasons include uncomfortable fit, difficulty with maintenance (changing battery, cleaning), attitude of hearing just as well without it, device factors (frequent battery change and service, whistling), financial reasons (cost of batteries and repair), psycho-social/situational factors (annoyance, lost, forget), health care professional's attitudes (not helpful), ear problems (ear wax, tinnitus), and appearance (cosmetic) (McCormack and Fortnum, 2013).

In addition to hearing aids, there are a number of hearing assistive technology systems (frequency modulation, infrared, visual) that mesh with the phone, doorbell, computer, TV, and others home appliances that may improve hearing with or without a hearing aid. Surgical implants (middle ear or cochlear) are effective but limited by their expense and surgical risk.

Proprioception Acuity Improves with Training Proprioception acuity is related to the level of physical activity. Physically active individuals exhibit greater proprioception sensitivity. Thus, *proprioception is modifiable with practice*. Interventions of yoga, tai chi, and balance training for 12 weeks were all effective in improving postural stability in the elderly (average age of 74 years) with a history of

---

Summary of Age Changes of the Sensory System

Vision
- Presbyopia: loss of accommodation (near vision) due to ↓ flexibility of lens
- ↓ color discrimination (blue, green, violet)
- Senile miosis → ↓ dark/light adaptation
- ↓ visual field: senile miosis
- Slowed visual processing, ↓ motion detection → deterioration of rods, changes in visual cortex
- ↓ acuity: thickened lens; corneal thickening (minor)

Audition
- Presbycusis (sensory, neural, metabolic, and cochlear conductive) → ↑ audible threshold; high frequencies (consonants) not heard; ↓ pitch discrimination

Olfaction
- ↓ olfactory receptor number
- ↓ smell perception → ↑ threshold of detection
- ↓ discrimination
- ↓ taste perception

Gustation
- ↓ taste perception → ↑ threshold of detection
- ↓ discrimination
- Worsens with ↓ production of saliva due to medications, oral diseases, systemic diseases, changed patterns of chewing

Proprioception
- Atrophy of skeletal muscle spindles
- ↓ acuity; ↓ discrimination of position/movement sense
- ↑ positional sway
- ↓ balance control
- Worsens with muscle fatigue, rapid movement, visual motion

Cutaneous mechanoreception
- ↓ Meissner's and Pacinian corpuscle number
- ↑ threshold/↓ discrimination for vibration, pressure, touch
- ↓ two-point discrimination for touch

Thermoreception
- No change in number
- Possible ↓ sensitivity to coolness and warmth

falls. A 9-month longitudinal study (Seco et al., 2013) of simple physical activity training (55 min biweekly, exercises of stretch, strength, and aerobics) with 3 months follow-up (two groups 65–74 and >74 years) significantly improve strength, flexibility, and balance. An RCT (Martínez-Amat et al., 2013) (44 individuals 61–90 years of age) targeting proprioception found that a 12-week focused program (using BOSU and Swiss balls) significantly improved measured parameters of proprioception acuity as well as balance and gait.

Somatosensory Receptors Preserved with Toxic Chemical Avoidance Prevention may be one of the best ways to preserve olfaction (and indirectly gustation). There is evidence to suggest that olfaction declines with repeat exposure to environmental abuse: toxins, volatile chemicals, tobacco smoke, and airborne pollutants. Additionally, chronic infections and inflammation (sinonasal disease) contribute to loss of the nasal epithelium and olfactory receptors. On the other hand, improvement in olfaction (threshold and discrimination) appears possible with prolonged olfaction practice.

SUMMARY

Perception of the environment occurs through specialized structures (sensory organs/receptors/afferent nerves/channels) for vision, audition, gustation, olfaction, temperature, pain, and touch/proprioception. These varied senses ensure individual safety and enhance quality of life. Explanations abound as to why these highly discriminating structures fail but few are well substantiated.

The stimulus is specific for each sense modality such as light waves (vision), sound waves (hearing), various small molecules (taste/smell/pain), pressure/deformation (touch, position), and muscle activity (proprioception). To initiate a perceptual response, the stimulus must reach a threshold (minimal quantity). In general, an increase in sensory threshold is characteristic of aging. Other characteristics are slowed processing time and reduced discrimination.

Aging of the eye brings with it presbyopia (loss of near vision), senile miosis (reduced dark/light adaptation), reduced spatial contrast (color) sensitivity in low light, slowed visual processing, and reduced motion perception.

Hearing loss in the elderly, called presbycusis, leads to the inability to hear high frequencies (consonants), loss of pitch discrimination, and/or elevation of speech threshold. Chronic exposure to noise and ototoxic drugs contribute to hearing loss.

Olfaction and secondarily gustation are reduced with age. Changes reduce interest in food and may lead to accidents (food poisoning; inability to detect gas).

The sense of proprioception that factors into balance and stance stability deteriorates with age and is postulated to contribute to positional sway and increased risk of falls. Threshold to pain, touch, vibration, and deep pressure may also increase with age, although data for this are limited.

Numerous interventions alleviate sensory loss but to different extents. Refractive lens (glasses, contact lenses, intraocular lenses, magnifying glass, and laser treatment) significantly improve vision, hearing aids, assistive devices and cochlear implants minimally compensate for age-related hearing loss, and physical and sensory-focused exercises hold promise to benefit declining proprioception.

CRITICAL THINKING

Why are sensory threshold and sensory discrimination important concepts to know?

What are the main visual changes experienced by the elderly? What problems might they cause?

What are the main age-associated changes in hearing? What problems might they cause?

How do changes in smell, taste, touch, proprioception, temperature, and pain affect QOL for the elderly?

Suggest as many ways as possible for the elderly to optimize their vision, hearing, proprioception, and sense of smell. Give an estimated monetary cost for each.

KEY TERMS

Accommodation contraction of the ciliary body to facilitate near vision (allows the lens to become more convex).

Afferents nerves that receive information from the environment and relay it to the brain.

Cochlea the structure of the inner ear containing hair cells (receptors for hearing) and the vestibular canals (for position sensing).

Dark adaptation reflex acclimation to the dark that dilates the pupils to allow more light in.

Decibel intensity level for sound; measurement of loudness.

Gustation sense of taste found in the special receptors dispersed among the taste buds of the tongue.

Hair cells specialized cells with small extension (cilia appearing as "hair"); found in the inner ear bathed in fluid.

Iris smooth muscles that form the pupil of the eye; contract (constrict) or relax (dilate) in response to changes in presence and absence of light respectively.

Lens encapsulated transparent spherical disc between the anterior and posterior chambers of the eye; refracts light for near and far vision.

Olfaction sense of smell due to receptors in the nasal mucosal epithelium that sense small airborne molecules.

Ototoxic harmful to audition; usually refers to drugs that damage the cochlear hair cells or auditory nerves.

Photoreceptors special receptors in the retinal nervous tissue that convert light to nerve activity. These are of two types: rods and cones.

Presbycusis age-related loss of hearing; primarily due to disappearance of hair cells.

Presbyopia age-related loss of accommodation.

Proprioception afferent nerves in muscles, joints, and tendons that sense position and movement together or separately

Rod and cones specialized receptors for sight; rods are involved in dark adaptation and cones participate in sensing light

Senile miosis age-related loss of dark adaptation—inability to fully dilate the pupils.

Spiral ganglion nerves nerves receiving information from the hair cells and relaying it to the auditory region of the cortex.

BIBLIOGRAPHY

Review

Barett KE, Barman SM, Boitano S, Brooks HL. 2012. *Ganong's Review of Medical Physiology*, 24th ed. New York: McGraw Hill, p. 178.

Barett KE, Barman SM, Boitano S, Brooks HL. 2012. *Ganong's Review of Medical Physiology*, 24th ed. New York: McGraw Hill, p. 188.

Boisgontier MP, Olivier I, Chenu O, Nougier V. 2012. Presbypropria: the effects of physiological ageing on proprioceptive control. *Age (Dordr.)* **34**(5):1179–1194.

Bromley SM. 2000. Smell and taste disorders: a primary care approach. *Am. Fam. Physician* **61**(2):427–436.

Costanzo LS. 2006. *Physiology*, 3rd ed. Philadelphia: Saunders Elsevier, p. 87.

Costanzo LS. 2006. *Physiology*, 3rd ed. Philadelphia: Saunders Elsevier, p. 76.

Hüttenbrink K-B, Hummel T, Berg D, Gasser T, Hähner A. 2013. Olfactory dysfunction: common in later life and early warning of neurodegenerative disease. *Dtsch. Arztebl. Int.* **110**(1–2):1–7.

Kidd AR, III, Bao J. 2012. Recent advances in the study of age-related hearing loss: a mini review. *Gerontology* **58**(6):490–496.

McCormack A, Fortnum H. 2013. Why do people fitted with hearing aids not wear them? *Int. J. Audiol.* **52**(5):360–368.

Murphy C, Schubert CR, Cruickshanks KJ, Klein BEK, Klein R, Nondahl DM. 2002. Prevalence of olfactory impairment in older adults. *JAMA* **288**(18):2307–2312.

Nash SD, Cruickshanks KJ, Klein R, Klein BE, Nieto FJ, Huang GH, Pankow JS, Tweed TS. 2011. The prevalence of hearing impairment and associated risks factors: the Beaver Dam Offspring Study. *Arch. Otolaryngol. Head Neck Surg.* **137**(5):432–439.

Owsley C. 2011. Aging and vision. *Vision Res.* **51**(13):1610–1622.

Pinto JM. 2011. Olfaction. *Pro. Am. Thorac. Soc.* **8**(1):46–52.

Proske U, Gandevia SC. 2012. The proprioceptive senses: their roles in signaling body shape, body position and movement, and muscle force. *Physiol. Rev.* **92**(4):1651–1697.

Spankovich C, Le Prell CG. 2013. Healthy diets, healthy hearing: National Health and Nutrition Examination Survey, 1999–2002. *Int. J. Audiol.* **52**(6):369–376.

Walling AD, Dickson GM. 2012. Hearing loss in older adults. *Am. Fam. Physician* **85**(12):1150–1156.

Experimental

Bao J, Ohlemiller KK. 2010. Age-related loss of spiral ganglion neurons. *Hear Res.* **264** (1–2):93–97.

Bisson EJ, Lajoie Y, Bilodeau M. 2014. The influence of age and surface compliance on changes in postural control and attention due to ankle neuromuscular fatigue. *Exp. Brain Res.* **232**(3):837–845.

Bruce MF. 1980. The relation of tactile thresholds to histology in the fingers of elderly people. *J. Neurol. Neurosurg. Psychiatry* **43**(8):730–734.

Cruickshanks KJ, Nondahl DM, Tweed TS, Wiley TL, Klein BE, Klein R, Chappell R, Dalton DS, Nash SD. 2010. Education, occupation, noise exposure history and the 10-yr cumulative incidence of hearing impairment in older adults. *Hear Res.* **264** (1–2):3–9.

Doty RL, Shaman P, Applebaum SL, Giberson R, Siksorski L, Rosenberg L. 1984. Smell identification ability: changes with age. *Science* **226**(4681):1441–1443.

Doumas M, Krampe RT. 2010. Adaptation and reintegration of proprioceptive information in young and older adults' postural control. *J. Neurophysiol.* **104**(4):1969–1977.

Gescheider GA, Bolanowski SJ, Hall KL, Hoffman KE, Verrillo RT. 1994. The effects of aging on information-processing channels in the sense of touch: I. Absolute sensitivity. *Somatosens. Mot. Res.* **11**(4):345–357.

Goble DJ, Coxon JP, Van Impe A, Geurts M, Van Hecke W, Sunaert S, Wenderoth N, Swinnen SP. 2012. The neural basis of central proprioceptive processing in older versus younger adults: an important sensory role for right putamen. *Hum. Brain Mapp.* **33**(4):895–908.

Heft MW, Robinson ME. 2010. Age differences in orofacial sensory thresholds. *J. Dent. Res.* **89**(10):1102–1105.

Hummel T, Rissom K, Reden J, Hähner A, Weidenbecher M, Hüttenbrink K-B. 2009. Effects of olfactory training in patients with olfactory loss. *Laryngoscope* **119**(3):496–499.

Hurley MV, Rees J, Newham DJ. 1998. Quadriceps function, proprioceptive acuity and functional performance in healthy young, middle-aged and elderly subjects. *Age Ageing* **27**(1):55–62.

Lin FR, Ferrucci L, Metter EJ, An Y, Zonderman AB, Resnick SM. 2011. Hearing loss and cognition in the Baltimore Longitudinal Study of Aging. *Neuropsychology* **25**(6):763–770.

Lord SR, Rogers MW, Howland A, Fitzpatrick R. 1999. Lateral stability, sensorimotor function and falls in older people. *J. Am. Geriatr. Soc.* **47**(9):1077–1081.

Maitre J, Jully J-L, Gasnier Y, Paillard T. 2013. Chronic physical activity preserves efficiency of proprioception in postural control in older women. *J. Rehabil. Res. Dev.* **50**(6):811–820.

Martínez-Amat A, Hita-Contreras F, Lomas-Vega R, Caballero-Martínez I, Alvarez PJ, Martínez-López E. 2013. Effects of 12-week proprioception training program on postural stability, gait, and balance in older adults: a controlled clinical trial. *J. Strength Cond. Res.* **27**(8):2180–2188.

Matsuda T, Doty RL. 1995. Regional taste sensitivity to NaCl: relationship to subject age, tongue locus and area of stimulation. *Chem. Senses* **20**(3):283–290.

Owsley C, McGwin G, Jackson GR, Heimburger DC, Piyathilake CJ, Klein R, White MF, Kallies K. 2006. Effect of short-term, high-dose retinol on dark adaptation in aging and early age-related maculopathy. *Invest. Ophthalmol. Vis. Sci.* **47**(4):1310–1318.

Seco J, Abecia LC, Echevarría E, Barbero I, Torres-Unda J, Rodriguez V, Calvo JI. 2013. A long-term physical activity training program increases strength and flexibility, and improves balance in older adults. *Rehabil. Nurs.* **38**(1):37–47.

Sindhusake D, Mitchell P, Smith W, Golding M, Newall P, Hartley D, Rubin G. 2001. Validation of self-reported hearing loss. The Blue Mountain Hearing Study. *Int. J. Epidemiol.* **30**(6):1371–1378.

Sturr JF, Zhang L, Taub HA, Hammon DJ, Jackowski MM. 1997. Psychophysical evidence for losses in rod sensitivity in the aging visual system. *Vision Res.* **37**(4):475–481.

Verrillo RT, Bolanowski SJ, Gescheider GA. 2002. Effect of aging on the subjective magnitude of vibration. *Somatosens. Mot. Res.* **19**(3):238–244.

Winn B., Whitaker D, Elliott DB, Phillips NJ. 1994. Factors affecting light-adapted pupil size in normal human subjects. *Invest. Ophthalmol. Vis. Sci.* **35**(3):1132–1137.

14

AGING OF THE ENDOCRINE SYSTEM

NEUROENDOCRINE SYSTEM

Overview

The classic definition of an *endocrine gland* is a tissue that produces *hormones* (proteins, steroids, small molecules) that exit the gland, circulate in the blood, and regulate diverse activities of growth, reproduction, metabolism, and response to stress, at distal sites throughout the body. In this regard, the endocrine system is a major supervisor of organ system function. Specific hormones, for example, parathyroid hormone, insulin and melatonin regulate levels of indispensible ions, nutrient availability and associated metabolic functions, and sleep/wake states, respectively. Others, such as adrenal steroids are crucial for the expression of the stress response, in which the characteristic "fight or flight" reaction quells a perceived threat. As hormone levels change with age, essential functions are either lost or dramatically impaired.

Endocrine glands fall into two categories: those within the hypothalamus–pituitary axis, such as the ovaries, testes, thyroid, adrenal glands, and the liver, and those outside this network. The latter respond to specific stimuli, such as glucose availability (pancreas), blood level of ionized calcium (parathyroid gland), and light/dark cycle (pineal). All hormones exert their diverse effects by binding to specific receptors.

Endocrine aging is gland specific. With one exception, age-related endocrine alterations are highly variable, as is the prevalence of age-associated endocrine-based disorders, for example, late-onset hypogonadism. The one exception is menopause,

Human Biological Aging: From Macromolecules to Organ Systems, First Edition. Glenda Bilder.
© 2016 John Wiley & Sons, Inc. Published 2016 by John Wiley & Sons, Inc.

the cessation of ovarian estrogen and progesterone production, a loss that terminates female reproductive function. The majority of age-dependent endocrine changes are characterized as reductions, for example, of concentrations, receptors, sensitivity, functions. In contrast, the prostate gland continues to grow producing a benign hyperplasia and the concentration of adrenal steroids remains elevated for a longer period than necessary. Elevated levels of cortisol are associated with a vulnerability to several diseases, including cancer.

Evidence of hormonal deficits in the elderly has suggested to some that hormone replacement is needed and as such could be antiaging. However, this has not been substantiated. Hormone therapy (HT) for relief of classic symptoms of menopause is safe and effective in the short term (several years) but long-term use remains controversial. Safety and efficacy studies on HT relating to growth hormone, testosterone, and melatonin changes have not been done and antiaging therapy with these hormones is considered premature.

Glands, Hormones, and Regulation

Components of the Neuroendocrine System: Hypothalamus, Pituitary, and Peripheral Glands

The *neuroendocrine system exerts exquisite control over reproduction, response to stress, growth, and metabolism.* The three components of this system are (i) the hypothalamus, (ii) the pituitary gland, and (iii) peripheral glands. *The hypothalamus is the "master gland"* and consists of a group of neurons emanating from the paraventricular nucleus located in the brain region. Hypothalamic nerves produce neuropeptides called releasing hormones (see Table 14.1 for specific hormones of the neuroendocrine system and associated gland). *Hypothalamic releasing hormones communicate with the pituitary gland* situated directly below it and stimulate anterior

TABLE 14.1. Glands and Hormones of the Neuroendocrine System

Gland	Hormone
Hypothalamus	Corticotrophin-releasing hormone (CRH), Thyrotropin-releasing hormone (TRH), Gonadotropin-releasing hormone (GnRH), Growth-hormone-releasing hormone (GHRH), Somatostatin
Pituitary	Adrenocorticotropic hormone (ACTH), Thyroid-stimulating hormone (TSH), Luteinizing hormone (LH), Follicle-stimulating hormone (FSH), Growth hormone (GH)
Peripheral glands	
Adrenals	Cortisol
	Aldosterone
	Androgens, for example, dehydroepiandrosterone (DHEA)
Thyroid	Thyroxine (T_4)
	Tri-iodothyronine (T_3)
Ovaries	Estrogen
	Progesterone
Testes	Testosterone
Liver	Insulin growth factor-1 (IGF-1)

pituitary cells to *produce and release tropic hormones. Tropic hormones enter the blood to act on peripheral endocrine glands (adrenal glands, thyroid gland, liver, ovaries in females, and testes in males)*. Peripheral endocrine glands in turn biosynthesize specific hormones that direct sexual maturation, reproduction, growth, metabolism, and response to stress. In all cases, hormones achieve their effects by binding to and activating specific membrane or intracellular receptors.

Hormone Levels Are Controlled by a Negative Feedback Inhibition Mechanism

Several different mechanisms operate to ensure hormone availability in amounts that match the needs of the organism. *One such mechanism is called negative feedback inhibition*. This mechanism works as follows: a releasing hormone (A) from the hypothalamus (gland 1) reaches the pituitary (gland 2) where it stimulates production and release of a tropic hormone (B) from the pituitary. Hormone B enter the blood and as the concentration of hormone B rises, it inhibits (negative feedback) production of hormone A in gland 1. Negative feedback inhibition causes the concentration of hormone A to fall, and with that, stimulation of gland 2 declines and the concentration of hormone B also falls. Figure 14.1 depicts negative feedback inhibition. *With age, the negative feedback inhibition lessens its regulatory control and in some cases, ceases completely*. Not surprisingly, tissues experience deviant levels of hormones that induce serious consequences leading to inefficiency, atrophy, damage, degeneration, and disease.

Other types of hormonal control include positive feedback in which a hormone enhances rather than inhibits its own production. In addition to stimulation/inhibition of hormonal synthesis and release, hormonal effects may be modulated by variations in release characteristics, for example, pulse, surge, tonic, cyclic, changes in their rate of metabolic inactivation (degradation), changes in binding proteins (increase or

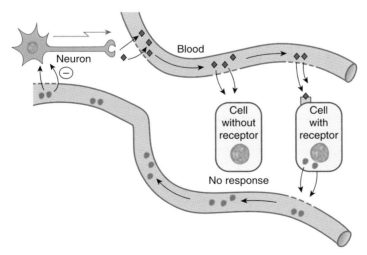

Figure 14.1. Negative feedback inhibition. Hypothalamic releasing hormone stimulates select pituitary cells (those with responsive receptors) to release the tropic hormone that inhibits release of the hypothalamic hormone (negative feedback loop). (Reprinted with permission from Patrick Bilder, PhD.) (See plate section for color version.)

NEUROENDOCRINE AGE CHANGES

Menstrual Cycle Is Regulated by the Hypothalamus–Pituitary–Ovarian Axis

Female sexual characteristics and reproductive potential develop during adolescence with the pulsatile release of gonadotropin releasing hormone (GnRH). This event in turn stimulates pituitary production and release of gonadotropins, luteinizing hormone (LH) and follicle-stimulating hormone (FSH). Subsequently, FSH and LH activate the ovaries to produce estrogen (17-β-estradiol, E2) and progesterone (P). Over the ensuing years from puberty to menopause, the menstrual period is tightly regulated by the hypothalamus-pituitary-ovary hormonal axis. The maturation of the ovum and preparation of the uterus for possible fertilized egg implantation is driven largely by LH and FSH surges and cycles that dominate the monthly fluctuations of E2 and P. Negative and positive feedback mechanisms are indispensable for normal cycles. Additional proteins, termed ovarian inhibins, contribute to neuroendocrine control.

Estrogen Stimulates Specific Intracellular and Membrane Receptors to Produce Diverse Effects

There is no question that E2 is a driving force for reproduction. However, as reviewed by Cui et al. (2013), E2 also affects activities in other tissues such as the heart, liver, bone, muscle, and brain. The role of E2 in bone density, cardiovascular health, lipid and energy metabolism, and cognition is presently under clinical study and is discussed below.

E2 acts by stimulating one of two nuclear receptors, E2 receptor alpha (ERα) and E2 receptor beta (ERβ). Once activated, these receptors bind to DNA as transcriptional factors and "turn on" specific genes, whose effects are experienced within hours or days. Receptors of the ERα type are found in reproductive tissues, brain, bone, liver, and in lesser amounts in many other tissues. Receptors of the Erβ type are expressed in the colon, bone marrow, brain, endothelium, lung, and bladder. E2 also exhibits rapid effects (within minutes) by stimulating membrane receptors or exerting direct effects on select enzymes. *Ovarian E2 is the main source of estrogen in premenopausal women*, synthesized from cholesterol through several enzymatic steps to the androgen, androstenedione, that is converted to testosterone and finally to E2.

Progesterone similar to E2 stimulates specific nuclear receptors, in this case, one of two forms termed PR-A and PR-B, to assist with reproductive function. P plays a prominent role in uterine homeostasis, stimulating and modifying cells lining the uterus and also exerts important effects on the brain such as neuroprotection in traumatic brain injury.

Menopause

Menopause is Unique to Humans Reproductive senescence in mid-life (around the age of 51 in humans) is a unique phenomenon that occurs in humans but not in closely related species. An analysis by Alberts et al. (2013) of five nonhuman primates (Madagascan prosimian, New World monkeys, Old World monkeys, and two Great Apes) with a human population (hunter-gathers in the Kalahari Dessert of South Africa devoid of effects of industrialization on lifespan) found that the accelerated rate of infertility and reproductive loss in mid-life in humans is distinct to them and not the norm in nonhuman primates whose fertility decline matches their mortality rate, that is, they reproduce at ages close to time of death. Why fertility in women declines at a faster rate than senescence in other organ systems is an enigma.

Dramatic Loss of Reproductive Function Due to Ovarian Failure

Menopause is the time at which reproductive function ceases for females. Clinically, the onset is defined as the time following one year of no menstrual cycles (amenorrhea). Figure 14.2 defines the peri- and postmenopausal changes.

According to Rance (2009), "Ovarian failure is the critical determinant of menopause in women." Women start life with between one-half to one million *primordial follicular cells that slowly degenerate over time*. With the disappearance of follicular cells, production of E2 and P ceases, causing a disruption of the negative feedback inhibition to the pituitary and hypothalamus. The consequence of this dysregulation is two-fold: *FSH, LH, and GnRH are no longer inhibited and their levels increase* and the loss of E2 and P bodes ill for E2- and P-dependent tissues, for example, all reproductive tissues as well as the brain, liver, bone, skin, and others.

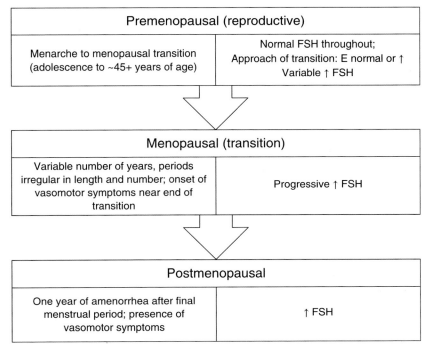

Figure 14.2. Stages of menopause. Estrogen (E); follicle-stimulating hormone (FSH). (Reprinted with permission from Soules et al. (2001).)

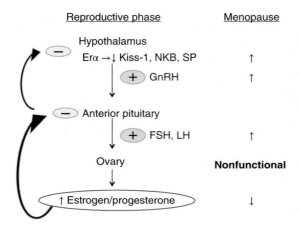

Figure 14.3. Regulatory changes in the hypothalamus–pituitary–ovarian axis during menopause. KiSS-1: kisspeptin; NKB: neurokinin B; Erα: estrogen receptor alpha; GnRH: gonadotropin-releasing hormone; FSH: follicle-stimulating hormone; LH: luteinizing hormone; SP: substance P. Hypothalamic mediators are discussed in Rance (2009). (See plate section for color version.)

The onset of menopause occurs in the perimenopausal phase. During this transition of several years, neuroendocrine control of ovulation and uterine preparation is slowly lost. Small increases in FSH signal waning inhibitory activity from the ovaries (decrease in inhibins and E2) tend to accelerate follicular pool exhaustion. In early transition, plasma E2 concentration is normal or elevated (FSH is elevated and inhibin is low) but as negative feedback inhibition lessens, there are fewer ovulatory cycles. Late transition brings amenorrhea (no menstrual bleeding), anovulatory cycles, variability in hormone levels, and cycle lengthening. Menopausal symptoms, for example, hot flushes, are related to erratic E2 levels.

The final menstrual period starts the postmenopausal phase in which there are no follicles in the ovaries, and no ovarian E2, a state of castration as if the ovaries had been surgically removed. Menopause does not diminish hypothalamic release of GnRH and their presence facilitates the increase in FSH and LH. A subset of hypothalamic neurons (infundibular nucleus) characterized by their lack of production of GnRH and expression of ERα receptors enlarge in menopause. It is proposed that the loss of E2 permits increased expression of various modulators (neurokinin B (NKB), substance P (SP), kisspeptin-1(Kiss-1)), and repression of others (dynorphin) that activate production of GnRH by neighboring neurons. These mechanistic changes of menopause are illustrated in Figure 14.3.

Effects of Menopause Include New Profile of Estrogens, Vasomotor Reactivity (Hot Flushes), and Atrophy of E-Dependent Tissues

Cessation of ovarian function reorders the availability of estrogens. Of the three forms of E—estrone (E1), 17-beta estradiol (E2 or estrogen), and estriol (E3)—the predominant and most potent E2 in premenopausal women is replaced with the less potent E1 in postmenopausal women produced by nongonadal tissues, for example, kidney, fat, skin, and brain. However, circulating levels of both E2 and E1 decline 90 and 75%, respectively. Tissues with the enzymatic machinery to convert cholesterol to E or at a minimum contain the aromatase, the obligatory enzyme for the final step in synthesis of E from testosterone (T) will continue to produce some E. Menopause does not prohibit production of E2 in cerebral neurons and astrocytes. In postmortem brain samples from individuals 50–97 years of age, E and also androgen levels remain constant (Rosario et al., 2011).

With the loss of circulating E/P in menopause, reproductive tissues that include the vaginal epithelium, the cervix, the uterus, oviducts, and ovaries undergo atrophy (reduction in size). The lining of the uterus (the endometrium) atrophies, but according to some studies retains for years the potential for hormonal "rejuvenation" and possible implantation of a fertilized egg later in life.

One of the most apparent effects of menopause, initially appearing in the perimenopausal phase, is the vascular or vasomotor dysfunction identified as "hot flashes", "night sweats," or "flushing". The word flushing is preferred since vascular irregularities are most frequently of a prolonged nature. Flushes are described as feelings of intense warmth, uncomfortable sweating, and/or sensations of unwanted body heat that may occur repeatedly throughout the day. Hot flushes are moderate to severe in 60% of the women, mild in 20%, and absent in 20% and may persist up to 5 years or longer. HT is generally effective in reducing or suppressing hot flushes but is recommended for short-term use due to an associated cancer risk. Some hot flush relief is experienced with weight loss, smoking cessation, and low ambient temperatures. Over-the-counter remedies such as evening primrose oil, soy, and black cohosh are not recommended as short-term efficacy is questionable and their long-term safety is unknown.

The mechanism of the vasomotor flushing remains a mystery, although some data show that the fluctuation of E2 (and other factors) in the hypothalamus narrows the normally sensed temperature range. Thus, a slight increase in body temperature (smaller than normal) provokes a sweating response associated with increased blood flow to the skin to facilitate heat loss. Table 14.2 summarizes the hot flush.

The second most frequent menopausal complaint is vaginal and vulvovaginal atrophy (VVA). It occurs in 50% of postmenopausal women, although it is considered underreported. Lack of E2 thins the vaginal epithelial lining, and decreases renewal rate such that exfoliation of cells eventually increase vaginal pH and permits bacterial infiltration. These changes lead to dryness, soreness, pruritus, painful intercourse, and infections. Therapy may include low-dose systemic or local HT.

TABLE 14.2. Additional Observations on Hot Flushes

- Prevalence—highly variable around world (0–80%)
- Occurrence—anytime of day or night; most likely with embarrassment, stress, sudden temperature change due to alcohol, caffeine, warm drink, and so on
- Starts in upper body and spreads downward
- Perceived duration 0.5–60 min (average: 3–4 min)
- Continues for 1 year; 5 years or more for 25% of women
- Interferes with work, activities, and sleep
- May lead to fatigue and depression
- General order: sweating, increased blood flow, sensation of warmth, and increase in heart rate
- Effective therapies—low dose daily estrogen of conjugated equine estrogens (0.3 mg), oral estradiol (0.5 mg), or transdermal estradiol 14 µg
- Other therapies—some efficacy with serotonin receptor uptake inhibitors (SSRI, for example, Prozac); generally ineffective are phytoestrogens, soy isoflavones, and acupuncture

Summarized from Sturdee (2008).

According to Ortmann and Lattrich (2012), other postmenopausal symptoms include sleep changes, urinary tract symptoms, mood changes, and sexual problems. These are less consistent and generally improve with diminution of hot flushes. Of these, only urinary tract symptoms improve with local HT.

Menopause Facilitates Bone Loss Estrogen is one of several factors required for normal bone formation (called osteogenesis) and, as expected, bone mineral density (BMD) declines following menopause (see Chapter 8). Clinically, significant bone loss termed osteoporosis is serious since this disease is associated with an increased risk of fractures. Fractures in the elderly lead to a vicious downward cycle of reduced mobility, dependence, and additional fractures. HT is approved to prevent osteoporosis in high-risk women and in osteoporotic women where other pharmacotherapies have failed.

Menopause Accelerates Aging of the Skin A deterioration of skin quality relating to thickness, wrinkles, and dryness appear in postmenopausal women. Observational and clinical trial results have shown that systemic E favorably benefits the skin by increasing thickness, elevating sebum production, lessening dryness, and erasing some wrinkles, effects attributed to E-induced increase and preservation of collagen and elastin synthesis. For example, in a prospective, randomized double-blind, placebo-controlled trial of 40 postmenopausal women who received 7 months of HT of E/P, epidermal hydration, vascularization, elasticity, thickness, and quantity of collagen increased and the number and depth of wrinkles decreased (Sator et al., 2007).

Controversial Use of HT to Prevent Cardiovascular Disease, Cognitive Decline, and Sexual Dysfunction It remains controversial as to whether HT can (i) reduce the risk of cardiovascular disease and (ii) preserve age-associated changes in cognition, libido, and mood. Several large randomized controlled trials, the Heart and Estrogen/progestin Replacement Study (HERS) and the Women's Health Initiative (WHI) attempted to investigate the role of HT in the prevention of cardiovascular disease and cognitive loss. One arm of the WHI sought to determine whether E/P HT in postmenopausal women protected women against heart disease. Accordingly, it enrolled more than 16,000 postmenopausal women, treating half with conjugated equine estrogen (0.625 mg) and medroxyprogesterone acetate (2.5 mg) and half with a placebo. All women had an intact uterus. The study was stopped three years prematurely because of the increased incidence of breast cancer observed in women taking the hormones. The study also found an increased incidence of coronary heart disease (CHD) events (nonfatal MIs), increased incidence of strokes, and increased incidence of venous thromboembolism. Additionally, there was also a *decreased* incidence of hip fractures and colorectal cancers in women taking E/P compared to placebo (see Table 14.3 for incidence values). The results of the HERS study completed prior to the WHI indicated that E/P therapy (same dose as used in WHI) had no effect on CHD in healthy postmenopausal women, but did *increase* CHD risk in the early phase of therapy in women with prior evidence of CHD (the bulk of the participants in HERS).

While the negative results of the WHI study resulted in a significant reduction in use of E/P among postmenopausal women, further scrutiny of the WHI findings

TABLE 14.3. Results of WHI: Estrogen/Progesterone to Women with Uterus, Postmenopausal

Clinical Outcome	Incidence (E/P versus Placebo)
CHD events	29% increase; 37 versus 30 per 10,000
Stroke	41% increase; 29 versus 21 per 10,000
Venous thromboembolism	Doubling; 34 versus 16 per 10,000
Breast cancer	26% increase; 38 versus 30 per 10,000
Colorectal cancer	37% decrease; 10 versus 16 per 10,000
Endometrial cancer; lung cancer	No change; 54 versus 50 per 10,000
Hip fractures; vertebral fractures	30% decrease; 10 versus 15 per 10,000
Mortality	No difference

Summarized from Rossouw et al. (2002).

revealed a flawed study design. Specifically, 66% of postmenopausal study participants were older than 60 years of age at the start of the study and 74% of them had never used E/P HT. Thus, *hormones were abruptly introduced to the majority of study participants more than 12 years after menopause.* A later review of a subgroup of WHI women at *ages closer to the menopause transition showed that HT provided cardiovascular protection*. Additionally, measurement of blood lipids in a small sampling of WHI women yielded a profile (high HDL and low LDL) consistent with cardiovascular protection. Furthermore, in women without a uterus, given only E therapy (another arm of the WHI) showed cardiovascular protective effects (less calcium deposition in arteries suggesting less atherosclerosis). In summary, the first major report on the findings of the WHI concluded that risks of E/P replacement in postmenopausal women for cardiovascular protection outweighed the benefits. Subsequent reanalyses point out a flawed experimental design; subset reanalyses suggest that HT during menopause transition may lower CV risk. Two additional trials (early versus late intervention with estradiol, the Kronos Early Estrogen Prevention Study) are underway to clarify these issues.

The effect of E and P on mood has not been firmly established. Depression is thought to be underdiagnosed in the elderly. However, although the incidence of depression in menopause has been reported in excess of 50% in the United States and ~42% in a foreign country, for example, Turkey, these rates appear to be similar to those reported for premenopausal women in the study sample. On the one hand, it is accepted that women may be vulnerable to depression as a result of hormone changes during the premenstrual or the period immediately preceding menstruation, pregnancy, postmiscarriage, postpartum, and the perimenopause phase but, on the other hand, it is generally concluded that most women do not suffer from depression during menopause. Some predisposing factors to depression in menopause are prior depression, postpartum depression, severe menopausal symptoms (flushes and insomnia), elevated body mass index (BMI), and low socioeconomic status.

Since E exerts effects on neuronal function in animal models and in cells in culture, there is considerable interest to determine whether cognition, for example, memory, processing speed, executive function, declines as a result of menopause and whether this decrement could be prevented with E therapy and by extension, and would E therapy prevent dementia. The majority of results show that *E therapy is ineffective in the prevention of dementia but whether small changes in cognition in fact occur with menopause and are prevented with E therapy is unresolved.* As discussed at an NIA symposium relating in part to defining cognitive complaints in the menopause transition, memory function and processing speed may decline during the transition in association with the presence of hot flushes. This group concluded that the data were limited as to whether E therapy reversed these changes or whether E/P might be in fact detrimental. Some reviews found that cognition decline is associated with mood disorders, including depression, anxiety, and panic, and treatment of these conditions should improve cognition, while other reviews noted that small transient changes in cognition are independent of hot flushes, mood disorders, and sleep disturbances. It seems reasonable to conclude that E could play a role in neuronal health, considering its role in brain development, its synthesis in the brain, and the presence of E receptors, but how that translates into a transient or permanent decline in cognition following menopause is yet to be determined.

Risk/Benefits of Hormone Therapy Current advice on HT for postmenopausal symptoms is detailed in Table 14.4. HT use is based on individual need and if deemed necessary, it is initiated at the minimally effective dose, which is generally much lower than that used in major clinical trials, and its need is reassessed on a regular basis.

TABLE 14.4. Recommendations for Hormone Therapy (HT) for Menopausal Symptoms

- Individualized evaluation of benefit/risk of HT is recommended
- Systemic estrogen therapy (ET) or estrogen/progesterone therapy (EPT) (FDA approved)—treatment of vasomotor symptoms (hot flushes) and associated consequences (sleep disturbances, irritability, difficulty concentrating, and decreased QOL)
- ET—short-term treatment of symptoms of genital and vaginal atrophy; local and systemic HT therapies effective and FDA approved. PT: not recommended
- ET—FDA approved for prevention of menopausal osteoporosis in women at high risk and as therapy in women with osteoporosis where other therapies have failed. Continuous use to maintain bone density
- To reduce risks from HT: initiate early in menopause; use of no more than 3–5 years for EPT, somewhat longer for E; the lowest effective dose (route a matter of choice as all are effective); EPT always if uterus intact
- HT—not recommended for the following: treatment of sexual dysfunction, for example, decreased libido; improvement in QOL issues (other than related to decrease in menopausal symptoms); improvement in physical strength; coronary artery protection; and treatment of cognitive aging or dementia
- Risks—increased cardiovascular risk with HT started late (>10 years postmenopause); no CV risk if started early in menopause; increased risk of thromboembolism, less if started early; increased risk of endometrial cancer if E used unopposed; increased breast cancer risk if EPT used >3–5 years

Based on the North American Menopause Society 2012 recommendations (The North American Menopause Society, 2012).

Andropause

Male Reproduction Is Regulated by the Hypothalamic–Pituitary–Testes Axis
The male reproductive system is regulated through the hypothalamic–pituitary–testes axis in a fashion that shares similarities with regulation of the female reproductive system already described. In particular, pulsatile release of GnRH from the hypothalamus stimulates the synthesis and release of pituitary gonadotropins, LH, and FSH, which stimulate cells of the testes. LH stimulates the production of testosterone (T) from the Leydig cells and FSH stimulates spermatogenesis and production of inhibin (protein participating with T in negative feedback inhibition) in the sertoli cells. As shown in Figure 14.4, both T and inhibin inhibit production of gonadotropins and GnRH.

Similar to regulation of female reproductive hormones, *T is regulated by negative feedback inhibition*. Additionally, blood levels of T are influenced by the presence of binding proteins, sex hormone binding globulin (SHBG), and albumin. T not bound to these proteins is "free" and, therefore, in the active form. Thus, changes in binding proteins influence the amount of free T. Another consideration is that T can be converted to dihydrotestosterone (DHT), which is more efficacious (effects at a lower concentration) than T. So, the enzyme required for this conversion indirectly regulates T-dependent activities. Finally, T can be converted to E by action of an aromatase and aromatase levels increase with enlargement of fat stores, thus potentially moderating T-dependent effects.

T and DHT are the main male reproductive androgens that are responsible for spermatogenesis and development of secondary sex characteristics. Additional actions include protein, lipid, and glucose metabolism, maintenance of skeletal muscle mass, and libido. Androgens play a role in bone maintenance (largely via conversion to E) and possibly in cognition, although this is controversial.

Modest Age-Related Changes in the Hypothalamic-Pituitary-Testis Axis
Most males experience modest changes in reproductive function with age. This is in marked contrast to the absolute infertility of menopause. The terms manopause and andropause have been coined, but unlike the precipitous drop in E/P concentrations in women, total T levels in males fall moderately and slowly at about a rate on average of *1% per year beginning about 35–40 years of age. Furthermore this change is highly variable*. The testicular Leydig cells, producers of T tend to disappear with time explaining in part the age-relate decline in T. As a result, LH and FSH levels increase (reduced negative feedback). Generally, the diurnal rhythm of T (peaking at 6–8 A.M.

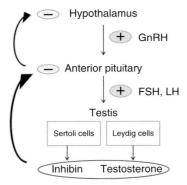

Figure 14.4. Regulation of the hypothalamus–pituitary–testes axis. GnRH: gonadotropin-releasing hormone; FSH: follicle-stimulating hormone; LH: luteinizing hormone.

and declining 35% later in the day) is well maintained into old age. Compared to changes in T, *serum androgen precursors, sulfated dehydroepiandrosterone (DHEAS), and androstenedione fall markedly with age*. Other changes reported in some men include an increase in SHBG (reducing free T) and variable changes in E. DHT levels appear to remain constant with age despite a reduction in T.

There is a debate regarding whether "low T" is a clinical entity. Clearly, a severe T deficiency independent of age (hypogonadism) with impaired reproduction, infertility, decreased libido, decreased muscle mass, and osteoporosis is an endocrine disorder that responds well to T replacement therapy. On the other hand, T therapy in the elderly with "low T" is generally not recommended unless accompanied by symptoms. The prevalence of low T (<300 pg/ml) in the elderly is 10–25%. Based on data from the Massachusetts Male Aging study (Araujo et al., 2004), 7–12% of men (40–67 years of age) in this study group had both low T and sexual symptoms. A larger study (European Male Aging Study) (Wu et al., 2010) of 3369 men (40–79 years of age) defined a syndrome of late-onset hypogonadism characterized by at least three sexual symptoms, for example, poor morning erection, low sexual desire, erectile dysfunction, and total T <320 pg/ml and free T <64 pg/ml. The prevalence of late-onset hypogonadism in this sampling was 0.1% in individuals 40–49 years to 5.1% in those 70–79 years.

T Therapy Remains Controversial for Late-Onset Hypogonadism

T levels are lowered by many factors other than age. They include smoking, severe psychological stress, alcohol, obesity, and type 2 diabetes (T2D). All are modifiable through lifestyle choices and changes should be instituted prior to election of T therapy. Results of short-term studies show that T therapy increases libido, muscle mass, and bone mineral density of the spine and decreases fat mass, thus producing favorable effects on sex drive and body composition. Improvement in erectile dysfunction with T therapy is less impressive and generally T therapy does not improve physical function or muscle strength. The most serious limitation to T therapy in the elderly at present is the lack of long-term efficacy and safety studies.

Dysregulation of the Hypothalamic-Pituitary-Testicular Axis Is One of Many Reasons for Benign Prostatic Hypertrophy

The prostate gland is a secondary sex gland, positioned around the urethra beneath the bladder. It provides about 30% of the seminal fluid that supports sperm survival. Continual growth of the gland (epithelial and stromal cells) creates a glandular enlargement, termed benign prostatic hyperplasia (BPH). BPH occurs in most men (prevalence > 80% at 80 years of age). In three-quarters of men, this size increase is benign but the remaining 25% suffer serious urinary problems to include: increased frequency of urination, difficulty starting/stopping urination, urinary retention that may lead to retrograde filling of ureters and inner part of kidney, painful urination; and urinary incontinence.

The cause(s) of BPH is(are) unknown. Since the prostate is a hormone sensitive gland, age-associated changes of the hypothalamic-pituitary-testes axis are considered responsible for BPH, at least in part. An elevation in the E/T ratio has been proposed. Other hormones, insulin and insulin-like growth factor-1 are additional possibilities. Risk factors of anabolic steroid use, environmental factors and genetics may also contribute to gland enlargement.

A four STEP (Simplified Treatment of the Enlarged Prostate) treatment of BPH has been proposed (Rosenberg et al., 2010). It begins with watchful waiting in patients with tolerable symptoms (Step 1) and progresses to use of medications (Steps 2,3) (one to relax the prostatic myocytes supplemented with another to reduce the level of the active form of T, dihydrotestosterone) in patients with intolerable symptoms. Step 4 encourages consultation with a urologist specialist.

The most serious concern with BPH is the possible progression to a malignancy. This may be assessed with an annual screening that includes a digital rectal examination (DRE) and measurement of the prostate specific antigen (PSA). If cancer is suspected, a biopsy is done. Treatment depends on the histological evaluation of the biopsy, comorbidities and the acceptance of adverse outcomes from the selected treatment. Generally for low and intermediate risk cancer in men with life expectancy of <10 years, active surveillance is reasonable which would include annual DRE and PSA. Where treatment is warranted, radical prostatectomy (removal of prostate, seminal vesicles, pelvic lymph nodes) is done. Other tested treatments include interstitial brachytherapy (local high dose irradiation delivered with implantation of radioactive pellets), and external beam radiotherapy. As adjuvant therapy, medical castration (with luteinizing hormone-releasing hormone agonist or anti-androgen) may be needed.

Hypothalamic–Pituitary–Adrenal Axis

The outer layer (cortex) of each adrenal gland produces several important hormones called adrenal steroids. Adrenal steroids include glucocorticoids (cortisol, corticosterone), mineralocorticoids (aldosterone), and androgens (dehydroepiandrosterone (DHEA), dehydroandrosterone). In particular the synthesis and release of glucocorticoids are stimulated by the pituitary adrenocorticotropic hormone (ACTH). ACTH secretion in turn is stimulated by the hypothalamic releasing hormone (corticotropin-releasing hormone (CRH)). Thus, *blood levels of adrenal corticoids are regulated by negative feedback inhibition*, as shown in Figure 14.5. Unique to the HPA is its responsivity to pain and emotional stress, for example, fear, anxiety. Adrenal steroids exert their diverse effects by combining with specific intracellular receptors, that act as transcriptional factors to stimulate gene expression.

The inner portion (medulla) of the adrenal gland is regulated by the sympathetic nervous system (SNS). With SNS activation, epinephrine, a vasoactive catecholamine is released into the circulation. It acts in concert with other effects of SNS activation.

Dysregulation of the Adrenal Glands Yields an Altered Response to Stress with Detrimental Consequences The concentration of cortisol is regulated by (i) a circadian or diurnal rhythm and (ii) the stress response. With regard to the circadian (24 h) clock, cortisol levels rise in the morning and fall at night. Evaluation of this pattern in young (~29 years), old (~80 years), and centenarians (100+ years) subjects revealed that the 80-year-old group experienced a reduction in nightly cortisol levels compared to the young group. Interestingly, the cortisol diurnal rhythm of centenarians matched that of the young. Other study results note that since diurnal rhythm in the elderly is influenced by many factors, for example, increase in body mass idex (BMI), socioeconomic level, adverse health, age-associated changes in diurnal rhythm are at present difficult to interpret.

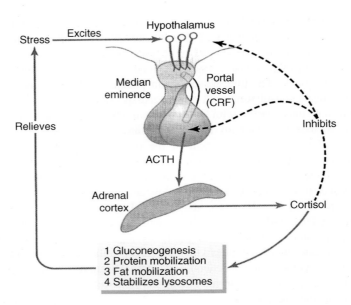

Figure 14.5. Regulation of the hypothalamic–pituitary–adrenal axis. (Reproduced with permission from Guyton and Hall (2006, p. 956).) (See plate section for color version.)

Higher order organisms react to a threat by a generalized but effective "flight or fight" response that includes (i) activation of the SNS (release of neurotransmitter, norepinephrine), (ii) release of adrenal steroids, mainly cortisol from the adrenal cortex, and (iii) release of epinephrine from the adrenal medulla. Together, the activated SNS and circulating hormones elevate blood sugar, improve blood flow to the skeletal muscles, dilate pupils, and stimulate brain and cardiovascular function. These changes enable humans and related organisms to stand ground or run away and hence generally survive environmental hazards. As observed in animals and subsequently confirmed in man, *age impacts the stress response by modification of the time course of adrenal steroids*. In aged animals, stress induces the "flight or fight" response as expected, but compared to young animals, the blood concentration of the *glucocorticoid remains elevated for a longer period of time*. Mechanistic studies show *that with age the hypothalamus and pituitary become less sensitive to the negative feedback of adrenal steroids, and specifically corticosterone in the case of rats*. Thus corticosterone persists longer than necessary.

Although data obtained in man regarding the hypothalamus-pituitary-adrenal axis are controversial and influenced by many confounding factors, Veldhuis et al. (2013) in critiquing this literature have emphasized several key findings. First as with animals, *the feedback inhibition of cortisol on ACTH lessens with age*. Therefore, in the elderly, the response to stress dissipates more slowly. On the one hand, elevated cortisol yields increased alertness and energy (increase blood glucose) but the *persistence of cortisol is detrimental because it is associated with reduced cognition, bone loss, immunosuppression, and sodium retention*. Repeated exposure to stress in the elderly correlate with the presence of proinflammatory mediators with potential for organ damage. Findings in animal models that relate abnormal adrenal stress responses to neurotoxicity and neurodegeneration additionally suggest that the age-associated aberrant hypothalamic-pituitary-adrenal response to stress is a possible risk factor for Alzheimer's disease.

Second, the release of ACTH/cortisol exhibits an age-dependent increased sensitivity to the effect of hypothalamic CRH, making it likely that smaller and smaller stresses induce higher and higher hormonal levels. This is especially prominent in women, an effect that can be dampened with HT. Overall, the HPA response to stress becomes more prominent (magnitude, time course) with age.

Circulating Levels of DHEAS and Androgens Decline with Age

The effects of DHEAS and other androgens are poorly understood. Their influence on male characteristics or protein metabolism is minimal unless converted to T or E. Hence, they are valuable sex steroid precursors. Production of DHEAS (main circulating form of DHEA) declines with age. DHEAS peak at around 30 years of age and then fall to levels 10% of that by age 80. Manufacturers of DHEA supplements claim that DHEA is antiaging because it restores muscle strength, physical vigor, and slows aging in general. However, the results of randomized placebo-controlled clinical trials, for example, DHEA and Well-Ness or DAWN trial, in *which subjects consumed 50 mg/day of DHEA for 1 year, showed no change in muscle strength, muscle mass, cognitive function, and bone mineral density (BMD) (men only) despite elevated levels of DHEA*, DHEAS, and metabolites generated from DHEA, for example, testosterone. However, an increase in BMD of the spine was observed in women. Although improved BMD would be an important effect of DHEA (if confirmed), this steroid failed as an antiaging therapy.

Effects of Aging on Aldosterone Are Unknown

The main action of aldosterone is to reabsorb sodium from the outgoing body fluids, for example, filtered blood and sweat, thereby regulating fluid volume and blood pressure. Aldosterone secretion is influenced not only by hypothalamic–pituitary hormones but also by renin, an enzyme from the kidney that generates angiotensin II, a potent vasoconstrictor and an inflammatory mediator, and by an elevation in blood potassium, sensed by the adrenals cortex.

Aldosterone does not appear to experience any particular effect of aging, although it is implicated in the development of cardiovascular (CV) diseases, such as hypertension and high levels of aldosterone in individuals with heart failure, CAD and myocardial infarction are associated with increased probability of death. One clinical trial found that pharmacological inhibition of aldosterone in patients with metabolic syndrome (collection of CV risk factors in one individual) already taking drugs that block angiotensin II experienced a reduction of left ventricular fibrosis. Additionally, recent studies have found aldosterone involvement in benign prostate hyperplasia in the presence of vitamin D deficiency. How age might have created this dependency remains to be determined.

Hypothalamic–Pituitary–Thyroid Axis

Thyrotropin-releasing hormone (TRH) activates the release of the pituitary thyroid-stimulating hormone (TSH) that subsequently causes the thyroid gland to synthesize and secrete thyroxine (T_4), which is converted in the thyroid and in peripheral tissues to the active thyroid hormone, tri-iodothyronine (T_3). Effects of T_3 are mediated by hormone binding to nuclear receptors that influence gene expression.

T_3 enhances the general rate of metabolic reactions (increases oxygen consumption) in most cells (except testes, uterus, lymph nodes, spleen, and anterior pituitary) and hence regulates resting or basal (baseline) metabolic rate. This major function exerts multiple secondary effects to accelerate fat, protein, and cholesterol metabolism and, influence cardiovascular and cognitive functions.

It is unclear whether the output of TSH increases or decreases with age as both results have been documented. One of the main problems contributing to this discrepancy is the presence of other diseases or conditions with potential to interfere with routine thyroid functions tests. For example, subclinical autoimmune thyroid disease, various chronic illnesses, such as atherosclerosis, certain medications, such as glucocorticoids for inflammation, lithium for psychiatric illness, and amiodarone for arrhythmia, malnutrition and endemic goiter skew thyroid test results. As reviewed by Peeters (2008) in studies that *excluded potential interferences, TSH and free T_3 levels decline with age, although T_4 remains unchanged*. An explanation for these changes is unknown.

Many factors determine whether age-associated changes in the hypothalamic-pituitary-thyroid axis warrant therapeutic intervention. Generally, if TSH is below 10 mIU/L, free T_4 is within normal range and there are no symptoms, therapy is not recommended. If with time, the concentration of TSH rises and hypothyroid symptoms, for example, fatigue, cold intolerance, weight gain, and depression develop, treatment with T_4 is considered reasonable. A puzzling observation that cautions the overuse of T_4 is that elderly with a profile of high TSH and low free T_4 appear to live longer than those with the opposite profile.

Although endocrine diseases of hypothyroidism and hyperthyroidism significantly depress and elevate basal or resting metabolic rate (RMR), respectively, the *small age-related reduction in thyroid hormones apparently contributes minimally to the well-known age-associated decline in RMR*. This is because at least two-thirds of the RMR is due to metabolism of fat-free body mass, for example, skeletal muscles and internal organs (see Chapter 7). As fat-free body mass disappears (due to sarcopenia, tissue atrophy), RMR also declines independent of thyroid status.

Hypothalamic–Pituitary–Liver Axis

Growth hormone (GH), synthesized in the anterior pituitary gland and secreted in pulses, is positively regulated by hypothalamic growth-hormone-releasing hormone (GHRH) and negatively regulated by growth hormone-inhibiting hormone (somatostatin). The secretion of GH is additionally influenced by diet, gender, adiposity, sleep, exercise, and age. The largest amount of GH is released during sleep and lesser quantities are evident following meals.

GH is a major player in skeletal muscle and skeletal systems. GH stimulates the production of the hepatic hormone, insulin growth factor 1 (IGF-1), and together as well as separately, these hormones support tissue growth and maintenance. In skeletal muscles, GH assists with the development of myofibers from satellite stem cells, enhances muscle protein synthesis, and prevents myocyte death or apoptosis to increase and/or maintain muscle mass. GH also increases and maintains bone mineral density by influencing proliferation of specialized bone cells, chondrocytes, and osteoblasts and by indirectly affecting calcium/phosphate metabolism via changes in kidney function. IGF-1 additionally controls GH levels through negative feedback inhibition of GHRH and stimulation of somatostatin.

Effects of GH Weaken with Age Due to Less Circulating GH and Reduced Tissue Sensitivity

GH and its induced companion, IGF-1, achieve maximal concentration at puberty and slowly decline thereafter. The time-dependent decay in circulating GH depends on aberrations in neuroendocrine regulation, for example, decrease in hypothalamic GHRH and an increase in somatostatin. However, *the extent of GH decline is primarily influenced by "controllable" factors such as adiposity (BMI), diet, and exercise.* Among these, elevated BMI is probably the most potent inhibitor of GH release. In general, older individuals exhibit GH and IGF-1 levels 5–20-fold lower than young individuals. In addition to a reduction in circulating amounts of GH and IGF-1, convincing results from animal models and limited results from human tissue biopsies and cultured cells suggest that the *response at the tissue level to these hormones also wanes with age*. Specifically, GH and IGF-1 receptors on skeletal muscle and bone slowly disappear with age. As these receptors decrease in number, so does the ability of GH and IGF-1 to induce beneficial tissue effects. These age-associated changes are termed somatopause.

The significance of somatopause is debated. In humans, somatopause is associated with sarcopenia and osteoporosis, conditions that could benefit from GH therapy. However, a meta-analysis of 18 clinical trials (Liu et al., 2007) with GH therapy (approximately 27 weeks) in the elderly (average age of 69 years) concluded that the risks, for example, onset of T2D and impaired fasting glucose, outweighed any benefits in body composition (increase muscle mass, decrease fat mass). Results of a recent clinical trial that used GHRH (subcutaneous injections of tesamorelin, an analogue of human GHRH) for 20 weeks found improvements in executive function, verbal memory, and visual memory in individuals with mild cognitive impairment and in healthy older individuals (55–87 years of age) (Baker et al., 2012). Nevertheless, a rush to use GH therapy to ameliorate age changes is premature until the role of GH/IGF-1 in aging is clarified and long-term safety studies with GH have been done.

ENDOCRINE GLANDS: SEPARATE FROM NEUROENDOCRINE AXIS

Endocrine glands that are not regulated by hypothalamic releasing hormones and pituitary tropic hormones include the parathyroid gland, the pancreas, and the pineal gland. They contribute to survival of the organism is different ways.

The pineal gland is located centrally at the base of the brain and is the locus for production of the hormone, melatonin. Melatonin production is circadian and levels increase in the dark and decrease in the light. Hence, melatonin serves as a marker of the sleep/wake cycle and disruption/dampening of this cycle is associated with insomnia, daytime sleepiness, and reduced alertness. A normal melatonin cycle is maintained primarily by sunlight impinging on the retinal ganglion photoreceptors in the eye that send neuronal information to the suprachiasmatic nucleus (SCN) and then to the pineal gland to regulate melatonin production. Melatonin in return circulates and stimulates specific melatonin receptors to produce brain activities related to sleep.

Optimal regulation of melatonin occurs when light illumination is high as in sunlight exposure. Light illuminance drops 200-fold with office or living room lighting and 2000-fold with nursing home lighting. Generally, when illuminance is extremely low, the "clock" cycling of the SCN takes charge but the consequences are reflected in poor quality sleep and cognitive difficulties. Melatonin acts to link the SCN function to the environmental light–dark cyclic changes.

Because some elderly experience poor quality sleep mainly characterized by a reduction in deep sleep time and an increase in light sleep time, changes that favor frequent awakenings and associated daytime sleepiness, an alteration in plasma melatonin is implicated. Most investigators have found an age-related decline in plasma melatonin. The 24 h plasma melatonin profile from a cross-sectional study of two groups of men (20–27 and 67–84 years of age) showed a significant 50% fall in melatonin concentration. A similar negative effect of age on melatonin levels has been reported for women.

Melatonin Levels Decline with Age in Large Part Due to Changes in the Eye

The age-associated drop in plasma melatonin appears related to several factors. Perhaps the most important is the aging eye, in particular the thickening of the crystalline lens that decreases light transmission and blocks entrance of the short wavelengths of violet and blue. It is the *blue portion of the light spectrum that is the most effective stimulator of retinal ganglion photoreceptors*. Second, senile miosis, weakened pupillary response to light, reduces the area of light penetration. These two alterations lessen photoreceptor stimulation and dampen the circadian rhythm of melatonin. Removal of the lens, as in cataract surgery and replacement with a intraocular lens, improves photoreceptor sensitivity and additionally, over a 9-month period following cataract surgery, enhances sleep quality. Moreover, reduced sunlight exposure as described above adversely affects melatonin levels. Other alterations such as calcification of the pineal gland and change in sensitivity of the SCN may also distort melatonin production.

Melatonin administration to animals reduces lipid peroxidation, inhibits expression of proinflammatory mediators, and in aged mice rejuvenates the atrophic thymus, a key motivator of immunosenescence. These changes rest partly on melatonin's antioxidant action, effective at pharmacological levels. Unfortunately, translating melatonin benefits in animals to humans has been difficult. At best, melatonin is an effective therapy to shorten the latency to sleep. Due to its short half-life (less than 1 h), stable analogues of melatonin (ramelteon in the United States and agomelatine in Europe) are approved for the treatment of insomnia. Whether melatonin preserves the immune system and prevents cancers and inflammatory diseases in man are yet to be determined.

Parathyroid Gland Senses Changes in Ionized Calcium

Parathyroid glands produce hormones, parathyroid hormone (PTH) and calcitonin that respond to disruption of calcium homeostasis. A fall in the level of ionized calcium in the blood stimulates the release of PTH that promotes calcium homeostasis in several ways: facilitates release of calcium from bone, stimulates vitamin D production, and prevents renal loss of calcium. In contrast, an elevation in ionized calcium signals secretion of calcitonin that counteracts the effects of PTH, lowers PTH levels, and ensures a reduction of blood calcium. *Maintenance of circulating free calcium is* critical for most activities in the body to include metabolic effects, neuronal function, muscle contraction, cardiovascular function, and many more.

Basal and Stimulated Parathyroid Hormone Levels Change with Age
Basal levels of PTH are influenced by serum phosphate and vitamin D. Meals high in phosphate elevate basal PTH, and vitamin D levels less than 30 ng/ml produce hyperparathyroidism. Basal PTH is *higher in the elderly than in the young possibly due to age-related changes of decreased renal and GI calcium absorption, increased prevalence of vitamin D deficiency, and reduced PTH calcium sensitivity.* Additionally, several studies in women and men observed a positive relation of age to the PTH maximal response, such that in the elderly, more PTH is produced to achieve the same level of ionized calcium compared to young subjects. Additionally, studies are needed to determine the significance of this change.

Pancreas: Nutrient Regulator

The pancreas is the chief regulator of nutrient distribution and utilization. It contains both endocrine and exocrine cells and is innervated by vagal neurons. Its two main hormones are insulin and glucagon produced in the islets of Langerhans by beta and alpha cells, respectively. The exocrine acinar cells supply the intestine with digestive enzymes. Exocrine enzymes and endocrine hormones work together to dispose of nutrients and although orchestrated in large part by the opposing actions of insulin and glucagon, there exists support from numerous assisting factors, for example, neuropeptides, incretins, and leptin.

It is well established that as plasma glucose levels increase, there is a concomitant increase in insulin that serves to reduce plasma glucose. In addition to this feedback, insulin inhibits release of its antagonist, glucagon. As reviewed by Chandra and Liddle (2013), in addition to plasma glucose, intestinal hormones (incretins) stimulated by ingested food play a major role in the stimulation of insulin release and the prevention of glucagon release. The hypothalamus by sensing glucose additionally helps and neurogenically stimulates insulin release. Leptin, a hormone from fat cells, decreases insulin secretion directly at the beta cell level or through effects on the hypothalamus. Other negative regulators of insulin secretion are melatonin, galanin (a ubiquitous neuropeptide), and endoplasmic reticulum (site of insulin formation) stress. It is hoped that a fuller understanding of the physiology of insulin and glucagon will lead to improved therapy for T2D and possibly delineate the role of aging in this disease.

Insulin Facilitates Uptake and Utilization of Sugar by Muscle/Fat Tissues and Enhances Storage in Liver
Insulin exerts its metabolic effect by binding to and activating the insulin receptor situated in the plasma membrane. This initiates the following changes: (i) accelerates the uptake of glucose and amino acids into muscle and increases the synthesis of muscle protein and glycogen, (ii) increases the uptake of glucose into adipose tissue and increases synthesis of fatty acids, and (iii) increases protein, fat, and glycogen synthesis in liver and decreases hepatic glucose production. Effects of glucagon oppose those of insulin and enhance the breakdown of hepatic glycogen (glycogenolytic), induce formation of glucose (gluconeogenesis), and accelerate breakdown of fats (lipolysis). Insulin itself is an effective inhibitor of glucagon secretion and vice versa, elevated glucagon blocks insulin secretion.

Prevalence of T2D Increases with Age The influence of age on the production and action of insulin are of considerable interest due to the current prevalence of T2D of ~26% (65 years and older) and the rising incidence with ~1.2 million diagnosed per year in individuals 45 years of age and older (National Diabetes Statistical Report, 2014). Unlike type 1 diabetes, in which there is a complete absence of insulin, *T2D is characterized by a normal, subnormal, or an elevated level of insulin and a resistance to the effects of insulin*. Initial compensation to elevated glucose leads to increased production of insulin that over time culminates in insulin exhaustion. However, the mechanism of insulin resistance is not known.

T2D displays definite manifestations that include elevated fasting glucose level >100 mg/ml (prediabetes), impaired glucose tolerance (inability to normally dispose of an oral glucose load), persistent hyperglycemia (elevated blood glucose), and subsequent microvascular damage that may cause retinopathy (blindness), neuropathy (pain), nephropathy (kidney failure), and limb ischemia (gangrene). Furthermore, the risk of bone fractures is elevated with T2D as is the risk for cardiovascular disease and its complications, for example, stroke.

Age and Lifestyle Choices Influence Insulin Sensitivity

A review of studies in man that probed the effect of age on insulin secretion and beta cell function found in general that *secretion of insulin declines with age independent of sensitivity changes*. There also appears to be an age-related *reduction in the response of beta cells to incretins* (GI hormones). Whether this is the initiating event to insulin resistance is not known. Complicating this are factors such as sarcopenia, lack of physical exercise, enlarged abdominal circumference, and dietary selections that contribute to impaired glucose tolerance and eventual insulin resistance. Insulin sensitivity can be improved with reversal of the aforementioned factors: replacement of a sedentary lifestyle with chronic aerobic and resistance exercise, reduced consumption of high carbohydrate diet, and reduction in weight, especially in abdominal fat. It is interesting to note that caloric restriction (CR), a procedure that extends the lifespan in animal models, improves insulin sensitivity and lowers fasting blood glucose. Comparable effects on insulin sensitivity and blood sugar are evident in studies of short-term (6 months to 1 year) CR in man.

T2D accelerates normal age changes and profoundly damages structural proteins such as collagen. The mechanism of vascular injury is largely the result of advanced glycation end product formation and continual oxidative stress. Reduced bouts of hyperglycemia significantly lessen microvascular damage, as shown by results of several clinical studies.

Management for T2D sets a glycemic control level at <7% glycated hemoglobin or the HbA_{1c} (American Diabetic Association). Individuals with prediabetes are encouraged to exercise, seek education on dietary options, and lose body fat. If these goals are not achieved, oral glucose lowering drugs, for example, metformin, and insulin sensitizing drugs, for example, rosiglitazone, may be added. Additional glycemic control is achieved with daily insulin injections.

Although insulin deficiency (relative or absolute) has been considered the sole cause of diabetes, convincing evidence in animals and emerging data from human studies point to a larger (and perhaps major) role of glucagon in diabetes. Glucagon opposes the action of insulin and promotes hyperglycemia through breakdown of liver glycogen and glucose formation. Study results suggest that *in T2D patients glucagon levels are high relative to insulin and could account for insulin resistance*. Results of physiological manipulations in animals are encouraging. For example, a 4-week infusion of glucagon in mice (Li and Zhuo, 2013) produced the T2D phenotype defined by elevation of fasted glucose level, impaired glucose intolerance without an increase in circulating insulin, and evidence of early kidney damage. These effects were inhibited with an antagonist of the glucagon receptor. As indicated by Li and Zhuo (2013), large-scale, long-term trials in man that investigate the role of glucagon in T2D are needed.

NEUROENDOCRINE THEORY OF AGING

The neuroendocrine theory of aging proposed by Dilman in a 1958 publication postulates that aging and related pathologies result from a progressive reduction of the hypothalamus–pituitary sensitivity. The consequence of this hypothalamic–pituitary dysregulation induces a loss of homeostasis, aging, and a vulnerability to disease. Specifically, abnormal endocrine states of menopause, elevation in glucocorticoids, dyslipidemia (increased cholesterol, fatty acids), obesity, hyperglycemia and hyperinsulinemia, immunodepression, thymic atrophy, hypertension, T2D, atherosclerosis, and cancer are attributed to an alteration in the hypothalamic–pituitary threshold. Menopause and andropause are considered initiating forces of neuronal and cognitive decline with age and set the stage for Alzheimer's disease. Furthermore, the reduction in GH/IGF-1 levels is responsible for altered body composition, sarcopenia and dynapenia, reduced physical function, frailty, and decreased cognition. Additionally, the reduced ability to recovery from stress favors continual damage to cells (immune, bone, neurons) and contributes to immunosuppression, cognitive decline, sarcopenia, and osteoporosis. *Overall, reduced hormone regulation and associated changes in tissue responsiveness put the internal milieu at risk*. Since homeostasis is more difficult to achieve, small stresses provoke disproportional detrimental effects eventually culminating in disease/disability and death.

To put all of these changes into a cause and effect scenario has not been possible in man, although strong associations exist, for example, with low GH/IGF-1 and muscle weakness, and with menopause and osteoporosis.

In animal models of aging, it is postulated that lifespan extension is mediated through neuroendocrine changes perturbed by CR. In response to a reduction in availability of food, gastrointestinal and associated hormones modulate specific neurons of the hypothalamus identified in rats as neuropeptide Y (NPY)-neurons, regulators of hunger. Importantly, the level of NPY rises with CR and is associated with a plethora of changes such as decreased body temperature, lowered levels of blood glucose, elevation of glucocorticoids, reduced fertility with decrease in LH/FSH secretion, decreased energy expenditure, decreased TSH and thyroid hormones, disease resistance, and, of course, life extension. The occurrence of lower fasting glucose and decreased TSH/thyroid hormones, also occur with CR in nonhuman primates but reproductive function and circulating adrenal steroids are similar to controls (food ad libitum). A study to assess the effects of long-term (2 year) CR in man is currently ongoing. CR effects on endocrine functions are eagerly awaited.

Summary of Age Changes of the Endocrine System

Hypothalamic–pituitary–ovarian axis
- Follicular cell disappearance → menopause
- ↓ E2, P, ↑LH, FSH, loss of NFI
- Atrophy of E2-dependent tissues: ovaries, oviduct, vagina, cervix, uterus; hot flushes; osteopenia
- Changes in cardiovascular system, brain are controversial

Hypothalamic–pituitary–testes axis
- Leydig cells disappear: minor (andropause?)
- Modest ↓ T, ↑LH, FSH
- Slight ↑ SHBP, slight↓ E
- Low prevalence of late-life hypogonadism
- High prevalence of benign prostatic hypertrophy

Hypothalamic–pituitary–adrenal axis
- Altered basal circadian rhythm: confounding factors
- ↓ sensitivity of hypothalamus/pituitary to NFI by cortisol: aberrant stress response affects bone, immune system
- ↑ sensitivity of pituitary/adrenal to CRH
- ↓ levels of DHEA and related androgens

Hypothalamic–pituitary–thyroid axis
- ↓ levels of T_3, TSH; ↔ levels T_4

Hypothalamic–pituitary–liver axis
- ↓ levels of GH and IGF-1; ↓ GHRH, ↑ somatostatin
- Changes modified by adiposity, diet, exercise
- ↓ tissue sensitivity (muscle/bone) due to disappearance of GH/IGF-1 receptor

Pineal gland
- ↓ levels of melatonin in part due to aging eye: thickening of lens/senile miosis
- Related to reduced sleep quality

Parathyroid gland
- ↑ basal levels of PTH; related to vitamin D deficiency, ↓ calcium absorption
- Exaggerated response to changes in calcium

Pancreas
- ↓ insulin secretion
- ↓ responsiveness of beta cells to incretins
- insulin resistance related to ↓ physical exercise, diet, abdominal circumference, sarcopenia
- ↑ levels of glucagon (?)

E2: 17-β-estradiol; P: progesterone; FSH: follicle-stimulating hormone; LH: luteinizing hormone; GH: growth hormone; GHRH: growth-hormone-releasing hormone; CRH: corticotrophin-releasing hormone; IGF-1: insulin growth factor-1; SHBG: sex hormone-binding globulin; T: testosterone; NFI: negative feedback inhibition; PTH: parathyroid hormone; TSH: thyroid-stimulating hormone; T_3: tri-iodothyronine; T_4: thyroxine.

SUMMARY

The consistency of the internal environment (homeostasis) is maintained in part by the endocrine system. The hypothalamic–pituitary–peripheral glands (ovaries, testes, thyroid, adrenals, liver) are largely regulated by negative feedback inhibition to ensure optimal concentrations of circulating hormones, mediators that act on tissues expressing specific hormone-sensitive receptors. Other endocrine glands maintain the homeostasis of nutrients (pancreas), calcium, and phosphate (parathyroid glands) and sleep–wake cycle (pineal).

The most dramatic endocrine change is the loss of female fertility (menopause). The production of E and P comes to an end and blood levels of FSH/LH and GnRH rise. The absence of E/P is responsible for hot flushes, atrophy of E/P-dependent reproductive tissue, and osteopenia (or possibly osteoporosis). In contrast, dysregulation of testosterone is less dramatic with a major influence on the development of benign prostatic hyperplasia and a minor impact on late-onset hypogonadism.

The second most notable age-associated endocrine dysfunction is the onset of insensitivity of the hypothalamus and pituitary to stress-induced elevation in cortisol level. Prolonged exposure of tissues to cortisol is associated with accelerated aging of immune, bone, and cardiovascular function. Basal and circadian rhythm of cortisol may also change with age but the identification of confounding factors suggests the need for additional studies.

Other endocrine changes include a impressive decline in circulating androgens (DHEA, dehydroandrosterone), variable fall in growth hormone/IGF-1 associated with loss of receptor sensitivity, slight increase in TSH and decline in T_3, elevation in basal and stimulated PTH, and reduction in melatonin. Insulin sensitivity may decline with age due to reduced beta cell function in conjunction with extrinsic lifestyle factors (abdominal adiposity, dietary choices, and inactivity). Loss of insulin sensitivity may lead to T2D.

The neuroendocrine theory of aging proposes that age changes are fostered by dysregulation of the HPA and these changes explain sarcopenia, osteopenia, cognitive decline, and immunosenescence that favor disease and disability. In man, loss of GH/IGF-1 effects, persistence of cortisol, and deficits in nutrient handling by insulin (and glucagon) appear to accelerate aging. Replacement of hormone deficiencies, for example, GH, T, and DHEA, with chronic hormone therapy is discouraged due to the lack of long-term safety assurance. The one exception is E/(P) short-term use to alleviate symptoms of menopause.

CRITICAL THINKING

Are there any hormones that could be of value for antiaging therapy? Support your answer with a benefits/risks assessment.

What are possible criticisms to the neuroendocrine theory of aging?

How might the incidence of T2D be mitigated? What hormonal changes are involved in this devastating disease?

What are the risks/benefits of estrogen/progesterone therapy in menopause?

Suggest ways to improve quality of sleep in the elderly?

Is "low T" a treatable condition"?

KEY TERMS

Adrenal steroids steroids produced by the outer layers or cortex of the adrenal glands. Examples are cortisol, androgen such as DHEA.

Andropause name assigned to purported loss of reproductive function in men.

Beta cells endocrine cells of the pancreas that secrete insulin in response to an increase in blood sugar.

Cortisol an adrenal steroid; hormone released during stress that optimizes the body's response to stress by facilitating a "fight or flight" response.

DHEA dehydroepiandrosterone—popular supplement thought to reverse aging; is a precursor to many androgens including testosterone.

Estrogen steroid synthesized from cholesterol by the ovaries; it is responsible for reproductive function, secondary sex characteristics, bone formation, and possibly cardiovascular and cognitive functions; exists in three forms.

Follicle-stimulating hormone (FSH) gonadotropic hormone produced by the pituitary gland to stimulate the production of estrogen/progesterone during the menstrual cycle in females and testosterone production in males.

Gonadotropic releasing hormone hormone of the hypothalamus that stimulates the release of LH and FSH.

Growth hormone hormone produced by the pituitary gland that stimulates production of hepatic IGF-1; with or without IGF-1, growth hormones stimulate skeletal muscle and bone growth.

Growth-hormone-releasing hormone hormone from the hypothalamus that stimulates the release of growth hormone from the pituitary gland.

Hormone substance produced by an endocrine gland that acts on distal tissues (tissues that express receptors to these hormones). Hormones are steroids, proteins, or small molecules.

Insulin protein hormone produced by the beta cells of the pancreas. It promotes transfer of sugar from the blood to the skeletal muscle and liver for use or storage.

Late-onset hypogonadism clinical syndrome of low testosterone and three symptoms of sexual dysfunction. Evident in ~5% of elderly.

Luteinizing hormone (LH) hormone produced by the pituitary that stimulates reproductive function in males and females.

Menopause cessation of reproduction function in women; onset around 50 years of age.

Neuroendocrine axis refers to the hypothalamus–pituitary–end organ (gonads, thyroid, adrenals, and liver). Hormones of these endocrine glands are tightly regulated by negative feedback inhibition.

Ovary site of maturation of the ovum and production of estrogen and progesterone.

Pancreas endocrine organ that produces insulin, glucagon, somatostatin, pancreatic polypeptide, and digestive enzymes.

Parathyroid hormone one of two hormones produced by the parathyroid gland that regulates calcium homeostasis.

Pineal gland endocrine gland that produces melatonin, regulator of the sleep–awake cycle.

Somatopause name given to aging phase in which growth hormone and IGF-1 levels are at their lowest levels.

Testes endocrine site for production and maturation of sperm; site for production of testosterone.

Testosterone steroid produced by the testes for maturation of sperm and other reproductive functions.

Thyroxine one of two hormones produced by the thyroid gland. Precursor to the active hormone tri-iodothyronine.

Tri-iodothyronine called T_3 (a hormone containing three iodine atoms). It is the "active" thyroid hormone that influences metabolic rate.

BIBLIOGRAPHY

Review

Aguilera G. 2011. HPA axis responsiveness to stress: implications for healthy aging. *Exp. Gerontol.* **46**(2–3):90–95.

Araujo AB, Wittert GA. 2011. Endocrinology of the aging male. *Best Pract. Res. Clin. Endocrinol. Metab.* **25**(2):303–319.

Archer DF, Sturdee DW, Baber R, de Villiers TJ, Pines A, Freedman RR, Gompel A, Hickey M, Hunter MS, Lobo RA, Lumsden MA, MacLennan AH, Maki P, Palacios S, Shah D, Villaseca P, Warren M. 2011. Menopausal hot flushes and night sweats: where are we now? *Climacteric* **14**(5):515–528.

Baker LD, Barsness SM, Borson S, Merriam GR, Friedman SD, Craft S, Vitiello MV. 2012. Effects of growth hormone-releasing hormone on cognitive function in adults with mild cognitive impairment and healthy older adults. *Arch. Neurol.* **69**(11):1420–1429.

Barnard RJ, Aronson WJ. 2009. Benign prostatic hyperplasia. Does lifestyle play a role? *Phys. Sportsmed.* **37**(4):141–146.

Chandra R, Liddle RA. 2013. Modulation of pancreatic exocrine and endocrine secretion. *Curr. Opin. Gastroenterol.* (5):517–522.

Cui J, Shen Y, Li R. 2013. Estrogen synthesis and signaling pathways during aging: from periphery to brain. *Trends Mol. Med.* **19**(3):197–209.

Dilman VM, Revskoy SY, Golubev AG. 1986. Neuroendocrine–ontogenetic mechanism of aging: toward an integrated theory of aging. *Int. Rev. Neurobiol.* **28**: 89–156.

Espino J, Pariente JA, Rodríguez AB. 2012. Oxidative stress and immunosenescence: therapeutic effects of melatonin. *Oxid. Med. Cell. Longev.* **2012**: 670294.

Guyton AC, Hall JE. 2006. *Textbook of Medical Physiology*, 11th ed. Philadelphia: Elsevier Saunders.

Hardeland R. 2012. Neurobiology, pathophysiology, and treatment of melatonin deficiency and dysfunction. *Sci. World J.* **2012**: 640389.

Heffner KL. 2011. Neuroendocrine effects of stress on immunity in the elderly: implications for inflammatory disease. *Immunol. Allergy Clin. North Am.* **31**(1):95–108.

Horstman AM, Dillon EL, Urban RJ, Sheffield-Moore M. 2012. The role of androgens and estrogens on healthy aging and longevity. *J. Gerontol. A Biol. Sci. Med. Sci.* **67**(11):1140–1152.

Junnila RK, List EO, Berryman DE, Murrey JW, Kopchick JJ. 2013. The GH/IGF-1 axis in ageing and longevity. *Nat. Rev. Endocrinol.* **9**(6):366–376.

Kushner JA. 2013. The role of aging upon β cell turnover. *J. Clin. Invest.* **123**(3):990–995.

Liu H, Bravata DM, Olkin I, Nayak S, Roberts B, Garber AM, Hoffman AR. 2007. Systematic review: the safety and efficacy of growth hormone in the healthy elderly. *Ann. Intern. Med.* **146**(2):104–115.

Li XC, Zhuo JL. 2013. Current insights and new perspectives on the roles of hyperglucagonemia in non-insulin-dependent type 2 diabetes. *Curr. Hypertens. Rep.* **15**(5):522–530.

MacBride MB, Rhodes DJ, Shuster LT. 2010. Vulvovaginal atrophy. *Mayo Clin. Proc.* **85**(1):87–94.

National Diabetes Statistical Report. 2012. Consensus document published by http://www.cdc.gov/diabetes/pubs/statsreport14/national-diabetes-report-web.pdf.

Ortmann O, Lattrich C. 2012. The treatment of climacteric symptoms. *Dtsch. Arztebl. Int.* **109**(17):316–323.

Pantalone KM, Faiman C. 2012. Male hypogonadism: more than just a low testosterone. *Cleve. Clin. J. Med.* **79**(10):717–725.

Peeters RP. 2008. Thyroid hormones and aging. *Hormones (Athens)* **7**(1):28–35.

Rance N 2009. Menopause and the human hypothalamus: evidence for the role of kisspeptin/neurokinin B neurons in the regulation of estrogen negative feedback. *Peptides* **30**(1):111–122.

Rosenberg MT, Miner MM, Riley PA, Staskin DR. 2010. STEP: simplified treatment of the enlarged prostate. *Int. J. Clin. Pract.* **64**(4):488–496.

Rossouw JE. 2014. Reconciling the divergent findings from clinical trials and observational studies of menopausal hormone therapy for prevention of coronary heart disease. *Semin. Reprod. Med.* **32**(6):426–432.

Sampson N, Untergasser G, Plas E, Berger P. 2007. The aging male reproductive tract. *J. Pathol.* **211**(2):206–218.

Sattler FR. 2013. Growth hormone in the aging male. *Best Pract. Res. Clin. Endocrinol. Metab.* **27**(4):541–555.

Singh P. 2013. Andropause: current concepts. *Indian J. Endocrinol. Metab.* **17**(Suppl. 3):S621–S629.

Soules MR, Sherman S, Parrott E, Rebar R, Santoro N, Utian W, Woods N. 2001. Executive summary: Stages of Reproductive Aging Workshop (STRAW). *Fertil. Steril.* **76**(5):874–878.

Sturdee DW. 2008. The menopausal hot flush-anything new? *Maturitas* **60**(1):42–49.

The North American Menopause Society. 2012. The 2012 Hormone Therapy Position Statement of the North American Menopause Society. *Menopause* **19**(3):257–271.

Turner PL, Mainster MA. 2008. Circadian photoreception: ageing and the eye's important role in systemic health. *Br. J. Ophthalmol.* **92**(11):1439–1444.

Veldhuis JD, Sharma A, Roelfsema F. 2013. Age-dependent and gender-dependent regulation of hypothalamic–adrenocorticotropic–adrenal axis. *Endocrinol. Metab. Clin. North Am.* **42**(2):201–225.

Experimental

Alberts SC, Altmann J, Brockman DK, Cords M, Fedigan LM, Pusey A, Stoinski TS, Strier KB, Morris WF, Bronikowski AM. 2013. Reproductive aging patterns in primates reveal that humans are distinct. *Proc. Natl. Acad. Sci. USA* **110**(33):13440–13445.

Araujo AB, O'Donnell AB, Brambilla DJ, Simpson WB, Longcope C, Matsumoto AM, McKinlay JB. 2004. Prevalence and incidence of androgen deficiency in middle-aged and older men: estimates from the Massachusetts Male Aging Study. *J. Clin. Endocrinol. Metab.* **89**(12):5920–5926.

Asplund R, Lindblad BE. 2004. Sleep and sleepiness 1 and 9 months after cataract surgery. *Arch. Gerontol. Geriatr.* **38**(1):69–75.

Greendale GA, Wight RG, Huang M-H, Avis N, Gold EB, Joffe H, Seeman T, Vuge M, Karlamangla AS. 2010. Menopause-associated symptoms and cognitive performance: results from the study of women's health across the nation. *Am. J. Epidemiol.* **171**(11):1214–1224.

Rosario ER, Chang L, Head EH, Stanczyk FZ, Pike CJ. 2011. Brain levels of sex steroid hormones in men and women during normal aging and in Alzheimer's disease. *Neurobiol. Aging* **32**(4):604–613.

Rossouw JE, Anderson GL, Prentice RL, LaCroix AZ, Kooperberg C, Stefanick ML, Jackson RD, Beresford SA, Howard BV, Johnson KC, Kotchen JM, Ockene J; Writing Group for the Women's Health Initiative Investigators. 2002. Risks and benefits of estrogen plus progestin in healthy postmenopausal women: principal results from the Women's Health Initiative randomized controlled trial. *JAMA* **288**(3):321–333.

Sharma M, Palacios-Bois J, Schwartz G, Iskandar H, Thakur M, Quirion R, Nair NP. 1989. Circadian rhythms of melatonin and cortisol in aging. *Biol. Psychiatry* **25**(3):305–319.

Sator PG, Sator MO, Schmidt JB, Nahavandi H, Radakovic S, Huber JC, Hönigsmann H. 2007. A prospective, randomized, double-blind, placebo-controlled study on the influence of a hormone replacement therapy on skin aging in postmenopausal women. *Climacteric* **10**(4):320–334.

Wu FC, Tajar A, Beynon JM, Pye SR, Silman AJ, Finn JD, O'Neill TW, Bartfai G, Casanueva FF, Forti G, Giwercman A, Han TS, Kula K, Lean ME, Pendleton N, Punab M, Boonen S, Vanderschueren D, Labrie F, Huhtaniemi IT, EMAS Group. 2010. Identification of late-onset hypogonadism in middle-aged and elderly men. *N Engl. J. Med.* **363**(2):123–135.

15

AGING OF THE IMMUNE SYSTEM

OVERVIEW

For individuals 65 years or older, malignant neoplasms and influenza/pneumonia are the second and fifth major causes of death, respectively. Although cause and effect have not been established, it is generally concluded that immunosenescence or dysregulation of the immune response plays a significant role in the age-associated prevalence of these diseases and furthermore may contribute to aging of other organ systems. What immunosenescence is all about is described in this chapter.

Humans are protected against microbial invasion, thanks to the following three interdependent systems: (i) assorted physical barriers/reflexes, (ii) the innate immune system, and (iii) the adaptive (acquired) immune system. Cooperation among the three levels is essential to detect and react to diverse pathogens: viruses, bacteria, parasites, and fungi, thereby thwarting infections. The most sophisticated of these is the adaptive immune system that additionally operates a type of surveillance to prevent cancers. Breakdown of regulatory controls, however, results in self-destruction called autoimmunity, which induces autoimmune diseases and generates insidious low-grade chronic inflammation, termed inflammaging. Figure 15.1 depicts the three levels of host defense and their interaction.

Innate immunity, formerly thought less important than adaptive immunity, is now considered a significant player in host defenses and more importantly in inflammaging. In particular, the dendritic cell of the innate system is a crucial link between innate and adaptive immunity, ensuring an optimal defense. Unfortunately, with age, the performance of dendritic cells declines and this connection weakens, to the detriment of the organism.

Human Biological Aging: From Macromolecules to Organ Systems, First Edition. Glenda Bilder.
© 2016 John Wiley & Sons, Inc. Published 2016 by John Wiley & Sons, Inc.

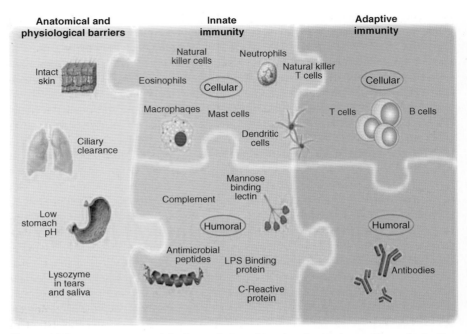

Figure 15.1. Components and the interrelation of barriers, innate immunity, and adaptive immunity. (Reprinted with permission from Turvey and Broide (2010).)

Physical barriers (if not broken) and innate immunity provide the first lines of immediate and effective defense against most pathogens. Adaptive immunity provided by T and B cells is the second line of defense and although not particularly rapid (taking days), adaptive immunity is expansive and efficacious. Its ability to recognize pathogens (and any foreign substance) is nearly limitless and on "recall" (second encounter) supplies a quick and complete pathogen elimination. This "memory" offers invaluable lifelong protection. However, while memory to childhood and adult pathogens remains more or less robust, generation of memory to new (not previously encountered) pathogens, slowly diminishes with age.

There are several basic cellular activities that are essential to a successful immune response. Many of them have been discussed in earlier chapters but are reviewed here because they significantly impact host defenses. Specifically, oxidative stress alters membranes through peroxidation reactions (see Chapter 4), disrupting membrane receptors and cell surface markers of innate and adaptive immune cells and rendering them dysfunctional. Secondly, clonal expansion (cell division) is the requisite step following immune cell activation. Repeated cell proliferation induces replicative senescence and the senescence-associated secretory phenotype (SASP) discussed in Chapter 5. The SASP is not only dysfunctional, and proinflammatory but it also cannot divide, negating clonal expansion. Thirdly, autophagy is another basic cellular activity (Chapter 5), that plays several roles in the host defense via adequate phagocytic pathogen killing, activation assistance to T and B cells and metabolic supplies for rapid cell proliferation and secretory protein production. A slowing of autophagy seriously impairs immune function. Fourthly, malnutrition, especially a deficiency in protein consumption, frequently present in many elderly and associated with sarcopenia (see Chapter 7), is a potent immunosuppressant. Finally, dysregulation of the hypothalamic–pituitary–adrenal axis with age (see Chapter 14) exposes immune cells to higher than normal levels of glucocorticoids for longer periods of time, changes that suppress immune reactions. Thus, many age-associated factors impact the responsiveness (quality and quantity) of host defenses against pathogens.

There is an ongoing debate as to what constitutes a disease-free (healthy) immune system in the elderly. Frequently, studies of "apparently healthy" subjects include many individuals with covert disease or conditions affecting immune activity, or with medication use that interferes with immunity. In an attempt to eliminate confounding factors and to define more specifically a disease-free norm, the Senieur protocol was developed for the study of European elderly. This protocol requires extensive data collection on each subject under three exclusion categories: clinical, for example, infections and inflammation, laboratory (values outside age-related average), and pharmacological (medications for immune system or specific disorder). The Senieur protocol was designed as the "first step in the study of ageing in a methodologically sound manner" (Ligthart, 2001). The important finding that fewer immunological deficits are evident in the elderly with this protocol suggests that the extent of immunosenescence *per se* in less serious than thought and that many factors (some mentioned above), some yet to be determined, negatively affect immune function of the elderly. The Senieur protocol is clearly valuable. To this, there is a need to add longitudinal studies that employ this protocol and evaluate large numbers of elderly subjects.

BARRIERS AND RELATED MECHANISMS

Components

The skin offers remarkable protection. The skin, if not broken, is impenetrable. Skin prevents pathogen entry by its blockade function, by virtue of its rapid cell turnover (sloughing), and/or by killing/retarding pathogenic growth with acidic sweat and sebum secretions. Resident cells, keratinocytes and sebocytes, produce an array of antibacterial peptides.

Beyond the skin, other antimicrobial mechanisms include (1) viscous and acidic secretions that prevent microbial entry and (2) reflexes and cilia that expel microbes. Secretions from the mouth, eyes, and nose contain antimicrobial proteins, for example, lysozyme. The sticky mucosal lining of the respiratory and gastrointestinal (GI) tract traps and smothers pathogens. Within the GI tract pathogens face the acidic stomach conditions, protein-destroying enzymes and intestinal flora that secrete bactericidal substances. Vomiting and diarrhea also eject large quantities of pathogens. Pathogenic invaders are expelled from the lungs by specialized cells with cilia (tiny hairs) of the trachea that beat upward and additionally are ousted by the sneeze and cough reflexes. Figure 15.1 illustrates these barriers.

Effects of Age

Reduced Barrier Quality Permits Pathogen Entry Intrinsic aging of the skin accelerated by the estrogen deficiency of menopause reduces the barrier quality allowing pathogens an easier entrance. Intrinsic aging is characterized by a thinner epidermis, by slower epidermal renewal (reduces sloughing), and reduced sebum production. Pruritus is a frequent outcome of intrinsic aging of the skin; its associated microcracks facilitate microbial access. Risk of infection is elevated with a composite of barrier deficiencies.

Reduced lung function enhanced by dynapenia contributes to a diminished cough reflex, thereby permitting bacterial residency and growth especially in under perfused lung tissue. In the GI tract of the elderly, stomach acidity trends to a more alkaline pH due in large part to increased use of antacids or proton pump inhibitors, for example, Prilosec, for the treatment of gastric reflux disease. To add to this, there is tendency for mucus production in trachea and GI tract to decline with advancing age. Refer to Summary Box for barrier aging.

INNATE IMMUNE SYSTEM

Components; Role in Immunity

Innate Immune System Provides the First Line of Cellular and Humoral Defense Against Pathogens As the front guard, innate immune components act rapidly against most pathogens. Innate cells include monocytes, macrophages, dendritic cells, neutrophils, eosinophils, mast cells, natural killer (NK) cells, NK T-cells (NKT), and epithelial cells of skin, GI, respiratory, and urogenital tracks. The non cellular component (humoral) consists of complement (extensive cascade of blood proteins/enzymes), C-reactive protein, endotoxin-binding proteins, and others, for example, defensins. These are shown in Figure 15.1.

Pattern Recognition Receptors/Phagocytosis Are Essential for Host Defense Innate immune cells express several different receptors called *pattern recognition receptors* (PRR) that distinguish and bind to *characteristic aspects of microbes called pathogen-associated molecular patterns (PAMPs) or damage-associated changes on cell membranes, or on cellular debris called damage-associated molecular patterns (DAMPs)*. Many of these PRRs have been identified: toll-like receptor (TLR), C-type lectin receptor (CLR), RIG-I-like receptor (RLR), and NOD-like receptors (NLR). The interesting and relevant aspect of the PRR family is that activation of PRRs on innate cells *yields secretion of proinflammatory mediators, chemokines, and antiviral mediators; maturation of dendritic cells; and with appropriate regulation, microbial suppression and beneficial interaction with adaptive immunity*. Activation of PRR is essential for rapid destruction of most pathogens, aided immensely by the proinflammatory nature of this first line of defense.

Many of the innate immune cells are phagocytic cells. Thus, they have the ability to engulf and digest pathogens (a process termed phagocytosis that is a type of autophagy). Polymorphonuclear neutrophils (PMNs), eosinophils, macrophages,

INNATE IMMUNE SYSTEM

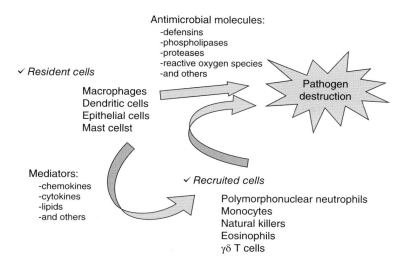

Figure 15.2. Innate immunity and inflammation—two facets of the same anti-infectious reaction. (Reprinted with permission from Si-Tahar et al. (2009).)

and dendritic cells are phagocytic cells. These cells either reside in tissues or circulate freely. To destroy some pathogens, a cascade of proteins (termed complement) is activated to facilitate phagocytic ingestion and destruction of bacterial membranes. Humoral components named above also participate in removing microbes.

The natural killer cell and NKT cells are lymphocytes that express the PRRs, for example, TLR that are also common to innate immune cells. These cells are especially useful because as "natural killers" they rapidly destroy tumor cells or cells infected with virus through cytotoxic cell lysis or induced apoptosis. Innate immunity interactions are shown in Figure 15.2.

Effects of Age

Signaling and Phagocytosis Decline with Age
Microbial invasion or tissue damage attracts PMNs and monocytes, cells with potential to eliminate intruders or remove debris. Other cell types, macrophages, dendritic cells, and NK cells join in. Dendritic cells process pathogenic components (degrade and re-present parts on their plasma membrane) to engage adaptive immune cells if necessary. *According to some studies, this initial innate immune response is deficient in the elderly.* Innate cell number remains constant with age but innate *cellular functions (signaling and subsequent activities) decline with age.* Specifically, the chemotaxis (migration) of PMNs to a site of infection lessens with age. PMN *phagocytic activity also declines.* This renders PMNs *incapable of a superoxide burst, thus reducing microbial killing.* The TLRs on PMNs, important bacterial detectors, fail to respond. Consequently, additional support cells are not recruited and a chance to obtain cooperation from adaptive immune cells is lost. Additionally, *conversion of monocytes to macrophages*

(inflammatory producers) rather than to dendritic cells is more prevalent in the elderly. This also hampers the interaction of innate cells with adaptive immune cells and reduces host protection and memory. Together, *the first line of defense exhibits aspects of dysfunction with age.*

An understanding of the aging of NK cells is limited. It is known that NK cells exert an important cytotoxic effect potent enough to totally annihilate aberrant cells, for example, cells modified by a virus, oxidative events, or malignancy and lacking markers of "self." Furthermore, as observed by Ogata et al. (2001) in 108 immunologically normal elderly in nursing homes due to physical limitations, *low NK cell activity was associated with increased infection and mortality in the 1-year follow-up.* There is some evidence to suggest that signal transduction, especially in response to a specific cytokine, interleukin-2 (IL-2), required for NK proliferation and cytotoxicity, is impaired with age. Other changes include an increase in mature (end state) NK cells, reduced production of mediators necessary for normal function, and loss of a key NK receptor. Anyone of these purported changes would be disadvantageous. Refer to Summary tabulation for age changes in innate immunity.

ADAPTIVE IMMUNITY

Adaptive immunity encompasses a complex highly evolved system of cellular and humoral responses against pathogens and modified cells. Furthermore, it has the capacity to "remember" initial encounters, called *immunological memory*, that generates a more rapid and destructive second response. The natural or artificial, for example, vaccines, exposure of an individual to a particular pathogen creates immunological memory for that pathogen. Over time each individual develops *a distinct repertoire of memory cells primed for the next encounter with "recall" pathogens.*

Any substance that evokes an adaptive immune response is termed an antigen. Antigens may be pathogens, portions of pathogens, foreign proteins, cell debris, or aberrant cells. The dendritic cell is one of the main processors of antigens, an activity that stimulates an adaptive immune response.

Cells of the immune system distinguish between natural "birth" cells of an organism and pathogenic/foreign or altered components. In other words, *immune cells differentiate between "self" (cells of the organism) and "non-self" (pathogens, foreign substances) by the presence of a specific marker set called the major histocompatibility complex (MHC) that is expressed on cells in the body.* Pathogens or foreign substances lack this marker set, are perceived as "nonself" and become targets for immunological destruction. Abnormalities of this exquisite detection system are manifest in the development of autoimmune disease, for example, multiple sclerosis, systemic lupus erythematosus, and certain types of arthritis. A disproportionate response of adaptive immunity leads to allergies, for example, rash, nasal congestion, respiratory distress, and anaphylactic shock.

Recall that the innate immune system also has a set of "antigens" that is recognized by the family of PRRs. While the PRRs are considered "ancestral," they nevertheless provide excellent frontline protection and are also essential for optimal adaptive immunity responses.

T Cell; B Cell; Role in Immunity

Two Adaptive Immune Cell Types Are the T Cell and the B Cell T and B cells originate from cell lineages in the bone marrow. T cells migrate to the thymus, an immune gland found in the upper chest, mature, for example, develop immunological "competence," and exit (hence, thymus or T-derived) into the circulation where some take up residence in the peripheral lymphoid tissue (nodes, spleen, peyers patches, and tonsils) or continue to circulate. B cells mature in a similar fashion, although the site of immunological maturation in man is unknown. In the chicken, B cells mature in the bursa, giving rise to the original name bursa-derived or B cells. In humans the intestine, liver or bone marrow are proposed bursa-homologous sites. This is shown in Figure 15.3.

T Cells Perform Cell-Mediated Immunity: Helper, Cytotoxic, and Regulatory Although T cells and B cells interact, their essential functions differ. *T cells perform cell-mediated immunity*. Numerous subsets exist to achieve cell-mediated immunity. Subsets are characterized by membrane receptors and markers. Four major subsets are (i) $CD8^+$ cytotoxic or killer cells, (ii) $CD4^+$ helper cells, (iii) regulatory cells, and (iv) memory cells. $CD8^+$ cytotoxic T cells target intracellular pathogens, such as viral-infected cells and tumor cells, and destroy them through release of several substances, for example, lymphotoxin. $CD4^+$ helper T cells assist B cells in production of antibodies (see B cells below), assist $CD8^+$ cell with cytotoxicity, support clonal expansion (proliferation to increase cell number), maximize phagocytic functions, and attract innate cells to sites of infection. Regulatory T cells dampen excessive immune reactions and limit the number of cytotoxic T cells. Memory T cells (antigen experienced) store information derived from

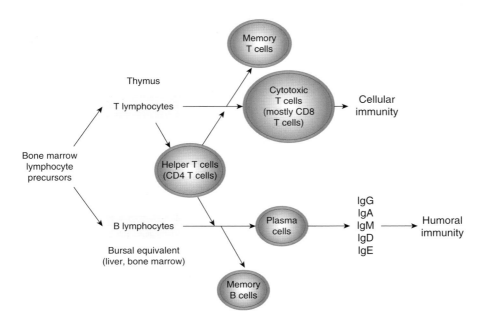

Figure 15.3. Adaptive immunity with cellular and humoral functions. (Reprinted with permission from Barrett et al. (2012).) (See plate section for color version.)

the initial antigen encounter so that on a second encounter with the same antigen, for example, a pathogen, a swift, efficient response is launched.

B Cells Mediate Humoral Immunity by Production of Antibodies

B cells mature into plasma cells to produce special circulating proteins called *immunoglobulins* or *antibodies and hence* mediate *humoral immunity*. Antibodies protect the organism by (i) binding to pathogens and causing lysis, (ii) preventing viruses from attacking normal cells, and (iii) neutralizing pathogen-secreted toxins. Antibodies also activate other immune cells to expand defense mechanisms. Immunoglobulins are classified into five types: *IgG* (most prevalent), *IgM, IgD, IgA*, and *IgE* (least prevalent). The abundant structural diversity within each type permits an almost unlimited defense against pathogens.

Similar to T memory cells, B memory cells are generated following the first encounter with an antigen. Subsequent encounters with the "recall" antigen induces a more effective response, for example, with *larger quantities of more avidly binding antibodies produced at a faster rate. Immunological memory is similarly generated with vaccination.* A vaccine contains a portion (antigenic) of a lethal pathogen in a nondetrimental form (heat killed or inactivated) that is introduced into an organism in a small dose. A primary humoral response ensues over the course of days culminating with the generation of B-cell memory cells and a measurable level of protective antibodies. In the future, exposure of a vaccinated (primed) individual to a lethal pathogen prevents death and to a nonlethal pathogen, shortens disease duration and associated tissue damage.

B-cell responses are enhanced with T-cell and antigen-presenting cell assistance, although B cells independent of accessory cells remain weakly immunologically competent. Cell-cell cooperation is important for maximal immunological utility. Deficiencies in the number or function of any one of the cell types (T-cell, B-cell, antigen-processing cell) positions the organism at a higher risk for infection, malignancies, or autoimmune disease. Figure 15.4 emphasizes key clonal expansion and functions of B cells and T cells.

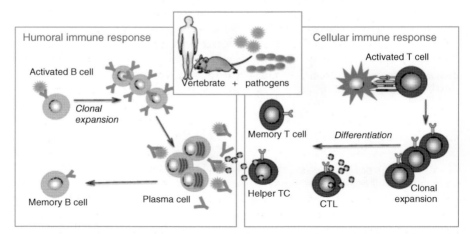

Figure 15.4. Clonal expansion is essential for immune response. (Reprinted with permission from Müller et al. (2013).) (See plate section for color version.)

Numerous Cytokines Facilitate Immunity Adaptive immunity is immensely effective not only because of the *diversity of T cells, B cells, and antibodies but also because of multitude of cytokines* secreted by adaptive immune cells. *Similar to innate immune cells, T cells release cytokines to direct immune responses and chemokines to attract support cells.* Cytokines influence distinct functions. For example, some of the interleukins are proinflammatory, for example, Il-6, and attract PMNs. Others are anti-inflammatory agents, for example, transforming growth factor beta (TGFβ), and prevent damage. Other relevant cytokines are tumor necrosis factors, interferons, colony-stimulating factors, and granulocyte–macrophage colony stimulating factors. Interferon exerts antiviral activity and the colony-stimulating factors influence proliferation and cell survival.

Immunosenescence

Thymus Gland Is One Determinant of T Cell Behavior *The thymus gland is the exclusive site for immunological maturation of T cells.* It is the locus of "education" where immature T cells learn self-tolerance, prepare for antigenic stimulation, and become immunocompetent T cells, joining one of the many subsets. *With age, the thymus dramatically shrinks, a phenomenon termed involution.* In humans, thymic involution begins almost at birth. By age 20, thymic size is reduced by approximately 50% and by 60 years of age, it is approximately one-sixth the birth size.

Concomitant with thymic involution, thymic function declines. Gradually, T-cell maturation (thymopoiesis) slows and thymocytes (future T cells) remain immature in the thymus. Importantly, *fewer immunocompetent T cells are released into the circulation, reducing T-cell diversity and, hence, responses to new antigens.* Eventually, a pattern emerges in the peripheral lymph nodes in which there exist *more old memory cells than new naïve T cells*.

Immunosenescence Is the Gradual Dysregulation of Immune Function with Age Immune dysregulation occurs with (i) an *accumulation of T-cell and B-cell memory cells* (to former antigens), many of which express the senescence-associated secretory phenotype (SASP); (ii) a *decrease in naive cells* (producing a vulnerability to new antigens, particularly intracellular pathogens); and (iii) a *shift in T-cell subsets* that favors an increased production of proinflammatory cytokines, decreased production of anti-inflammatory cytokines, less B-cell assistance, and less control over abnormal responses to self-antigens. These changes contribute to *immunosenescence*.

Thymic involution is held responsible for immunosenescence. The reason for progressive thymic involution is unknown. *Adding to the impact of thymic involution is the chronic mitotic pressure experienced by immune cells from persisting antigens*, for example, *cytomegalovirus (CMV)*. Recent data suggest that chronic CMV stimulation augments the progression of immunosenescence because it targets the $CD8^+$ cell. A seminal study (Fagnoni et al., 2000) that examined 120 subjects (18–105 years of age) found that *naive T-cell number (antigen-inexperienced) dramatically decreases with age in the $CD4^+$ subset* and is *nearly absent in the $CD8^+$ subset*. Concomitant with the decline in naive T cells, there is an increase in experienced T cells (memory or antigen experienced). Furthermore, some of antigen-experienced $CD8^+$ T cells express CMV markers and lack the CD28 receptor, the linchpin to a normal immune response. These findings suggest that *diversity of the $CD8^+$ T-cells repertoire in the elderly is dramatically limited* as it consists primarily of chronically stimulated cytotoxic cells and at the same time these antigen-experienced cells are poorly functional, most likely senescent and resistant to apoptosis (just occupying space). Thus, with the paucity of naive immune cells, the elderly are helpless against new antigens, for example, constantly mutating influenza microbes, and the aberrant tumor cell.

Similarly, the number of *helper T cells ($CD4^+$) existing without the essential CD28 coreceptor increases with age. Receptor loss brings about poor T-heper cell performance, for example, slowed response to antigen and retardation of clonal expansion.* Moreover, CD28 receptor disappearance limits adaptive immunity in several other subsets of immune cells.

Regarding aging of B-cell function, evidence shows that the *quality, that is, specificity and quantity of antibodies declines*. There is an age-related reduction in naive B-cell generation and associated parallel decline in response to new antigens and not unexpectedly there is also an increased presence of old antigen-experienced B-cell memory cells. *Dysfunctional $CD4^+$ cells (helper cells) contribute in part to poor B-cell performance*. There is also an *increase in autoantibodies* (to self components). Immunosenescence is summarized in the Summary Box.

CONSEQUENCES OF IMMUNOSENESCENCE

Increased Prevalence of Infectious Diseases Due in Part to Immunosenescence

In general terms, the *aged organism responds poorly to new pathogens*. Reduced T-cell helper activities coupled with deficiencies of dendritic cell antigen-processing function contribute to a decrease in *B-cell antibody production and specificity*. Cytotoxic T-cell activities are also limited and innate barriers and reflexes more readily permit pathogen entry. Thus, bacterial and viral infections that are effectively checked with impenetrable barriers or are suppressed with antibodies and cytotoxicity in the young become serious threats to the elderly. The increased incidence of infections in the elderly from pathogens such as *Escherichia coli* (certain strains), *Streptococcus pneumonia*, *Mycobacterium tuberculosis*, *Pseudomonas aeruginosa*, and the family of herpes viruses is attributed in part to immunosenescence. Similarly, immunosenescence appears responsible for the elevated frequency of gastroenteritis, bronchitis, and influenza in the elderly.

In addition to immunosenescence, development of infectious diseases, for example, pneumonia, is encouraged by several other factors common to the elderly. For example, age changes in lung structure and function (especially reduced tissue oxygenation that supports pathogen growth), delay in diagnosis due to atypical symptoms, for example, absence of fever, malnutrition or undernutrition, nursing home residency, and comorbidities, for example, chronic obstructive pulmonary disease and cardiovascular disease exacerbate common infections.

Incidence of Most Cancers Increases with Age

A second consequence of immunosenescence is the *reduced ability to recognize and destroy malignant cells*. The initiation of a malignancy is multifactorial. Among the many causal factors, the SASP is thought to play a significant role (see Chapter 5). Regardless of the origins, immunosenescence reduces proactive immunological surveillance, thereby permitting tumor formation. Many immune cells act to destroy tumors but the age-associated deficiencies in the cytotoxic T-cell populations, macrophages, NK cells, and PMNs summate in favor of tumor formation and growth. Although it seems reasonable to conclude that these aspects of immunosenescence contribute to the reported increased prevalence of malignancies with age, Fülöp et al. (2007) cautions that "a direct demonstration of a causal link between immunosenescence and infections or tumors is still missing."

Response of the Elderly to Vaccines Is Suboptimal

Protection from influenza vaccine is reduced by ~30–50% or more in the elderly. Vaccines rely on a complete immune response that includes antigen processing by dendritic cells, activation of helper T-cells, and release of appropriate cytokines to stimulate B-cell maturation and production of immunoglobulins and memory cells or expansion of cytotoxic function. *Most of these processes, as reviewed above, are suboptimal in the elderly* and so *vaccinations provide less protection* for this age group.

It has been reported from results of two clinical trials that among the vaccines (influenza, pneumococcal, tetanus, pertussis, and herpes zoster) approved for use in the elderly, only tetanus–pertussis–diphtheria combination provides satisfactory protection, albeit at a level less than observed in young subjects (Goronzy and Weyand, 2013). The other vaccines have marginal effects. These findings question the use of annual influenza shots for the elderly even when the vaccine is "manipulated" with an adjuvant, used in higher doses, or followed with booster shots to supposedly yield a stronger response.

Delayed Hypersensitivity Response Is Diminished

Cutaneous delayed-type hypersensitivity response (DTH) is decreased in the elderly. This is a T-cell-mediated response and includes, for example, cutaneous reaction to poison ivy and tuberculin skin test. The skin reaction to the latter is much weaker in the elderly than in the young. This change has practical implications not only for routine screening in the elderly of "older, recall" antigens but has also been linked to the increased prevalence of skin cancers and skin infections with age. Studies to define the mechanism for reduction in the DTH response have *attributed it to reduced phagocytosis by cutaneous macrophages* that poorly process the antigen and hence fail to cooperate with the T cell to produce the DTH response. Defective phagocytosis of cutaneous macrophages from elderly subjects is reversed *in vitro* suggesting that the *microenvironment of the skin itself contributes to this problem*. Recall aging of the skin (Chapter 6).

Presence of Autoantibodies Increases with Age

The appearance of autoantibodies (antibodies to self-components, for example, DNA or proteins) increases as a function of age. Why this occurs is not clear, but a recent hypothesis suggests that clearance of apoptotic bodies slows with age. Apoptotic bodies are the tiny remnants left from a cell that has undergone apoptosis. This type of death (see Chapter 5) is valuable because it generally does not induce inflammation. Apoptotic bodies are readily engulfed by phagocytic cells. However, if apoptotic bodies persist with their residual components, for example, DNA, RNA, which may have already been altered by oxidative stress or epigenetic changes (hence the original initiator of apoptosis), antigenic stimulation could occur and the resulting antibodies would be anti-DNA, anti-RNA, and harmful to the self. Other possible hypotheses include a shift in T-cell subsets that removes certain suppressive signals or cellular changes induced by oxidative stress that modify plasma membrane lipids and proteins, thereby creating neoantigens. Despite an age-associated increase in autoantibodies, the elderly do not endure an increased incidence of autoimmune diseases. However, this benefit comes with a cost as it is proposed that the presence of autoantibodies enhances regulator cell number and function, an effect that exacerbates existing dysregulation of helper and cytotoxic cells.

Reemergence of Latent Virus Is Common in the Elderly

An additional consequence linked to immunosenescence is the *phenomenon of reemergence of a latent virus*. This is illustrated by the high incidence of shingles in the elderly. Shingles is the reactivation of varicella-zoster virus. Exposure to this virus in early childhood produces the chicken pox infection. After first exposure, the varicella virus retreats within the nerve terminals of the central nervous system and later under stress and/or immune dysfunction, reemerges and produces the infection called shingles. Reemergence is characterized by flu-like symptoms and a skin rash that identifies the nerve termini of viral emergence. These symptoms may be followed by potentially severe complications such as persistent postherpetic neuralgia (chronic nerve pain and itching). Other viruses such as Epstein Barr and cytomegalovirus may exhibit a latent reemergence profile and potentially could contribute to the development of certain malignancies.

Immunosenescence Is Associated with Chronic Inflammation

Inflammatory components, cells, and cytokines have been found in every age-associated disease and are considered major players in the development of cardiovascular disease, malignancies, T2D, dementias, arthritis, and osteoporosis. Additionally, chronic low levels of inflammatory mediators (inflammaging) are

TABLE 15.1. Consequences of Immunosenescence

- Poor immune response to "new" antigens of pathogens (influenza) → associated with increased incidence of infections
- Adequate response to "old" antigens
- ↓ Immunological surveillance → associated with increased incidence of malignancies
- Poor to absent response to vaccines
- Reduced delayed type hypersensitivity → weak skin tests, ↑ skin cancers
- ↑ Autoantibodies → tissue damage
- Onset of low-level chronic inflammation

evident in the elderly and associated with organ aging. The source of inflammatory markers is not clear. *Deficiency of innate immunity favors macrophage activation and secretion of proinflammatory factors*. Second, inflammatory markers (Il-1, Il-6, tumor necrosis factor-alpha, C-reactive protein) are *elevated in obesity, smoking, and psychological stress and decline with weight loss, smoking cessation, and stress reduction*. While there are data to suggest that aberrant immune responses of immunosenescence induce high levels of proinflammatory mediators, convincing findings also indicate that lifestyles strongly influence the inflammatory state. Additionally, secretory products produced by replicative senescent cells as well as chronic CMV stimulation in the elderly may add to inflammaging as well. Consequences of immunosenescence are summarized in Table 15.1.

Preservation Strategies

Modification of Immunosenescence Is Possible in Animal Models

Several groups (Franceschi and Bonafè, 2003; Fülöp et al., 2007) have suggested strategies to combat immunosenescence and associated inflammation (inflammaging). Although controversial, it is proposed that immunosenescence accelerated by chronic viral stimulation, for example, CMV, could be prevented with early childhood vaccination. Another approach recommends rejuvenation of the thymus. Successful procedures used only experimentally, include immunological restoration by stem cell transplants, autologous T-cell banking (removing young T cells, freezing and later re-infusing), infusion of thymic peptides/thyroid hormones, and administration of cytokines (IL-2, Il-6, Il-12, interferon-gamma). It is well established that caloric restriction (CR) with adequate nutrition preserves immune function in animal models but the effect of CR on immune function in man has yet be evaluated. In contrast, the immunosuppression evident with protein malnutrition, fairly common in the elderly, can be alleviated with adequate nutrition.

IMMUNOLOGICAL THEORY OF AGING

The immunological theory of aging was originated by Walford (1964) following extensive observations on changes in immune function in aged rodents. Initially, this theory proposed that progressive age-dependent dysregulation in immune function

TABLE 15.2. Support for the Immunological Theory of Aging

- Correlation of thymic atrophy to gradual immune dysfunction and increase in incidence of autoantibodies, infectious diseases and malignancies
- Reduced ability of the thymus to replenish T cells; replicative senescence of T cells adds to this; associated with an increased incidence of malignancies and poor response to new antigens
- Decrease in the number of naive T-cell helpers and cytotoxic T cells; associated with an increased incidence of infectious diseases, cancers, and cardiovascular disease
- With age there is quantitative and qualitative decline in ability to produce antibodies (due to dysfunction of T-cell helper); correlated with an elevated susceptibility to infections, presence of autoantibodies and; poor response to vaccines
- In animal models of aging, caloric restriction (CR) reverses age-related decline in immunological function and extends the lifespan
- Maladaptive response of immune cells to ongoing tissue damage contributes to aging phenotype of inflammatory-mediated degeneration

leads to an increased vulnerability to major life-threatening diseases, for example, infectious disease, cancers, and autoimmune diseases. In particular, Walford defined aging as the failure of immunological surveillance in which the loss of self-recognition damages tissues in an "auto-immune disease-like" fashion. Support for this theory (see Table 15.2) points to thymic involution, decreased T- and B-cell function, and associated susceptibility to infections, malignancies, and autoimmune disease.

The immunological theory of aging assumes that all life-threatening diseases have an immunological basis. Initially very little evidence was available to reinforce this. However, as functions of T-cell and B-cell subsets and their secreted cytokines were revealed, evidence relating aberrant immune function to the development of major mortality-inducing diseases, such as atherosclerosis and diabetes, accrued. These diseases are now generally included with the more obvious immune-dependent maladies.

Setting disease apart, whether the immunological theory of aging as a theory of organ-system failure fully explains aging remains under review. One aspect of immunosenescence that might account for aging, at least in part, *relates to the interaction between immune dysregulation and low-grade chronic inflammation.* Although poorly understood, a vicious cycle has been noted in which immunological dysregulation induces inflammation and proinflammatory mediators induce immunological deficits. *Low-grade inflammation fosters organ damage and disease vulnerability.* In this regard, the following scenario has been proposed: (i) The immune system is a source of ROS, an inducer of cellular and subcellular destruction via epigenetic influence or direct structural damage; (ii) these effects initiate immune cell infiltration; (iii) with repeated episodes, proinflammatory situations persist or anti-inflammatory effects are overwhelmed by proinflammatory factors; and (iv) proinflammatory cytokines suppress normal immune function and damage tissues by encouraging apoptosis or necrosis. All of these changes could contribute to aging. Longitudinal studies with cause and effect evidence are presently absent.

Summary of Aging of the Immune System (Immunosenescence)

Barriers	Innate Immunity	Adaptive Immunity
• Skin thins, ↓ renewal • ↓ Sebum • ↓ Antimicrobial defense substances in tears and secretions • ↓ Stomach acid • ↓ Function of respiratory cilia • ↓ Epithelial mucus of GI, respiratory, urogenital tracts • ↓ Strength of cough/sneeze reflex	• ↓ Phagocytosis of neutrophils, macrophages, dendritic cells (DC) • ↓ Response of neutrophils to chemokines and cytokines • ↓ Maturation of DCs; ↓ number; ↓ function → ↓ antigen presentation to adaptive cells → poor or no response • ↓ Cytotoxic function of NK cells	• Thymic involution → slowed maturation of T and B cells → deficiency of peripheral naive T and B cells • Abundance of antigen-experienced memory cells of $CD8^+$ and $CD4^+$ subset • Poor T-cell helper function → ↓ response to new antigens • ↓ Immunosurveillance • ↑ Autoantibodies → ↑ T-regulatory function

SUMMARY

The immune system provides host defense against pathogens and malignancies. Its success is derived from it numerous physical barriers and reflexes, the phagocytic and cytotoxic cells of the innate immune system and the immense diversity and recall ability of T and B cells of the adaptive immune system. The first line of defense is the barriers/reflexes that block microbial entry. This is followed by a rapid response from innate immune cells to clear the pathogen and to activate the slower but more powerful response from the T and B cells. Both systems are driven by a plethora of plasma membrane receptors, cytokines, and chemokines.

Immunosenescence is the descriptive term for dysregulation of the immune system. Cell numbers remain constant but cellular activities are reduced. The most obvious age-related change is thymic involution that decreases renewal of peripheral T cells, an effect that limits T-cell diversity. Restricted cell-mediated and humoral-mediated immunity creates a vulnerability to new pathogens and poor immunological surveillance of tumors. Additionally, vaccination, a traditional artificially-induced defense enhancement is largely ineffective in the elderly.

Defects in phagocytic activity of innate immune cells further diminish immune responses and contribute to chronic inflammation. Barrier and reflex protection also deteriorate. Associated with these age changes is an increased prevalence of infectious diseases and malignancies.

The immunological theory of aging proposes that immunosenescence is the cause of aging. Immunological dysfunction stemming largely from thymic involution and associated with an elevated incidence of relevant diseases supports this theory. Second, aspects of immune dysfunction leads to chronic low-grade inflammation, a type of "autoimmune" condition. Together, immunosenescence destroys organ system function and creates a vulnerability to all life-threatening diseases. The theory is supported by associative data; it lacks cause and effect.

CRITICAL THINKING

What are the advantages of rejuvenating the thymus?

In what ways do innate immunity and adaptive immunity cooperate?

How does the immunological theory of aging explain aging? Does it explain organ failure?

How does immunosenescence differ from immunodeterioration?

What are some important cells of innate/adaptive immunity? How do their activities protect the organism?

What are some concerns with data regarding aging of the immune system?

KEY TERMS

Adaptive immunity characteristic response of the immune T cells and B cells that enables adjustments to environmental insults, for example, microbes and aberrant cells. Unique characteristics are unlimited pathogen recognition and production of immunological memory.

Antibody special proteins (immunoglobulins) produced by a plasma cell (mature B cell). Antibodies have the capability of halting progression of bacterial and parasitic infections and possibly tumor growth.

Autoimmune disease loss of ability by immune cells to recognize "self." Immunological attack on self produces unwanted tissue destruction manifest, for example, as rheumatoid arthritis, lupus.

B cell one of the major cell types of the adaptive immune system; mediator of humoral immunity (production of antibodies).

Cell-mediated immunity adaptive immunity attributed to actions of T cells, for example, cytotoxicity, helper function, and regulatory function.

Chemokine molecule with ability to attract cells to a specific locus; effective in "calling in" cells to site of damage or infection.

Clonal expansion stimulation of a cell to keep dividing and produce many daughter cells; required for an adequate immune response.

Complement a humoral component of innate immunity that is comprised of multiple proteins/enzymes that bind to and lyse pathogens, facilitate uptake of antigenic components by phagocytic cells, liaison with B cells, and can induce inflammation.

C-reactive protein circulating protein produced by the liver in response to inflammation.

Cytokine molecules released by innate/adaptive immune cells that stimulate proliferation, cell–cell interactions, and other assorted activities pro- or anti-inflammatory. Examples are interleukins and interferons.

Cytomegalovirus (CMV) member of the herpes family of viruses (another member is the chickenpox virus) and displays "latent" characteristics (reemerging at a later time); widely distributed.

Defensins a family of host antimicrobial peptides produced by innate immune cells, for example, granulocytes (PMN), epithelial cells of the GI tract, and skin.

Dendritic cell special phagocytic cell that has the capacity to engulf an antigen and convert it to a form recognizable by the T-helper cell and B cell to elicit an antibody response.

Endotoxin the lipopolysaccharide (LPS) and lipoprotein components of the bacterial cell wall that is recognized by the PRRs of innate immune cells. Inducers of fever.

Eosinophil a circulating white cell (bilobed nucleus) member of innate immune system. Cell induces inflammation and antimicrobial activity.

Immunosenescence state of dysregulation in which there is dysfunction due to failure of some but not all immune components. Refers to the decrease in T-cell diversity that appears to reduce protection against microbes and cancers and destroys the self with an increase in autoantibodies.

Innate immunity one-dimensional defense against pathogens; there is no memory. It includes barriers, pH, movement via epithelial-cilia, enzymes, phagocytic cells, and natural killer cells.

Lymphocyte general name given to cells of the immune system that include the T cells and B cells.

Macrophage monocyte-derived cell that resides in tissues. It engulfs and degrades particulate matter, a function termed phagocytosis (type of autophagy). Cell can process antigen and assist with immune responses.

Major histocompatibility complex (MHC) specialized proteins (marker) on cell membranes that supply information to the immune system that marker-bearing cells are "self." Pathogens lack MHC and are recognized as "non-self" and hence elicit an immune response through this and other mechanisms.

Mast cell specialized cell residing in tissues that on activation releases substances inducing allergic inflammation, both acute and chronic.

Natural killer (NK) cell lymphocyte with cytotoxic function, classified as part of the innate immune system.

Neutrophil innate immune cell that arrives first to the site of tissue damage. Has phagocytic function, releases lytic substances from cell granules, and influences the nature of the response.

Pattern recognition receptors (PRRs) receptors found on innate immune cells that respond to common molecular motifs of all pathogens (pathogen-associated molecular pattern) or molecular motifs of damaged tissue (damage-associated molecular patterns)

Phagocytosis process of engulfing particulate matter, for example, pathogen, foreign debris, and destroying it.

T cell cell that gains immune competence by prior residence in the thymus gland; mediates cell-based immunity.

T-cell repertoire includes subpopulations of T lymphocytes with specialized functions. Major subpopulations include T-helper cells (assist the B cell during antibody production), T-cytotoxic cell (cells with capacity to directly kill foreign cells), T-regulatory cells (regulate extent of immune response), and T-memory cell (cells primed with information of an earlier antigenic encounter).

Thymus major gland of the immune system. It serves as locus for "training" immature T cells.

BIBLIOGRAPHY

Review

Barrett KE, Barman SM, Boitano S, Brooks HL. 2012. *Ganong's Review of Medical Physiology*, 24th ed. York: McGraw Hill pp. 67–95.

Camous X, Pera A, Solana R, Larbi A. 2012. NK cells in healthy aging and age-associated diseases. *J. Biomed. Biotechnol.* **2012**: 195956.

Chandra RK. 2002. Nutrition and the immune system from birth to old age. *Eur. J. Clin. Nutr.* **56** (Suppl. 3):S73–S76.

Dowling DK, Simmons LW. 2009. Reactive oxygen species as universal constraints in life-history evolution. *Proc. Biol. Sci.* **276**(1663):1737–1745.

Franceschi C, Bonafè M, Valensin S. 2000. Human immunosenescence: the prevailing innate immunity, the failing of clonotypic immunity, and the filling of immunological space. *Vaccine* **18**(16):1717–1720.

Franceschi C, Bonafè M. 2003. Centenarians as a model for healthy aging. *Biochem. Soc. Trans.* **31**(2):457–461.

Fülöp T, Larbi A, Hirokawa K, Mocchegiani E, Lesourds B, Castle S, Wikby A, Franceschi C, Pawelec G. 2007. Immunosupportive therapies in aging. *Clin. Interv. Aging* **2**(1):33–54.

Fülöp T, Witkowski JM, Pawelec G, Alan C, Larbi A. 2014. On the immunological theory of aging. *Interdiscip. Top. Gerontol.* **39**: 163–176.

Goronzy JJ, Weyand CM. 2013. Understanding immunosenescence to improve responses to vaccines. *Nat. Immunol.* **14**(5):428–436.

Gupta S, Agrawal A. 2013. Inflammation & autoimmunity in human ageing: dendritic cells take a center stage. *Indian J. Med. Res.* **138**(5):711–716.

Ligthart GH. 2001. The SENIEUR protocol after 16 years: the next step is to study the interaction of ageing and disease. *Mech. Ageing Dev.* **122**(2):136–140.

Luciani F, Valensin S, Vescovini R, Sansoni P, Fagnoni F, Franceschi C, Bonafè M, Turchetti G. 2001. A stochastic model for CD8(+) T cell dynamics in human immunosenescence: implications for survival and longevity. *J. Theor. Biol.* **213**(4):587–597.

Man AL, Gicheva N, Nicoletti C. 2014. The impact of ageing on the intestinal epithelial barrier and immune system. *Cell Immunol.* **289** (1–2):112–118.

Macaulay R, Akbar AN, Henson SM. 2013. The role of the T cell in age-related inflammation. *Age (Dordr.)* **35**(3):563–572.

McElhaney JE, Effros RB. 2009. Immunosenescence: what does it mean to health outcomes in older adults? *Curr. Opin. Immunol.* **21**(4):418–424.

McElhaney JE, Zhou X, Talbot HK, Soethout E, Bleackley RC, Granville DJ, Pawelec G. 2012. The unmet need in the elderly: how immunosenescence, CMV infection, co-morbidities and frailty are a challenge for the development of more effective influenza vaccines. *Vaccine* **30**(12):2060–2067.

Meredith PJ, Walford RL. 1979. Autoimmunity, histocompatibility, and aging. *Mech. Ageing Dev.* **9** (1–2):61–77.

Müller L, Fülöp T, Pawelec G. 2013. Immunosenescence in vertebrates and invertebrates. *Immun. Ageing* **10**(1):12.

Nakatsuji T, Gallo RL. 2012. Antimicrobial peptides: old molecules with new ideas. *J. Invest. Dermatol.* **132** (3 Part 2):887–895.

Pawelec G, Derhovanessian E, Larbi A, Strindhall J, Wikby A. 2009. Cytomegalovirus and human immunosenescence. *Rev. Med. Virol.* **19**(1):47–56.

Sambhara S, McElhaney JE. 2009. Immunosenescence and influenza vaccine efficacy. *Curr. Top. Microbiol. Immunol.* **333**: 413–429.

Si-Tahar M, Touqui L, Chignard M. 2009. Innate immunity and inflammation: two facets of the same anti-infectious reaction. *Clin. Exp. Immunol.* **156**(2):194–198.

Turvey SE, Broide DH. 2010. Innate immunity. *J. Allergy Clin. Immunol.* **125** (2 Suppl. 2): S24–S32.

Vadasz Z, Haj T, Kessel A, Toubi E. 2013. Age-related autoimmunity. *BMC Med.* **11**: 94.

Ventevogel MS, Sempowski GD. 2013. Thymic rejuvenation and aging. *Curr. Opin. Immunol.* **25**(4):516–522.

Walford RL. 1964. The immunological theory of aging. *Gerontologist* **4**(4):195–197.

Weinert BT, Timiras PS. 2003. Physiology of aging invited review: theories of aging. *J. Appl. Physiol.* **95**(4):1706–1716.

Experimental

Agrawal A, Agrawal S, Cao J-N, Su H, Osann K, Gupta S. 2007. Altered innate immune functioning of dendritic cells in elderly humans: a role of phosphoinositide 3-kinase-signaling pathway. *J. Immunol.* **178**(11):6912–6922.

Agius E, Lacy KE, Vukmanovic-Stejic M, Jagger AL, Papageorgiou A-P, Hall S, Reed JR, Curnow SJ, Fuentes-Duculan J, Buckley CD, Salmon M, Taams LS, Krueger J, Greenwood J, Klein N, Rustin MH, Akbar AN. 2009. Decreased TNF-alpha synthesis by macrophages restricts cutaneous immunosurveillance by memory CD4$^+$ T cells during aging. *J. Exp. Med.* **206**(9):1929–1940.

Fagnoni FF, Vescovini R, Passeri G, Bologna G, Pedrazzoni M, Lavagetto G, Casti A, Franceschi C, Passeri M, Sansoni P. 2000. Shortage of circulating naive CD8(+) T cells provides new insights on immunodeficiency in aging. *Blood* **95**(9):2860–2868.

Ogata K, An E, Shioi Y, Nakamura K, Luo S, Yokose N, Minami S, Dan K. 2001. Association between natural killer cell activity and infection in immunologically normal elderly people. *Clin. Exp. Immunol.* **124**(3):392–397.

Weston WM, Friedland LR, Wu X, Howe B. 2012. Vaccination of adults 65 years of age and older with tetanus toxoid, reduced diphtheria toxoid and acellular pertussis vaccine (Boostrix(®)): results of two randomized trials. *Vaccine* **30**(9):1721–1728.

Wu J, Li W, Liu Z, Zhang YY, Peng Y, Feng DG, Li LH, Wang LN, Liu L, Li L, Liu J. 2012. Ageing-associated changes in cellular immunity based on the SENIEUR protocol. *Scand. J. Immunol.* **75**(6):641–646.

INDEX

acetylation, histone, 86-7
adipose tissue, 110, 111
  age changes, 110–111, 123, 131, *Fig.* 6.3
advanced glycation end products(AGEs), 63
  arterial stiffness, 177–8
  cardiac fibrosis, 168–9
  skin, 108
  type 2 diabetes (T2D), 66, 148, 151, 294
advanced lipid end products (ALEs), 65
adverse drug reaction (ADR)
  role of cytochrome P450 enzymes, 211
  role of kidneys, 217
age spots, *see* senile lentigo
aging
  accelerated, *see* progeroid syndromes
  characteristics, 6–9, *Table* 1.1
  commencement, 8
  evolutionary theory, *see* antagonistic pleiotropy, disposable soma, mutation accumulation
  evolutionary side effect, 39
  face, 115–6, *Fig.* 6.5
  molecular fidelity, loss of, 7–8
  organelle, *see* mitochondria, lysosome, nucleus, peroxisome
  programmed, 35–36
  rates species, 9
  role in disease, 163–4
    atherosclerosis, 94, 179, 181–2
    heart failure, 172–4
    infections, 312–3
    malignancy, 94–95, *Fig.* 5.7, 313
    osteoporosis, 151–3, *Table* 8.4
    Parkinson's disease, 95
    systolic hypertension, 178–9
    type 2 diabetes(T2D), 66, 294–5
  senescence phenotype, human, 8–9
  sensory, *see* audition, gustation, olfactory, somatosensory, vision
  theories of, overview, 12–13, *Table* 1.2, *see also* individual theory
alopecia, senescent, 107
antagonistic pleiotropy theory, 39–40
apoptosis, 40, 79
autophagy, 81, 82–3, 93 *see also* lysosome

Baltimore Longitudinal Study of Aging, 27
  findings, 27, *Table* 2.1
baroreflex, 153, 182
  gender, 182
basal metabolic rate (resting metabolic rate), 132
  age changes, 133, 290
base excision repair (BER) function, 60–1, 66, 67
benign prostatic hyperplasia (BPH), 286
  malignancy, 287
  role of endocrine system, 286
  treatment, 287
beta-adrenergic receptors
  disappearance, 176, 197–8
  role in exercise, 175–6
biogerontology, 3
  goals, 4
  milestones, 4
bone, *see* skeletal system

Caenorhabditis elegans, see round worm
Calcium
  recommended daily allowance, 150
  regulation, endocrine, 292–3
  role in
    bone function, 150
    calcification, cartilage, 171, 178, 196
    muscle contraction, 124, 181

caloric restriction, 21
  effect on
    animal models of aging, 21–2, 184
    cardiovascular system, 184
    endocrine system, 294
    humans, 21–2
    immune system, 315
    mechanism, 22, 87, *Box* 2.1
  relation to disposable soma theory, 42–43
  significance, 23, 42–43
cardiovascular system, 168–9, *Fig.* 9.1, *Fig.* 9.2, *Fig.* 9.3, *Fig.* 9.5, *Tables* 9.2, 9.3
  aerobic exercise, benefits, 174–5, 183–4
  aging, heart
    AGEs, 168–169
    consequences, 174
      atria, hypertrophy, 173
      heart failure, diastolic, 173
      response to exercise, 174
    diastole/twist, 172–4, *Fig.* 9.4
    fibrosis, 168
    preconditioning, 176–7
    renewal of cardiac myocytes, 170–1
    remodeling, 169–70
    valvular changes, 176
    ventricular hypertrophy, 169
    ventricular stiffness, 168–9
      gender difference, 169, 170
    ventricular twist, 172–4
  aging, vasculature
    arterial stiffness, 178–9
    consequences, 179, 180, 182–3, *Fig.* 9.6
    endothelial dysfunction, 181
    intimal-medial thickening, 179–80, *Fig.* 9.7
    smooth muscle dysfunction, 181–2
  caloric restriction in animals, 184
  diseases, 165, *Table* 9.1
  homeostatic measurements, *Table* 9.2
  risk factors for cardiovascular disease, 28–9, 182–3
catalase
  function, 66, *Table* 4.1
  location, 66–7, 84
  manipulations, effects of, 68–9, 80, 85
cells, 50, *Fig.* II.4, *Table* 5.1
  aging
    fibroblast, 108, 168

innate immune cells, role in oxidative stress, 58, 304, 306–8
mitotic, 89
  replicative senescence or senescence-associated secretory phenotype (SASP), 89–90, 94, *Fig.* 5.6, 170, 181, 304
postmitotic, 92, 124–5, 229, 230
stem, 90–1
cell cycle, 88–9, *Box* 5.5
germline, 27, *Fig.* 3.1
metabolism, general, 49
organelle, *see* lysosome, mitochondria, nucleus, peroxisome
role in disease, 94–95, 182–3, 312–14 *Fig.* 5.7
somatic, 37, 88, *Table* 5.1, *Fig.* 3.1
central nervous system, 225, 228 *Fig.* 12.1, *Fig.* 12.2, *Fig.* 12.3, *Box* 12.2, *Table* 12.2
  aging of
    brain structure, 229–31
      connectivity loss, 234
      microglia, 234
      neurotransmitters, 231–3, *Box* 12.3, *Table* 12.3
    cognition, 235
      dedifferentiation, 239
      interventions, 240–5
      reserve, 239
      reversibility, 238
      time of onset, 238
    exercise benefit, 241–2, *Table* 12.5
    maintenance, 240, *Table* 12.4
    preservation, 240–2
  neuroimaging tools, 226, *Box* 12.1
  neuroplasticity, 245–6
  research issues, 225–7, *Table* 12.1
  tests, cognitive, 235, *Table* 12.3
  training, cognitive, 241–2
cytomegalovirus (CMV), 311–12

data
  types, correlative or cause-effect, 17–18
dehydration, 215–16
delayed hypersensitivity, 313–14
deoxyribonucleic acid (DNA), 58–9, 85, *Fig.* 4.3
  composition of, *Box* 4.1, *Fig.* II.3
  damage response(DDR), 89–90, 93
  epigenetic modification, 85–6, 87

telomeres, 86–7, *Fig.* 5.5
types
    euchromatic, 85
    heterochromatic, 85
    telomeric, 86
dermis, *see* integument system
dietary restriction, *see* caloric restriction
disposable soma, theory, 39, 41–3, *Fig.* 3.3, *Fig.* 3.4
    relation to
        caloric restriction, 42
        fecundity and longevity, 42–3
        hydra, 43
Drosophila melanogaster, *see* models of aging, fruit fly
dynapenia, 126–8
    consequences, 132–4
    excitation-contraction coupling, 130
    interventions, 134–7
    mitochondrial dysfunction, 127–8
    physical inactivity, 128–9
    protein consumption, 129–30
    respiratory muscles, 196

elastocalcinosis, 178
endocrine system, 275–7
    hypothalamic-pituitary-adrenal axis, 287
        aging, 287–9
            gender sensitivity, 289
        aldosterone, 289
        feedback inhibition, *Fig.* 14.5
        dehydroepiandrosterone sulfate (DHEAS), 289
    hypothalamic-pituitary axis (HPA), general
        feed-back inhibition, negative, 277, *Fig.* 14.1
        hormones, *Table* 14.1
    hypothalamic-pituitary-liver axis, 290–1
        aging, 291
        somatopause, 291
    hypothalamus-pituitary-ovary axis, 278
        menstrual cycle/estrogen, 278
        menopause, 278–84, *Fig.* 14.2
            controversial issues, 282–4, *Table* 14.3
            effects, 111, 280–2
            hormonal mechanism, *Fig.* 14.3
            hormone therapy, 284, *Table* 14.4
            hot flush, 280–2, *Table* 14.2

    hypothalamic-pituitary-testes axis, 285
        benign prostatic hyperplasia, 286–7
        late-onset hypogonadism, 286
            testosterone therapy, 286
        negative feedback inhibition, 285–6, *Fig.* 14.4
    hypothalamic-pituitary-thyroid axis, 289–90
        aging, 290
        resting metabolic rate (RMR), 290
    pancreas, 293
        glucagon, 294–5
        insulin, 293
            resistance, 133, 294–5
        type 2 diabetes, 294
    parathyroid glands, 292–3
    pineal gland(melatonin), 291–2
    neuroendocrine theory, 295
epidermis, *see* integument System
evolutionary theory, 35–6
    aging, explanation of, *see* antagonistic pleiotropy theory, disposable soma theory, mutation accumulation theory
    genetics, role of, 37, *Fig.* 3.1
    tenets, Darwin, 36, *Fig.* 3.2
exercise, 128–9
    aerobic, 135
    effects on
        cardiovascular system, 183–4
        cognition, 242–4, *Table* 12.5
        skeletal muscle system, 134–5
        skeletal system, 154–5, *Table* 8.5
    four-pronged program, 136–7
    guidelines, 183–4
    resistance, 134

falls
    causes
        baroreceptor aging, 182
        cognitive aging, 237
        fractures, 153
        dynapenia, 136
        sensory aging, 259, 263–5
    prevention, 135, 136–7, 153, 266
fibroblasts
    aging of
        cardiac, 168–9, *Fig.* 9.2
        dermal skin, 104, 108, 111, *Fig.* 6.2
frailty syndrome, 132
fruit fly, *see* models of aging

genes, *see also* deoxyribonucleic acid
  epigenetic, 85–6
  expression, 58, 60
  function of,
    germline, 37
    soma, 37, *Fig.* 3.1
  manipulation, *Fig.* 2.1
    effects, 69–71
    knock-in, 23–4
    knock out, 23–4
gastrointestinal system, 207–9, *Fig.* 11.1, *Box* 11.1
  age-associated disorders
    constipation, 212
    malnutrition, 212
  age changes
    hypochlorhydria, 209
    liver, 211
    microbiota diversity, 209–11
    salivary glands
      xerostomia, 211
    transit time, colonic, 209
glands,
  endocrine, 275, *see also* endocrine system
  exocrine, 104, 207–9, *see also* prostate
glutathione peroxidase, 66–7
  manipulations, 69
glutathione, *see* redox

hair, *see* alopecia, senescent
Hayflick's number, 89
homeostasis, 8, 11, 48, 71, 77, 79, 83, 85, 113, 129, 133, 135, 150, 168, 181, 182, 199, 234, 247, 266, 278, 292, 295, Table 1.2, *Fig.* 4.5, *Fig.* 5.2
hormesis
  caloric restriction, 23, 79
  mitochondrial ROS, 79, *Fig.* 5.2
hormone therapy(HT)
  effect on
    bone, 149
    estrogen-dependent tissues, 118, 153, 280–4
    skin, 111, 118
Huntington's disease, 36
  mutation accumulation example, 40
Hutchinson–Gilford syndrome, *see* progeroid syndromes
hydrogen peroxide, 55
  effects, 55, 58, 70–1, 84, 92
  generation of, 66–8

hypodermis, *see* integument system
hypotheses, *see* oxidative stress, redox stress, mitotic clock
inflammaging, 90, 303, 315
  consequences, 315, *Table* 15.1
  inflammation, 179
    role of endothelium, 180–181
    prevention strategies, 183–4, 314–5
immune system, 303–305, *Fig.* 15.1. *Fig.* 15.2, *Fig.* 15.3
  adaptive immunity, 308–11, *Fig.* 15.3, *Fig.* 15.4
  age changes
    barriers, 306
    immunosenescence, 311–12
      consequences, 312–15, *Table* 15.1
        autoantibodies, 314
        chronic inflammation, 315–16
        delayed hypersensitivity, 313–14
        infectious diseases, 312–13
        latent virus onset, 314
        malignancies, 313
        vaccination suboptimal, 313
    innate immunity/phagocytosis, 306–7, *Fig.* 15.2
    prevention, 315
    thymic involution, 311
  barriers, 305
  immunological theory, 315–16, *Table* 15.2
  innate immunity, 306–7
  pattern recognition receptors, 306
  Senieur protocol, 305
  T/B-cell, 309–10, *Fig.* 15.3, *Fig.* 15.4
integumentary system (skin), 103–105, 110, *Fig.* 6.1
  age changes, *Fig.* 6.2, *Fig.* 6.3, *Fig.* 6.7
    consequences, 111–15
      barrier dysfunction, 105–7
      hair characteristics, 107
      photoaging, 104, 108
      photocarcinogenesis, 105–6, 114–15
      pruritus, 112
      temperature dysregulation, 113
      vitamin D production, 113
      wound healing, 113–114
    extrinsic, 104
      dermis, 107
        solar elastosis, 108
      epidermis, 105–6
      hypodermis, 110, Fig. 6.3

intrinsic, 104
  dermis, 108
  epidermis, 106–7
  keratinocytes, 105, 106
  Langerhans cells, 105, 107, 113
  melanocytes, 105, 106
  menopause effects, 111, 282, 306
  smoking effects, 103, 116, 118
  sweat glands, 108–9
aging face syndrome, 115–16, *Fig.* 6.5
preservation
  fillers, 108, 117
  hormone therapy, 118, 290
  retinols, 116–17
  sunscreens, 116
kyphosis, 194, *Fig.* 10.3
  gender differences, 194

life expectancy, 4–6, *Fig.* 1.1
  issues, 6
  reasons for increase, 5
lifespan
  extension, 21, 25–27
  manipulations, experimental, 69–71
  maximal, 8, 13, 21
lipids, 49, *Box* 4.3, *Fig.* 4.3, *Fig.* II.2
  peroxidation, 65
  saturation, 64
    longevity, 65, 68
lipofuscin, 83
longevity
  centenarians, phenotype, 11
  contribution, genetic 9–10, *Fig.* 1.2
  determinants (also assurance genes), 9–10
  gerontogenes, 10–11, 24
  relation to
    environment, 9–11
    fecundity, 41–2
    maintenance, soma, 41
lysosome, 82
  autophagy, 82–3, *Box* 5.2
    pathways of, *Fig.* 5.3
  lysosomal-associated membrane protein (LAMP-2A), 83
  mitochondrial-lysosomal axis, 83, 93
  mitophagy, 83

Macaca mulatta, see models of aging, nonhuman primate
macrophage
  role in,
    apoptosis, 93
    immunity, 209, 306, 307–8
malignancies
  photocarcinogenesis, 105–6
  role of
    immune system, 313
    SASP, 94–5
  types, skin, 114–15
malnutrition, *see* gastrointestinal system
matrix metalloproteinases (MMPs), 108
menopause, *see* endocrine system, hypothalamic-pituitary-ovary axis
methionine sulfoxide reductase, 63, 67, *Fig.* 4.4
mitochondria, 78–9, 81, *Fig.* 5.1, *Box* 5.1
  age changes,
    humans, 80–1
  fission/fusion, 80–1
  role in
    apoptosis, 79
    cell dysfunction, 81, 83, 91
    disease, 95
    ROS production, 79, *Fig.* 5.2
  theory,
    mitochondrial-lysosomal axis, 83, 93
    mitochondrial free radical (MFRTA), 80
mitotic clock, 89 *Fig.* 5.5
models of aging
  laboratory animal
    fruit fly, 26
    mouse, 20, 26
    nonhuman primate, 26–7
    round worm, 25
    yeast, 24–5
  significance, 23–4, *Fig.* 2.1
Mus musculus, see models of aging, mouse
mutation accumulation theory, 40–1

necrosis, 93–4
neuroimaging, *Box* 12.1
  issues, 226
  patterns, 238–40
neuroplasticity, *see* central nervous system
neurotransmitters/receptors, Table 12.2,
  aging of, 176, 231–4
nitric oxide, 57, 177, 181, *Table* 4.1
  radical, 55
nocturia, 215, 216
nucleotide excision repair(NER), 60, 66, *Fig.* 4.4

nucleus, 85–6 *see also* deoxyribonucleic acid, progeroid syndromes
  aging, 86–8, *Box* 5.4
  histones, 85–6
  lamins, 87–8

organelles, 50, 77 *see also* mitochondria, nucleus, lysosome, peroxisomes
organism
  cell components, 50–1, *Fig.* II.4
  hierarchy, 47–8, *Fig.* II.1, II.2
  structure/function, 49–50, *Fig.* II.3
osteoarthritis
  diagnosis, 155, *Fig.* 8.5
  factors, 94–5, 156–7, *Fig.* 8.6
  incidence, 156
  pharmacotherapy, 158
osteopenia, 152–3
osteoporosis, 152–3
  pharmacotherapy, 153, 291
oxidation, 54–7
  biomarkers of oxidation, *Table* 4.1
  countermeasures, 66–8, *Fig.* 4.4
    membrane composition, 67
    Msr, 67
    NER/BER, 67
    redox pairs, 67
    superoxide dismutase (SOD), 10, 66
      manipulations, experimental, 69–70, 71
  effects on
    lipids, peroxidation, 64–5
    nucleic acids, breakage, 58–61
    proteins, cross-linkage, 61–4, 108, 168, 178
  free radical, 53,54–55, 68, *Table* 4.1
  initiators, 57–8, 63, 65, *Fig.* 4.2
  measurements, experimental, 71
  non radical oxidant, 55, *Table* 4.1
  principles, oxidation/reduction, 54–5, *Fig.* 4.1
  source, 57–8, *Fig.* 4.2
  targets, 58, 61, 64, *Table* 4.2
oxidative stress, 57, *Fig.* 4.5
  hypotheses, 53–54
    free radical theory (oxidative stress), 53–4, *Fig.* 4.5
      caloric restriction, effect on 69
      strengths, 68–9
      trials, clinical, 70
      weaknesses, 69–70

    redox stress hypothesis, 53–54, *Fig.* 4.5,
      caloric restriction, effect on, 71
      strengths, 70–1
      weaknesses, 71–2
  role of,
    reactive oxygen species(ROS), 55, 57, *see also* superoxide anion, hydrogen peroxide, nitric oxide
oxygen, 55

pacemaker cells, 167, *Fig.* 9.2
  role in heart rate aging, exercise, 175
pancreas
  endocrine, *see* endocrine system
  exocrine, 209
Parkinson's disease, 95, 232
peroxisome, 77, 84, *Box* 5.3,
  aging of, 84–5, *Fig.* 5.4
phenotype, senescence
  cells, 89–90, 94, *Fig.* 5.6, 170, 181, 304
  humans, 8–9
    factors, 18
phospholipids, 65
  peroxidation, 65
photoaging, 104, 105
  prevention of, 116–18
pollution,
  effects, 104, 196
progeroid syndromes
  Hutchinson–Gilford syndrome, 29
  Werner's syndrome, 29
proteins, 49, 60–1, 63, 64, *Fig.* II.2, *Box* 4.2, *Fig.* 4.3, *Table* 4.2
pruritus, 112
pulmonary system, 193–5, 197, *Fig.* 10.1, *Fig.* 10.2, *Fig.* 10.4
  aging of
    airways, 197–8
    airway lining, 198
    chemoreceptors, 199
    consequences, 201
    forced expiratory volume (FEV), 200
    forced vital capacity, 200
    gas diffusion, 198–9
    interventions to improve function, 201–2
    maximal inspiratory pressure, 197
    maximal oxygen consumption (VO2max), 200–1
      gender differences, 201

prevention, 201–3
residual volume (RV), 200, 201, *Fig.* 10.4
thoracic cavity, 195–7
calcification, 196
dynapenia, 197
hyperkyphosis, 195–6, *Fig.* 10.3
pulse wave velocity (PWV), 178
calculation, *Fig.* 9.6
consequences, 179

radiation, ultraviolet
effects on
skin, 103, 104, 109, 111, 113
stochastic, 11
reactive carbonyl compounds (RCC), 57, 60, *Table* 4.2, *Fig.* 4.4
reactive nitrogen species (RNS), 57, 60, 65, 67, *Fig.* 4.4
nitric oxide, endothelial, 91, 181
reactive oxygen species (ROS), 53–4, 57, 60, 63, 65, 67, *Table* 4.1, *Fig.* 4.4
experimental, 68–9
levels, 78–90, 83
signaling, 53–4, 57, 79
redox (reduction/oxidation)
glutathione, 54, 57, 64, *Fig.* 4.4
pairs, 66–7
potential, cellular, 70–1
stress hypothesis, 53–4, 70, 71
thiol (sulfur) switch, 55, *Table* 4.2
oxidation, cysteine, 64
oxidation, methionine, 63
renin-angiotensin system (RAS), 184
inhibition/longevity, 215
replicative senescence, *see* cells, mitotic, aging
retinoids, 116–17, *Fig.* 6.6
ribonucleic acid, 58, *Box* 4.1, *Fig.* 4.3
noncoding, 86, *Box* 5.4

Saccharomyces cerevisiae see models of aging, yeast
sarcopenia, 124–126
basal metabolic rate, 132–3
consequences, 132–4,
consumption, protein, 129–30
insulin resistance 134
interventions, 134–7
physical inactivity, 128–9
resting metabolic rate, 133
scientific method, 17

senescence-associated secretory phenotype (SASP), *see* cells, mitotic, aging
senile lentigo, 106
sensory system
age changes
consequences/compensation, 266–9, *Table* 13.2
cornea, 256, 259
olfaction/gustation, 261–2
gender difference, 261
presbycusis, 260–1
presbyopia, 257–8, Fig. 13.2
relation to falls, 263
retina, 258–9
senile miosis, 258
modalities,
audition, 259, *Fig.* 13.3
decibels, relevant, *Table* 13.1
gustation, 262
olfaction, 261–2
somatosensory
mechanoreceptors, 264–5, *Fig.* 13.4
nociceptors, 265
proprioception, 263–4
thermoreceptors, 266
vision, 256, *Fig.* 13.1
principles, 255–6
sirtuins, 87, *Box* 2.1, *Box* 5.4
skeletal muscle system, 123–4, 127 *Fig.* 7.1
aging (senescent) phenotype, 124–8, 129
consequences, 123, 132–4
dynapenia, 126–8
hormone decline, 131
interventions, 134–7
motoneuron dysfunction, 130
myocyte changes
dihydropyridine receptor, 128
mitochondrial dysfunction, 127–8
physical inactivity, role of 128–9
reduced protein consumption, role of, 129–30
risk factors, 136, *Table* 7.1
sarcopenia
factors, causative, 128–31
gender difference, 125
prevalence, 125
stem cell dysfunction, 131
myocyte, 124, 126, 134, *Fig.* 7.1

skeletal system, 143–4, *Fig.* 8.1, *Fig.* 8.3 *Table* 8.1, *Table* 8.2
  age changes
    bone cell dysfunction, 147
    bone loss patterns, 147–9, *Fig.* 8.4
      gender differences, 148, *Fig.* 8.4
      exercise benefit, 154–5, *Table* 8.5
      factors, contribution *Table* 8.4
        adrenal steroids, role of, 151
        estrogen, role of, 149–50, 280
        growth hormone/insulin, role of, 151
        mechanical strain reduction, 146–7
        smoking, 151
        vitamin D/parathyroid hormone, role of, 150
      fracture prevention, 153
      modeling/remodeling, 146–9
    cells, 145–6, *Fig.* 8.2
    customary strain stimulus, 144–5, *Table* 8.3
    mechanical stimulation benefit, 155
    mechanostat, 144–145
    modeling/remodeling, 144–5, *Table* 8.3
    osteopenia, 152
    osteoporosis, 152–3
skin, *see* integumenatary system
smooth muscle cell phenotypic change, 181–2
study design
  cross-sectional, 19
  longitudinal, 19–20
    Baltimore Longitudinal Study, 27, *Table* 2.1
    Framingham Heart Study, 28–9
  randomized control trial, 20
  meta-analysis, 20

superoxide anion, 55, 58, 63, Table 4.1, *Fig* 5.1
superoxide dismutase (SOD), 10, 63, 66, *Table* 4.1

thermoregulation
  effects of sarcopenia, 133–4
  skin, 110, 113
  thermoreceptors, 266
transcriptional factors, 86, 278, 287

urinary System, 207–9 *Fig.* 11.2
  age-associated conditions
    adverse drug reaction, potential for, 217
    dehydration, 215–6
    nocturia, 216–7
    urinary incontinence, 217–8
      gender differences, 217–8
  age changes
    glomerular filtration rate (GFR), 214
    glomerulosclerosis, 214
    ion imbalance, 214–5

vitamins
  vitamin A, *see* retinoids
  vitamin C, 66, 67, 70, *Table* 4.1
  vitamin D, 113, *Fig.* 6.4,
    deficiency, 113
    role in bone, 150–1
    RDA, 150
  vitamin E, 66, 67, 70

Werner's syndrome *see* progeroid syndrome
wound healing, 113–14

Xerosis, 111, 112